Introduction to PLASMAS AND PLASMA DYNAMICS

Introduction to PLASMAS AND PLASMA DYNAMICS

With Reviews of Applications in Space Propulsion, Magnetic Fusion and Space Physics

THOMAS M. YORK

York Scientific Consultants

HAI-BIN TANG

Beijing University of Aeronautics and Astronautics (BUAA), Beihang University

ELSEVIER

Amsterdam • Boston • Heidelberg • London
New York • Oxford • Paris • San Diego
San Francisco • Singapore • Sydney • Tokyo

Academic Press is an imprint of Elsevier

AP

Academic Press is an imprint of Elsevier
125 London Wall, London EC2Y 5AS, UK
525 B Street, Suite 1800, San Diego, CA 92101-4495, USA
225 Wyman Street, Waltham, MA 02451, USA
The Boulevard, Langford Lane, Kidlington, Oxford OX5 1GB, UK

Library of Congress Cataloging-in-Publication Data
A catalog record for this book is available from the Library of Congress

British Library Cataloguing-in-Publication Data
A catalogue record for this book is available from the British Library

ISBN: 978-0-12-801661-9

For information on all Academic Press publications
visit our website at http://store.elsevier.com/

Publisher: Joe Hayton
Acquisition Editor: Hayley Gray
Editorial Project Manager: Cari Owen
Production Project Managers: Pauline Wilkinson and Julie-Ann Stansfield
Designer: Greg Harris

Typeset by TNQ Books and Journals
www.tnq.co.in

Printed and bound in the United States of America

In dedication:

To our wives, Mary and Yan Liu:

They shared the vision and the journey.

CONTENTS

Preface	*xi*
Acknowledgments	*xiii*
1. The Plasma Medium and Plasma Devices	**1**
Introduction	1
Plasmas in Nature	2
Plasmas in Laboratory/Device Applications	3
References	5
2. Kinetic Theory of Gases	**7**
Introduction	7
Basic Hypotheses of Kinetic Theory	7
Pressure, Temperature, and Internal Energy Concepts	8
Kinetic Theory and Transport Processes	13
Mathematical Formulation of Equilibrium Kinetic Theory	20
References	32
3. Molecular Energy Distribution and Ionization in Gases	**33**
Introduction	33
Molecular Energy	33
Ionization in Gases	50
References	63
4. Electromagnetics	**65**
Introduction	65
Electric Charges and Electric Fields—Electrostatics	65
Electric Currents and Magnetic Fields—Magnetostatics	68
Conservation of Charge	69
Faraday's Law	71
Ampere's Law	72
Maxwell's Equations	72
Forces and Currents due to Applied Fields	73
Plasma Behavior in Gas Discharges	76
Illustrative Applications of Maxwell's Equations	81
References	84

5. Plasma Parameters and Regimes of Interaction **87**

Introduction 87

External Parameters 87

Particle (Collision) Parameters 87

Sheath Formation and Effects 87

Plasma Oscillations and Plasma Frequency 90

Magnetic Field Related Parameters 91

Electrostatic Particle Collection in (Langmuir) Probes 92

References 98

6. Particle Orbit Theory **99**

Introduction 99

Charged Particle Motion in Constant, Uniform Magnetic ($\overrightarrow{B}$) Field 100

Particle Motion in Uniform Electric and Magnetic Fields 102

Particle Motion in Spatially Varying (Inhomogenous) Magnetic Fields 105

Particle Motion with Curvature of the Magnetic Field Lines 107

Particle Motion in Time-Varying Magnetic Field 108

Particle Trapping in Magnetic Mirrors 110

Adiabatic Invariants 111

References 113

7. Macroscopic Equations of Plasmas **115**

Introduction 115

Electromagnetic Energy and Momentum Addition to Plasmas 115

Conservation Equations of Magnetofluid Mechanics 120

Single Fluid Equations of Magnetofluid Mechanics 125

The MHD Approximations 132

Similarity Parameters 134

References 136

8. Hydromagnetics—Fluid Behavior of Plasmas **137**

Introduction 137

Basic Equations of Continuum Plasma Dynamics 137

Transport Effects in Plasmas and Plasma Devices 138

Kinematics (and Dynamics) of Magnetic Fields in Plasmas 143

Magnetohydrostatics 147

Hydromagnetic Stability 152

Waves in Plasma—Propagation of Perturbations 158

Fluid Waves and Shock Waves in Plasma 164

References 192

9. Plasma Dynamics and Hydromagnetics: Reviews of Applications 195
Introduction 195
Plasma Acceleration and Energy Conversion 197
Introduction 197
References 216
Plasma Thrusters 217
Introduction 217
References 246
Magnetic Compression and Heating 249
Introduction 249
References 263
Wave Heating of Plasmas 264
Introduction 264
References 277
Magnetic Fusion Plasmas 279
Introduction 279
References 303
Space Plasma Environment and Plasma Dynamics 305
Introduction 305
References 323

Appendix A: Conversion between MKS and Gaussian System 325
Appendix B: Definite Integrals – Maxwellian Distribution Functions 327
Appendix C: Nomenclature 329
Appendix D: Problems 331
Index 339

PREFACE

The study of plasmas—ionized gases which are generally electrically neutral—emerged as an important topic because of the importance of the subject in energy, communications, space exploration, and defense applications. Intense interest in the subject emerged from astrophysics[1] and study of thermonuclear processes[2] in the 1950s. While the subject matter foundation is inherently physical science, development and construction of devices of a broad variety require interpretation to the engineering applications. This process was assisted in the 1950s with the publication of two comprehensive volumes on *Gas Discharge Physics*.[3]

The material presented here has been organized and found useful in instruction and research over a period of many years. One of the authors (*TMY*) first began dealing with the unique aspects of high temperature and high energy gases as a result of the work on reentry theory and experiments with shock tunnels in the 1960s. This was not an academic endeavor per se, but the study grew out of the need to build physical devices that had to meet real needs. Following periods of research were motivated by problems of space propulsion, magnetic fusion, laser fusion, and space physics. The second author (*HBT*) has been similarly motivated in the need to understand the physical interactions in real devices. As with the study of all fluids, behaviors of plasmas are complex, and without simple observational models, understanding comes with the combination of precise experimental evidence and appropriate theoretical and computational models. Experience has taught that the conception of principles and their application in directed research efforts based on anecdotal results from either experiment or theory have proven to be ineffective. Therefore, this material is presented in the context that there is a need for a framework of knowledge that can guide the student and researcher in the examination and exploration of the intricate and exquisite behaviors that occur in gases which are influenced by high temperatures and electric and magnetic fields.

[1]Alfven, H., 1950. Cosmical Electrodynamics. International Monographs on Physics. Clarendon, Oxford.

[2]Proceedings of the 2nd United Nations Conference on Peaceful Uses of Atomic Energy, Geneva, 1958.

[3]Flugge, S. (Ed.), 1956. Handbuch der Physik, Gas Discharges I: vol. 21; and II, vol. 22. Springer, Berlin.

It is clearly intended that this serves as introductory text for those approaching the study of ionized gases. It is not intended as a text in plasma physics or as a reference for gas discharge applications; there are a number of excellent works on those subjects, and they are given as references. It is intended to provide an introduction based on physical concepts and straightforward mathematical treatment so that the reader will gain a comprehensive exposure to the basis, techniques, and problems encountered in plasma studies and applications. Physical understanding is paramount; the work always points to further study and research on any subject of interest. For the students, there are a number of new areas of physics that need a basic foundation for the engineering applications. This work presumes an undergraduate degree involving fluid and thermal engineering or in physics, and the text attempts to extend this into the introductory domains of atomic physics, electricity and magnetism, and quantum mechanics. This background is necessary in order that the ultimate effort of applications of plasma principles does not remain in the framework of simple substitution in available equations. The coverage of kinetic theory is extended into regimes of transfer and transport of internal particle energies. Electricity and magnetism coverage emphasizes not only Maxwell's equations, but the application and effects of those equations to physical experiments and devices that utilize plasmas. The equations of fluid mechanics are extended to include electromagnetic energy and momentum components, but a serious attempt is made to develop understanding of the complex fluid mechanical behaviors that result when interactions include plasma physics and transport processes. This complex behavior is made more intractable by the occurrence of both collisional and collisionless behavior in plasma devices. The authors believe that sound preparation for work with plasmas involves detailed consideration of specific plasma devices and phenomena. The applications and examples are taken from plasma accelerators/thrusters, compression/heating devices including magnetic fusion, and space physics descriptions of magnetospheres/ionospheres. The solution of numerous problems in the future involving energy, electronics, communications, and transportation fields will involve understanding plasmas and plasma dynamics. We hope this work will assist those who will face these challenges.

ACKNOWLEDGMENTS

A work of this type and extent has drawn its integrity from a number of contributors in a number of ways. For both authors, each of us owes a debt to some exceptional teachers who opened our vision to understanding thoughts, concepts, and goals that became a driving force. We have gained immeasurably from coinvestigators on research projects, colleagues in research laboratories and in universities. We have gained insights from the unique relationships with our students in the process of defining and executing their research accomplishments. These individuals are too numerous to name and recognize here.

For the first author (TMY), it is appropriate to recognize the contribution of his academic affiliation with Professor Bob Jahn at Princeton University; in the formative period of his PhD work, he was encouraged to pursue a broader academic inquiry into the scientific foundations of his research activity.

For the second author (HBT), it is real pleasure to acknowledge the idea and understanding of plasma and plasma propulsion from Professor Yu Liu at Beihang University who not only gave encouragement but also shared his keen insight into the best way to present difficult concepts at the beginning of the author's research.

In the preparation of the specific document, the authors are appreciative of the help at BUAA of Dr. Chaojin Qin, who transformed written text equations into precise document form. An extensive contribution was made by Mengdi Kong (MS candidate), who prepared numerous drawings and confirmed the details of the mathematical developments that are presented.

Finally, in a work of this extent, the reader will find the inevitable error; for this the authors assume complete responsibility.

CHAPTER 1

The Plasma Medium and Plasma Devices

INTRODUCTION

The world in which we function is consistent with our physical characteristics defined by mass, volume, and energy. Our natural environment is benign—a gaseous atmosphere of nitrogen and oxygen at pressures of 10^5 N/m^2, temperatures of 0–40 °C, and particle densities of 10^{25} m^{-3}. We are continuously receiving radiant energy from the Sun at a rate of about 300 W/m^2, in a 24-h cyclical pattern due to the Earth's rotation, which is modified by the annual cycle of the Earth's orbital motion around the Sun.

In the course of history, we have observed in our local environment exceptional natural displays of energy that demonstrate the existence of forces and energies well beyond our control. The Sun itself is clearly of a very high temperature and is capable of transient, powerful eruptions. Storms in the atmosphere display enormous wind power; electrical lightning strikes generating shock waves and creating local temperatures that can ignite combustion. Polar latitudes evidence dynamic geophysical scale displays of light that inspire awe and require understanding. All these natural events demonstrate and testify to the high-energy excitation of our gaseous atmosphere in response to geophysical electric and magnetic field-based mechanisms. In fact, in the total physical world, with the exception of the near-Earth environment, the medium we exist in is composed of high-energy particles with electric charges, and they are in incessant motion, sometimes directed and sometimes random. In short, the physical universe is largely composed of plasma.

This work is an introduction to the properties and behavior of that electrically active medium and of some of the devices that have been developed to utilize the characteristics of energy and force transfer with the plasma. Plasma is a medium that includes species of charged particles, and plasma dynamics is the description and analysis of force generation and energy transfer with that medium. The important characteristic of gaseous plasmas is their physical makeup, which allows reaction to electric and

Introduction to Plasmas and Plasma Dynamics
ISBN 978-0-12-801661-9

magnetic fields, particularly and including the conduction of current. There is a conceptual similarity of plasmas with solid electrical conductors whereby flowing electrons and electromagnetic waves move through static ions in response to electric and magnetic fields. The charged plasma particles develop organized (collective) behavior due to interaction with large numbers of nearby charged particles. Due to the energy equilibrium but mass differences of plasma component species, there is the occurrence of local electric field generation, which is the beginning of a complex interplay of particle motion and electric and magnetic fields. These behaviors are the ingredients that allow unique device performance using plasmas.

With our relatively recent discovery (and still developing knowledge) of atomic structure, electrical charges and currents, electric and magnetic fields, and electromagnetic radiation, we have begun the process of defining and controlling particle behavior to develop new devices to serve our needs. Particularly in the last 50 years, we have seen the application of such knowledge to create devices with enhanced capability in light and power generation, communications, scientific diagnostics in the physical and biological sciences, and space exploration (National Research Council, 1995). This work introduces the student and researcher to the basic mechanics of the particle interactions inherent in devices that utilize charged particles and presents the framework for understanding their further application in new devices.

PLASMAS IN NATURE

General Description

A general representation of plasmas that are observed in nature is presented in Figure 1.1.

The plasma regions are identified by their properties of particle density and particle temperature.

The Solar Plasma

It can be identified that gases in the solar system occur over the range of 10^{33} p/m^3 and 10^7 K in the solar core to 10^9 p/m^3 and 10^5 K in the Earth's aurora (Kivelson and Russell, 1993). Both these extremes in properties represent plasmas that have important physical characteristics and if produced in the laboratory can be utilized in practical devices. It can be seen that lightning, which occurs at atmospheric pressure conditions, is typified by temperatures of 10,000 K or more.

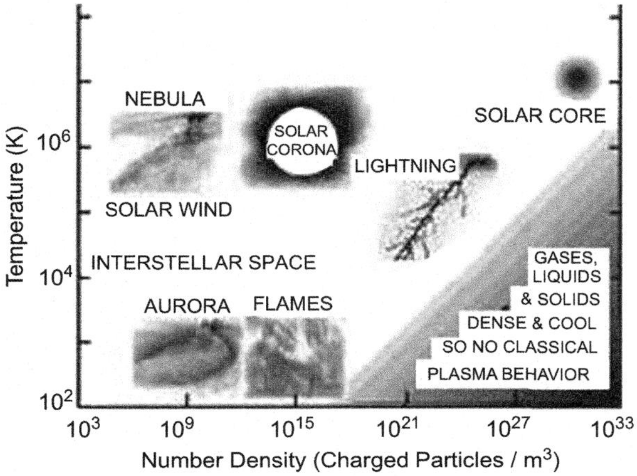

Figure 1.1 Property domains of plasmas occurring in space and natural environment. *Adapted from web site: http://www.cpepphysics.org/fusion_chart_view.html, Contemporary Physics Educ. Project (2010), with permission.*

As the solar plasma and its energies are so significant in our environment, it is useful to identify as a reference the orders of magnitude of a set of specific properties and parameters relative to the Earth. The plasma in the interplanetary system originates from the Sun. The Sun has a mass of 2×10^{30} kg, diameter of 1.4×10^6 km, and a composition of 75% hydrogen and 25% helium. The thermonuclear fusion of hydrogen to helium produces a core temperature of 1.6×10^7 K and a corona temperature of 5×10^6 K. This plasma of the Sun escapes in all directions and expands into all regions of the solar system. At the Earth radius from the Sun the particle proton and electron densities are about 10 cm^{-3}, with proton temperature of 4×10^4 K and electron temperature of 1.5×10^5 K, and most importantly a solar wind flow speed of about 400 m/s. The interaction of this flowing plasma with the Earth's magnetic field produces the hypersonic flow field of the asymmetric magnetosphere (Bothmer, 1999), as shown in Figure 1.2.

PLASMAS IN LABORATORY/DEVICE APPLICATIONS

General Description

Because of the potential for application in new revolutionary devices that can extend our capabilities in a number of technologies (Charles, 2009), the behavior of ionized gas plasmas has been explored over a broad range of densities and temperatures, steady state and transient conditions, small and

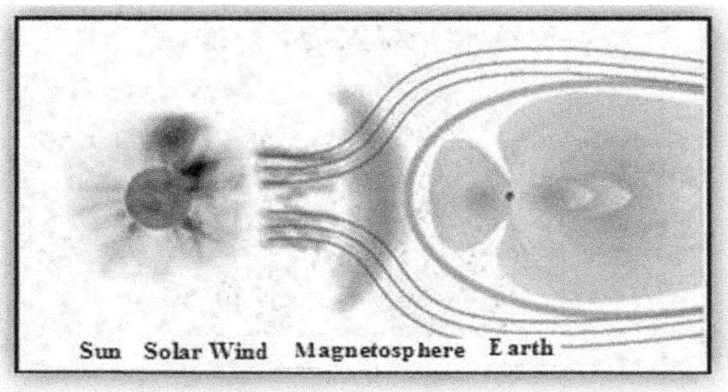

Figure 1.2 Schematic of the solar plasma and the Earth's magnetosphere structure. *Adapted from European Space Agency, ESA (2006) with permission.*

large size scales, power levels and sources, and geometries. Laboratory devices have been constructed for basic scientific research studies (McCracken and Stott, 2005) and as test beds for product development (Cappitelli and Gorse, 1992). As with any new technology, the identification of operating principle is basic and the definition of scalability of the principle is critical to expand the operating range. A schematic display of some of the general types of plasma devices that have been developed are presented in Figure 1.3. General indications of plasma length scales are

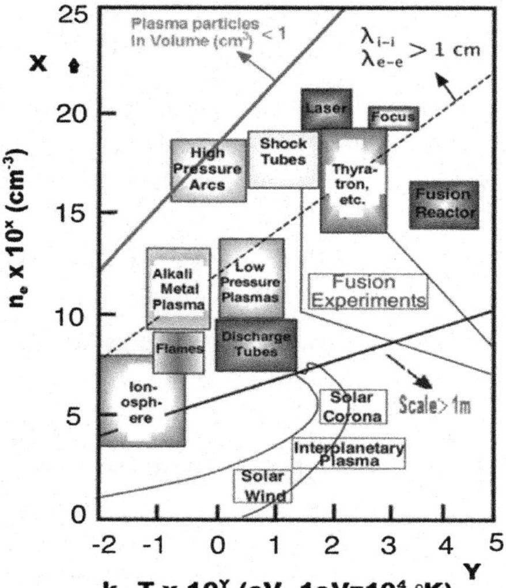

Figure 1.3 Schematic of plasma density and temperature in various types of plasma devices. *Adapted from Sheffield (1975).* Plasma Scattering of Electromagnetic Radiation. *Academic, New York.*

shown with respect to plasma charge separation (upper left), particle mean free path (λ), and geophysical size (lower right).

Categories of Device Plasmas

There are a number of ways to classify the different types of devices that generate and utilize the unique characteristics of plasmas. Historically, devices for generating light were most basic, and fluorescent discharge tubes have been in use for over 100 years. Gas discharge vacuum tubes (Cobine, 1957) for voltage and signal modification in communication devices enabled advances that changed society. However, perhaps the most effective criteria for classifying devices are that shown in Figure 1.3: the density and temperature ranges of the plasma are as follows:

1. Relatively low-temperature, higher pressure plasmas
   a. Flames
   b. Gas dynamic plasmas (incl. reentry)
   c. Shock tube plasmas
   d. Laser-target plasmas
   e. Electric arc plasmas
2. Lower pressure, higher temperature plasmas
   a. Discharge tubes
      - Fluorescent lights
      - Plasma screen displays
      - Laser source plasmas
      - High-power switching devices
   b. Space propulsion thrusters
      - Ion, Hall, MPD thrusters
3. High-density, high-temperature plasmas
   a. Magnetic fusion power experiments
   b. Laser target implosion experiments

REFERENCES

Bothmer, V., 1999. Solar corona, solar wind, structure, and solar particle events. In: Proceedings of ESA Workshop on Space Weather-1998, ESTEC. Noordwijk, The Netherlands, pp. 117–126.

Cappitelli, M., Gorse, C., 1992. Plasma Technology: Fundamentals and Applications. Plenum, New York.

Charles, C., 2009. Plasmas for spacecraft propulsion. J. Phys. D. Appl. Phys. 42, 163001.

Cobine, J.D., 1957. Gaseous Conductors: Theory and Engineering Applications. Dover, New York.

Contemporary Physics Education Project, CPEP, 2010. Characteristics of Typical Plasmas. Retrieved from: http://www.cpepphysics.org/fusion_chart_view.html (accessed 15.06.15).

European Space Agency, ESA, 2006. Solar Wind Buffets Earth's Magnetic Field. Retrieved from: http://www.esa.int/spaceinimages/Images/2003/05/Solar_wind_buffets_ Earth's_magnetic_field (accessed 15.06.14.).

Kivelson, M., Russell, C., 1993. Introduction to Space Physics. Cambridge Univ. Press, Cambridge.

McCracken, G.M., Stott, P.E., 2005. Fusion: The Energy of the Universe. Elsevier, London.

National Research Council, 1995. Plasma Science: From Fundamental Research to Technological Applications. National Academies Press, Washington, DC.

Sheffield, J., 1975. Plasma Scattering of Electromagnetic Radiation. Academic, New York.

CHAPTER 2

Kinetic Theory of Gases

INTRODUCTION

In the study of the mechanics and energetics of fluid flow, normally the fluid is considered to be a continuous medium (continuum), describable by properties such as density, temperature, pressure, and viscosity. For example, energy is defined as $C_V T_0$. Since the basic problem is that of the interchange of a large amount of energy in and out and fluid systems, we must look at what a fluid is "in the small" (microscopically) as well as "in the large" (macroscopically) so that we can understand what energy "is" (what its forms are), and how it can change when added to or removed from a fluid. The energy exchange is central, and the effects of the energy exchange are secondary.

Kinetic theory originated in an attempt to explain and correlate the familiar physical properties of gases on the basis of molecule behavior (perfect gas law as stated for imperfect gases, viscosity, conduction, and diffusion).

BASIC HYPOTHESES OF KINETIC THEORY

Basic Hypotheses (Present, 1958)

1. Molecule hypothesis—"matter is composed of small discrete units known as molecules: that the molecule is the smallest quantity of substance that retains its chemical properties, that all molecules of a given substance are alike, and there are three states of matter which differ in the arrangement and state of motion of molecules."
2. "The interaction of gas molecules in collisions with each other and the walls of the container obey the laws of classical mechanics (conservation of momentum and energy)"; the collisions are elastic.
3. Gas properties are described by statistical methods. A large number of molecules imply that average behavior can be determined by statistics; dynamical method implies that initial condition (such as position and speed) and forces acting determine behavior. Statistical method implies that behavior is independent of initial conditions, and we seek proper averages, that is, the average over all molecules at one instant.

Introduction to Plasmas and Plasma Dynamics
ISBN 978-0-12-801661-9

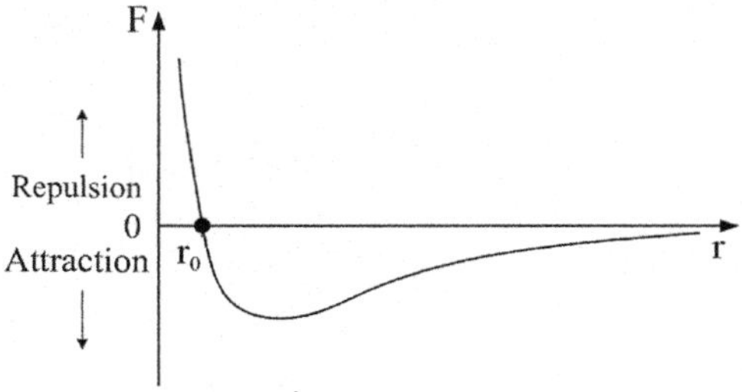

Figure 2.1 Forces of interaction between molecules as a function of separation distance.

Secondary Hypotheses

Molecules are always in motion—incessant, translating motion. Molecules possess only kinetic energy (neglect internal modes for now). The size of molecules is small compared with the separation of particles, and particles interact only on colliding (ideal gas law). We have described a *billiard ball model* of molecules. The real force interaction of molecules is shown in Figure 2.1.

PRESSURE, TEMPERATURE, AND INTERNAL ENERGY CONCEPTS

Consider the behavior of a group of gas molecules inside a fixed control volume (Figure 2.2).

Size of volume:

$$V = xyz;$$

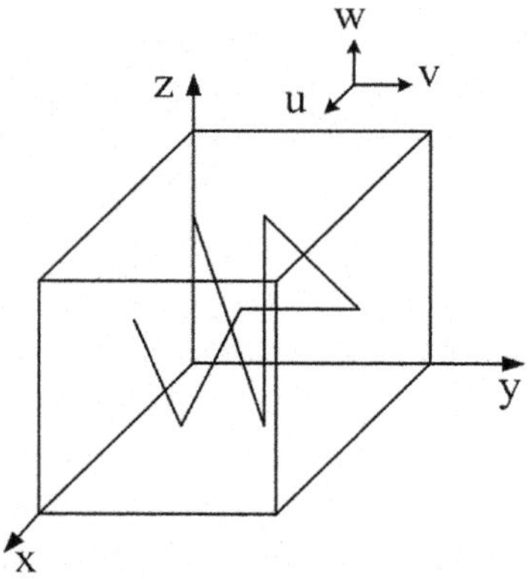

Figure 2.2 Control volume for molecule motion.

Total number of molecules in V:

$$N = n\left(\frac{\text{molecules}}{\text{vol}}\right)V;$$

Mass of single molecule: m;
Speed of molecules: c, with:

$$c^2 = u^2 + v^2 + w^2, \text{ where } u = \text{speed in } x \text{ direction, etc.;}$$

Kinetic energy of molecule:

$$\frac{1}{2}mc^2.$$

Now, assume equilibrium (no directed motion), random motion. Then, $\overline{u} = \overline{v} = \overline{w}$, or average values of speed are equal, and since $\overline{u^2} \approx \overline{u}^2$ in equilibrium, we get: $\overline{c^2} = \overline{u^2} + \overline{v^2} + \overline{w^2} = 3\overline{u^2}$. Effectively, 1/3 of all molecules move in the x direction, 1/3 in the y, and 1/3 in the z direction. Consider particle motion along the x axis (→) (Figure 2.3), where the y–z plane is at $x = 0$, and a second plane is at $x = x$ (which is the average distance between molecular collisions).

Average number of collisions/time that a single molecule will hit the y–z plane, is:

$$\frac{\text{Collisions}}{\text{time}} = \frac{\text{velocity}}{\text{dist./coll.}} = \frac{u}{2x}.$$

Also, the momentum change on particle collision with the y–z wall (mom. in: mu←, out: mu→), so:

$$\frac{\Delta\text{Mom}}{\text{Collision}} = 2mu.$$

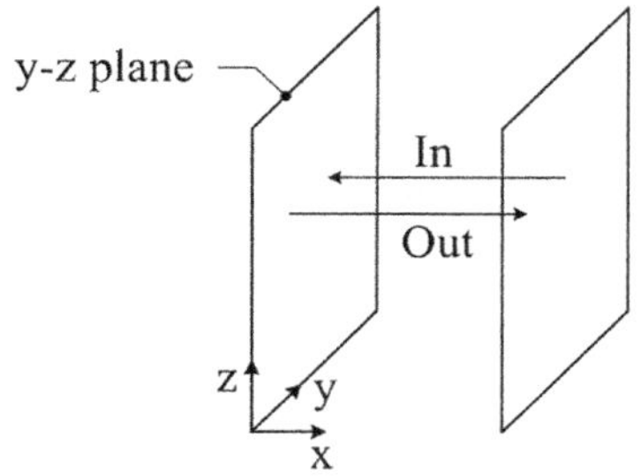

Figure 2.3 Particle motion along the *x* axis.

Now, force on y–z wall (per molecule) = Rate of change momentum (per molecule), as:

$$\left(\frac{\text{Collisions}}{\text{time}}\right)\left(\frac{\Delta\text{Mom}}{\text{Collision}}\right) = \frac{u}{2x}\cdot 2mu = \frac{mu^2}{x},$$

and the force on the y–z wall from *all* molecules in the volume is: $\frac{Nmu^2}{x}$, ($-x$ direction).

The pressure on the y–z wall from all molecules is force/area, or:

$$p_{yz} = \frac{Nmu^2}{x}\cdot\frac{1}{yz} = \frac{Nm\overline{u^2}}{V} = \frac{1}{3}\cdot\frac{Nm\overline{c^2}}{V}, \qquad \text{and} \qquad p_{yz} = \frac{1}{3}nm\overline{c^2},$$

where n is the number density $\left(\frac{\text{molecules}}{\text{vol}}\right)$. In fact, this is independent of direction. This statement mathematically defines the macroscopic property, pressure, using kinetic theory concepts. In the kinetic particle description of this macroscopic property, pressure is related to momentum transfer due to particle collisions.

Now, recall from experiment: $pv = RT$ (from macroscopic thermodynamics), or:

$$p\frac{V}{M_t} = RT \text{ where } M_t = m\left(\frac{\text{mass}}{\text{particle}}\right)N(\text{total number of particles}), \text{ and:}$$

$$p\frac{V}{mN} = p\frac{1}{mn} = RT \quad \text{or} \quad p = mnRT.$$

We now have a further microscopic definition of pressure. With p, the pressure defined, let us consider other properties.

The kinetic energy per particle (this is random kinetic energy only) is:

$$E = \frac{1}{2}m\overline{c^2} \quad \text{but} \quad \frac{1}{2}m\overline{c^2} = \frac{3p}{2n},$$

and with

$$\frac{p}{n} = mRT \quad \left\{\textit{Note that: } E\left(\frac{\text{energy}}{\text{particle}}\right)n\left(\frac{\text{particles}}{\text{vol}}\right) = \frac{3p}{2}\right.$$

$$\left.\therefore\ E\cdot n\left(\frac{\text{energy}}{\text{vol}}\right) = p\left(\frac{3}{2}\right)\right\}, \quad E = \frac{1}{2}m\overline{c^2} = \frac{3}{2}mRT.$$

But,

$$mR = m\left(\frac{\text{mass}}{\text{particle}}\right)\cdot\frac{\overline{R}(\text{universal gas const.})}{M(\text{mass/mole})} = \frac{\overline{R}}{N_A(\text{Avogadro's number})}$$
$$= k\ (\text{Boltzmann const.})$$

then,

$$E\left(\frac{\text{Energy}}{\text{particle}}\right) = \frac{1}{2}m\overline{c^2} = \frac{3}{2}kT,$$

which defines temperature in terms of microscopic properties.

Therefore, temperature is a property indicative of random, translational energy. Also, since $\frac{\text{Energy}}{\text{particle}} = \frac{3}{2}\frac{p}{n}$, then $p = nkT$ is the kinetic form of equation of state.

Let us now consider some orders of magnitude of the kinetic properties that have been defined; we know that pressure and temperature are related to molecular motion. Let us look at the speed involved in this motion. Take air at room temperature and substituting values, we get:

$$\bar{c}(\text{average speed}) \approx (\overline{c^2})^{\frac{1}{2}} \approx \left(\frac{3kT}{m}\right)^{\frac{1}{2}}$$
$$= \left\{\frac{3\left(1.38\times 10^{-23}\frac{J}{K}\right)(300\ \text{K})}{(28)\left(\frac{1}{6\times 10^{26}}\right)}\right\}^{\frac{1}{2}} \approx 500\ \text{m/s},$$

where, $\frac{\text{mass}}{\text{molecule}}\cdot\frac{\text{molecules}}{\text{mole}}$ Molecular weight (air) $\approx$ 28 kg mole. Note that this speed is on the order of speed of sound (pressure perturbations) in the medium.

We now examine the energy relationships. We have seen that the average translational energy per particle is $E_{tr} = \frac{3}{2}kT = \frac{3}{2}mRT$.

So for *all* the particles:

$$e_{tr}\left(\frac{\text{energy}}{\text{mass}}\right) = \frac{\text{energy}}{\text{molecule}}\cdot\frac{\text{molecule}}{\text{mass}} = \frac{3}{2}mRT\cdot\frac{1}{m} = \frac{3}{2}RT.$$

Now, evaluating this energy using the definition of a specific heat at constant volume:

$$C_V \equiv \left(\frac{\partial e}{\partial T}\right)_V = \frac{3}{2}R(tr), \text{ and since } C_p - C_V = R, \text{ therefore,}$$
$$C_p \equiv \frac{5}{2}R(tr).$$

This result is in good agreement for monatomic gases at moderate temperature.

As noted above, random kinetic energy is related to pressure. Now, we can observe that for pressure: $p = \frac{1}{3}nm\overline{c^2} = \frac{2}{3}nE_{tr} = \frac{2}{3}n\left(\frac{\text{particles}}{\text{volume}}\right) \times E_{tr}\left(\frac{\text{Energy}}{\text{particle}}\right) = \frac{2}{3}nE_{tr}$, which has units of $\left(\frac{\text{Energy}}{\text{volume}}\right)$. So, just as we have defined mass per unit volume as mass density, now we have recognized pressure as an *energy density*, the *random thermal energy density*.

We have been considering molecules of single species. Now we can look further at some aspects involving gas mixtures. Since we are dealing with plasmas, it is interesting to look at plasmas as gas mixtures composed of electrons and ions, and note that these particles have largely different masses.

Gas Mixtures

The effect of combining different gases in a given volume at a uniform temperature, T, can be evaluated by considering energy, an extensive property, as:

$$\frac{\text{Energy}}{\text{volume}} = n\left(\frac{\text{molecules}}{\text{volume}}\right) \cdot E_{tr}\left(\frac{\text{Energy}}{\text{molecule}}\right).$$

Adding gases together in the same volume gives:

$$\left(\frac{E}{V}\right)_{tot} = \sum nE_{tr,i} = n_1E_1 + n_2E_2 + \cdots.$$

But E/Volume is related to p, and so $p_{mix} = p_1 + p_2 + \cdots = \sum p$; this is Dalton's law of partial pressures. Gases with p_1 in V will contribute to the total pressure of the mixture: when gases are mixed, $p = \sum p_i$ (partial pressures).

Further, since we know that pressure is equal to nkT, we can see:

$$nkT \text{ (for a mixture)} = n_1kT_1 + n_2kT_2 + \cdots,$$

and if $T = T_1 = T_2 = \cdots$, i.e., we have temperature equilibrium, then $n = \sum n_i$.

As $E_{tr1} = E_{tr2}$, for in equilibrium they must have equal energy, we can see that, $kT_1 = kT_2$ and $m_1\overline{u_1^2} = m_2\overline{u_2^2}$. This must be true in general, so if we have particles where $m_1 \ll m_2$, for example, the electron's mass is much less than that of the ion, that is, $\overline{u_1^2} \gg \overline{u_2^2}$. The mass of a proton is $m_p = 1.38 \times 10^{-24}$ g, while the mass of an electron is $m_e = 0.91 \times 10^{-27}$ g, so this results in a very large difference between the speeds of electrons and ions in a plasma.

What we have accomplished is to express basic continuum properties as a function of molecular properties: m, n, $\overline{c^2}$; what we must now consider in detail is the particle velocity, what affects it, and how it is determined. The primary events affecting particle velocities are collisions!

Let us now develop the phenomenology, terminology, and behavior of gases related to collisions.

KINETIC THEORY AND TRANSPORT PROCESSES

Particle Collisions

In the above developments, we considered particle collisions with boundaries, and this resulted in our definition of pressure. Clearly, in the random motion of a large number of particles, most collisions that occur are between particles.

We treat molecules as smooth, rigid, elastic spheres of diameter, d. For like molecules, when collision takes place, centers of molecules are a distance, d, apart. In Figure 2.4, two molecules of diameter d are shown in collision, and the distance between centers on collision is d.

The sphere of radius d shown in the figure is called the sphere of influence—no other molecule can have its center within the sphere.

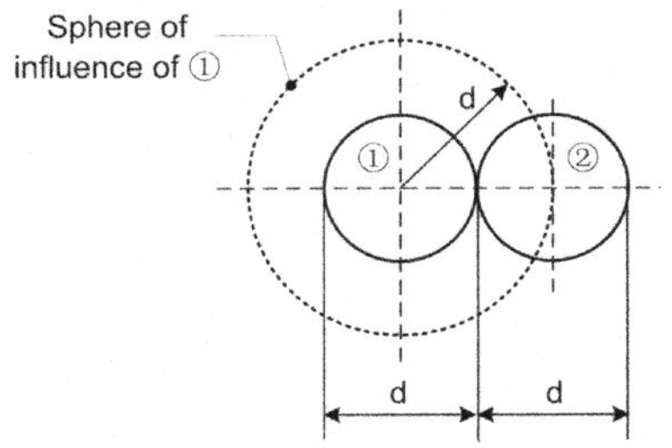

Figure 2.4 Two molecules of diameter d in collision.

For a simple model, assume that all molecules in a volume are at rest except one molecule which moves with speed, c. Assume that static molecules are held fixed so that the moving particle velocity, c, is maintained. Referring to Figure 2.5, which represents a unit time (on the average), there are two circles, left and right, representing spheres of influence of moving (L) and static (R) particles. The large circle on the left is the end of a cylinder of length $l = c \cdot 1$ (this is in a unit time), i.e., length = speed·time, so we have a cylindrical volume of end area πd^2 and length $c \cdot 1$.

We define this area πd^2 as the cross-section, with symbol, σ; therefore, any molecule with a center in the volume would have a collision with the molecule that is moving. The number of molecules in the volume swept in a unit time is:

$$n\left(\frac{\text{molecules}}{\text{volume}}\right) \cdot \pi d^2 c = \frac{\text{collisions}}{\text{time}}.$$

In actuality, the cylinder for a real molecule can have kinks and bends, but we neglect such effects.

If we consider that all particles are moving and that c_r represents average relative velocity, an exact solution gives $\frac{\text{collisions}}{\text{time}} = n \cdot \pi d^2 c_r$. We will consider this fact again, as it is one of the important points in kinetic theory: what is the difference between a solution for a model with an average particle moving at speed c where the other particles are static, and the solution where a particle moving with speed c has to be taken with respect to other particles also moving at speed c? Therefore, we write:

ν_c or $\Theta\left(\frac{\text{collisions}}{\text{time}}\right) = n \cdot \pi d^2 \bar{c} \equiv n \cdot \sigma \bar{c}$, with the definition that, σ (cross-section) $= \pi d^2$.

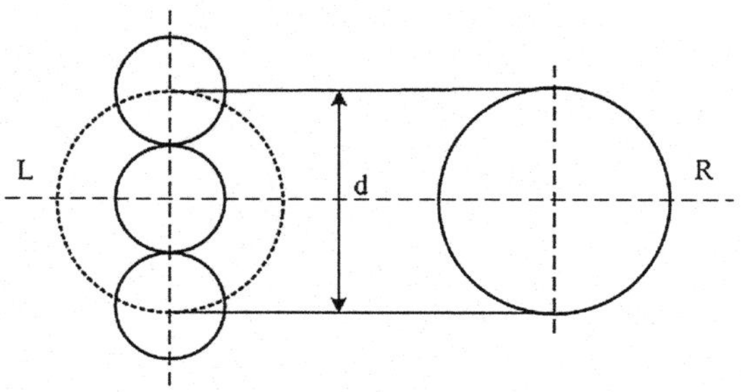

Figure 2.5 Collision model.

Above, we considered, $\bar{c}$, average speed $\left(\frac{\text{distance}}{\text{time}}\right)$ and Θ, average $\left(\frac{\text{collisions}}{\text{time}}\right)$; therefore, we can now define, $\left(\frac{\text{distance}}{\text{collision}}\right) \equiv \lambda = \frac{\bar{c}}{\Theta}$. Here λ is the mean free path, the average distance between collisions:

$$\lambda = \frac{\bar{c}}{\Theta} = \frac{\bar{c}}{n \cdot \pi d^2 \bar{c}} = \frac{1}{n \cdot \pi d^2} = \frac{kT}{p(\pi d^2)}, \quad \begin{array}{l} T\uparrow \Rightarrow \lambda \text{ increases} \\ p\downarrow \Rightarrow \lambda \text{ increases} \end{array}.$$

If a gas is expanding, then $p\downarrow, T\downarrow$ downstream and there are two competing effects which influence λ. Now, with our basic understanding of gas mixtures and collisions, we can utilize these concepts to consider the transport of molecular gas properties by random thermal motion.

Transport Phenomena (Viscosity, Conduction, and Diffusion)

By observation:

Fluid with pressure difference $(\Delta p \Rightarrow \overline{\Delta \text{v}})$ has friction (viscosity)—momentum loss.

Fluid with temperature difference $(q \sim \Delta T)$ conducts heat—energy loss.

Fluid with density differences $(n \sim \Delta \rho)$ has diffusion—mass loss.

In each case noted above, a difference in a macroscopic property (Δp, ΔT, $\Delta \rho$) will result in the molecular transport of properties from one region to another. The molecules will carry information from one region to another; this is a nonequilibrium process. In all three cases, there is a gradient of physical property and related molecular transport, and that transport is proportional to the gradient and in the opposite sense.

The normal macroscopic way of expressing these three different transport behaviors is as follows:

For viscosity: $\tau\left(\frac{\Delta\text{momentum}}{\text{area}-\Delta t}\right) = -\mu \frac{du}{dy}$. This is Newton's law of viscosity.

For heat transport (conduction): $\left(\frac{\Delta\text{energy}}{\text{area}-\Delta t}\right) = -K\frac{dT}{dy}$. This is Fourier's law of heat transfer.

For (self-) diffusion: $\left(\frac{\Delta\text{number}}{\text{area}-\Delta t}\right) = -D\frac{dn}{dz}$. This is Fick's law of diffusion.

Let us consider an analysis of transport based on mean free path methods. Let $A(z)$ denote a molecular property which varies in the z direction. Consider the net transport of A across an imaginary plane located at $z = z_0$ shown in Figure 2.6 with axes x to the right, y forward, z up. Now, with the gradient in the z direction, let us consider transport between two planes of area S at $z_0 + \Delta z$ and $z_0 - \Delta z$.

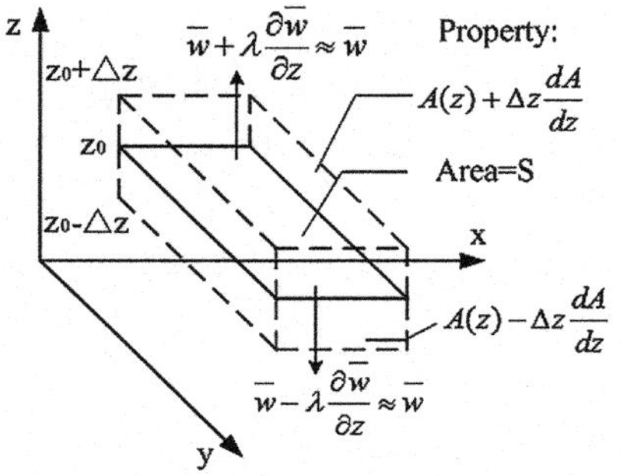

Figure 2.6 Transport between two regions of a fluid.

The particles that reach S from above or below carry properties that derive from their last collisions in those regions, i.e., $\Delta z = \lambda$ (the mfp). Now the number of particles crossing S (per unit time) is:

$$\frac{\text{particles}}{\text{time}} = n\left(\frac{\text{particles}}{\text{vol}}\right) \cdot S\ (\text{area}) \cdot \overline{w}\ (\text{avg. speed perp. to } S).$$

So for any property A, the flux of A across the area S is: $\frac{\text{A}}{\text{unit time}} = \frac{\text{A}}{\text{particle}} \cdot nS\overline{w}$.

A diagram (Figure 2.7) shows the z and x axes with $-\overline{w}$ carrying property from above, flux down is $nS(-\overline{w})\left[A + \lambda \frac{dA}{dz}\right]$, negative in our reference frame; velocity from below carries flux up $+\overline{w}$, $nS(\overline{w})\left[A - \lambda \frac{dA}{dz}\right]$.

So the net transport of A across area S per unit time is expressed as:

$$\Gamma\left(\frac{A}{\text{time}}\right)_{\text{net}} = -nS\overline{w}\left[A + \lambda\frac{dA}{dz}\right] + nS\overline{w}\left[A - \lambda\frac{dA}{dz}\right] = -2nS\overline{w}\lambda\frac{dA}{dz}.$$

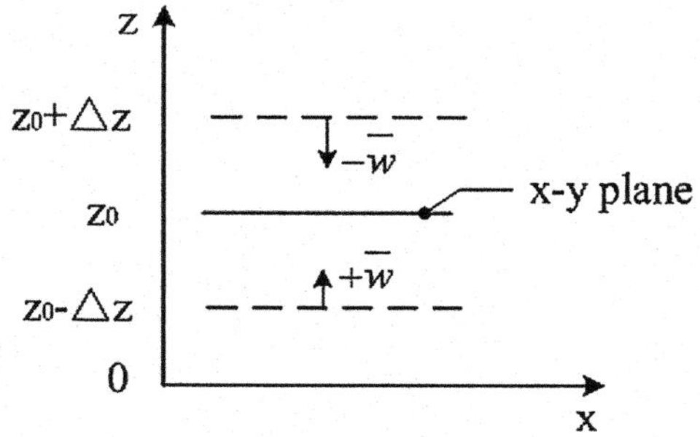

Figure 2.7 Transport of property flux.

Flux of A per unit time and area is $\frac{\Gamma}{s}\left(\frac{A}{\text{time}\cdot\text{area}}\right) = -2n\overline{w}\lambda\frac{dA}{dz}$ (if A incr. with z, net flux is negative). We know that $\lambda = \frac{1}{n\pi d^2}$, so $\frac{\Gamma}{s}\left(\frac{A}{\text{time}\cdot\text{area}}\right) = -\frac{2\overline{w}}{\pi d^2}\frac{dA}{dz}$.

But how is $\overline{w}$, for this very specific case, related to $\bar{c}$? We have to try to understand how to calculate a value for $\overline{w}$. This is at the heart of kinetic theory and an exact calculation is complex. It requires the definition and application of a velocity distribution function. We will do this analysis below. We will now simply state (we will prove this later) that $\overline{w} = \bar{c}/4$; this is one of the most important results of kinetic theory. It cannot come from intuition (recall intuition implies that $\overline{c^2} = 3\overline{w^2}$), it must be calculated from mathematics. We now simply state that $\overline{w} = \bar{c}/4$, then the net transport of A per unit area per unit time is:

$$\frac{\Gamma}{s}\left(\frac{A}{\text{time}\cdot\text{area}}\right) = -\frac{1}{2}n\bar{c}\lambda\frac{dA}{dz}.$$

For the different properties which can be given for A, we now look at the predictions from kinetic theory.

Viscosity (Momentum Transport)

$A = mu$, where u is directed flow velocity along the x axis. The transport is:

$$\frac{\text{Momentum}}{\text{time}\cdot\text{area}} = -\frac{1}{2}n\bar{c}\lambda\frac{d(mu)}{dz} = -\frac{1}{2}mn\bar{c}\lambda\frac{du}{dz},$$

and as Figure 2.8 shows, higher velocity at larger z and lower velocity at lower z mean that there is a net momentum transfer from higher to lower z.

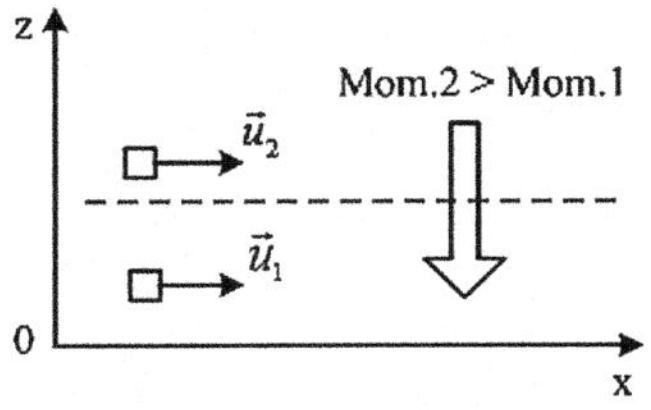

Figure 2.8 Momentum transfer.

But macroscopically, $\frac{\text{Momentum}}{\text{time}\cdot\text{area}} \equiv \tau = -\mu \frac{du}{dz}$, a phenomenological law. These two relationships are equal, and $\mu = \frac{1}{2} mn\bar{c}\lambda$, but $\bar{c} \approx (\overline{c^2})^{\frac{1}{2}} = \left(\frac{3kT}{m}\right)^{\frac{1}{2}}$, so:

$$\mu = \frac{1}{2} mn\lambda \left(\frac{3kT}{m}\right)^{\frac{1}{2}} = n\lambda \left(\frac{3mkT}{4}\right)^{\frac{1}{2}}, \text{and since } \lambda = \frac{1}{n\pi d^2} \text{ get}$$
$$= \frac{1}{\pi d^2} \left(\frac{3mkT}{4}\right)^{\frac{1}{2}}.$$

This predicts that μ, viscosity, is proportional to $T^{1/2}$, and it is not proportional to p (pressure) and ρ (density); the smaller the cross-section, $\sigma = \pi d^2$, the higher the viscosity, μ.

Thermal Conduction (Energy Transport)

$A = \frac{1}{2} mc^2 = \frac{3}{2} kT = \frac{3}{2} mRT = mC_{V_{tr}} T$, as $C_{V_{tr}} = \frac{3}{2} \frac{k}{m} = \frac{3}{2} R$, the specific heat for thermal energy. Then the transport of thermal energy is:

$$\frac{\text{Energy}}{\text{time}\cdot\text{area}} = -\frac{1}{2} n\bar{c}\lambda \frac{d(mC_{V_{tr}} T)}{dz} = -\frac{1}{2} n\bar{c}\lambda m C_{V_{tr}} \frac{dT}{dz},$$

but, macroscopically, $\frac{\text{Energy}}{\text{time}\cdot\text{area}} = -K \frac{dT}{dz}$, where K is the coefficient of thermal conductivity.

Therefore: $K = \frac{1}{2} nm\bar{c}\lambda C_{V_{tr}} = \mu C_{V_{tr}}$.

So the coefficient of conduction is the product of viscosity and specific heat of random translational energy; it is proportional to $T^{1/2}$, it is not proportional to p (pressure).

Diffusion (Mass Transport)

The formalism of diffusion requires care in definition; to clarify, consider that some molecules of the same species are tagged by color. The transport of tagged molecules (n_a) through similar molecules of same species ($n - n_a$) is called self-diffusion. What is transported, then, is the probability of the occurrence of a tagged molecule, that is, n_a/n, the ratio of n_a to total n. Then transport is:

$$\frac{\text{number of molecules}}{\text{time}\cdot\text{area}} = -\frac{1}{2} n\bar{c}\lambda \frac{d\left(\frac{n_a}{n}\right)}{dz}$$
$$= -\frac{1}{2} \bar{c}\lambda \frac{dn_a}{dz}, \text{ with units } \left(\frac{\text{number of particles}}{\text{s}\cdot\text{m}^2}\right),$$

which is by definition the same as G, the macroscopic flux. So with:

$$\left(\frac{\text{number of particles}}{\text{s}\cdot\text{area}}\right) \equiv G = -D\frac{dn}{dz}, \quad \text{then:}$$

$$D_{11}(\text{self} - \text{diffusion coefficient}) = \frac{1}{2}\bar{c}\lambda = \frac{1}{2}\bar{c}\cdot\frac{1}{n\pi d^2} = \frac{1}{2}\cdot\frac{1}{n\pi d^2}\cdot\left(\frac{3kT}{m}\right)^{\frac{1}{2}}.$$

Since $D_{11} = \frac{1}{2}\cdot\frac{(3mkT)^{\frac{1}{2}}}{nm\pi d^2}$, the self-diffusion coefficient is proportional to $T^{1/2}$ and to $\frac{1}{\rho}\sim\frac{1}{\text{density}}$.

Now we have seen how the three transport coefficients (viscosity, conductivity, and diffusion) are related to kinetic properties, but we can also see that they are all related to each other. The coefficients are measurable macroscopic quantities, and because of these kinetic relationships we can use them to determine molecular properties. Consider the relationships between different transport coefficients and the meaning of this. We have:

$$\left.\begin{aligned} \mu &= \frac{1}{2}nm\bar{c}\lambda \\ K &= \frac{1}{2}nm\bar{c}\lambda C_V \\ D &= \frac{1}{2}\bar{c}\lambda \end{aligned}\right\} \Rightarrow \begin{aligned} \mu &= \rho D_{\|} \\ K &= C_V\rho D_{\|} \\ \frac{\mu}{\rho} &= D_{\|} \end{aligned}.$$

So all transport is conceptually similar to diffusion.

First, $\frac{\mu}{\rho}$ is referred to as viscous diffusivity because from this type of billiard ball kinetic theory it is actually equal to the diffusion coefficient. If we look at $\frac{K}{\rho C_V} = D_{11}$, this is referred to as thermal diffusivity, because $\frac{K}{\rho C_V}$ is once again equal to the kinetic relationship for diffusion coefficient. Also, the Prandtl number is defined as: $\frac{\mu C_p}{K}$, and this is equal to $\frac{\rho C_V}{K}.\frac{\gamma\mu}{\rho} = \frac{\mu/\rho}{K/(\rho\cdot C_V)}\cdot\gamma$, with $\gamma = C_p/C_V$, so the Prandtl number expresses: $\left(\frac{\text{viscous diffusivity}}{\text{thermal diffusivity}}\right)\cdot\gamma$.

As we primarily have been considering physical relationships, let us briefly examine the mathematics of diffusion behavior in the partial differential equation of the process. To emphasize process, Figure 2.9 shows a side view of a control volume of height Δz with the area for flux at top and bottom, S; the volume of this cube is $S\cdot\Delta z$. Now macroscopically we

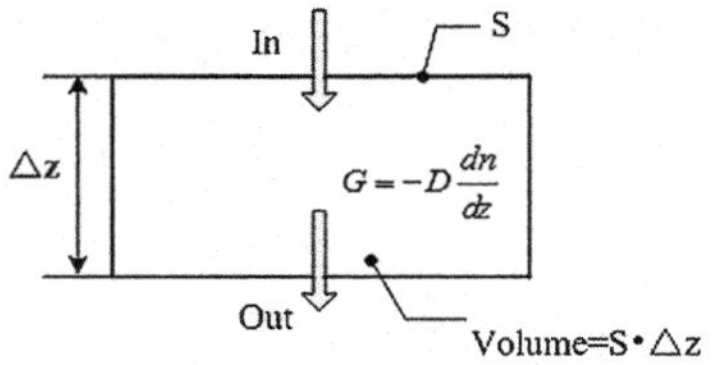

Figure 2.9 Side view of control volume.

write $G = -D\frac{dn}{dz}$. The net flow of particles (particles per unit time) through the volume when the terms in brackets are cleared, is:

$$\frac{\Delta n}{\Delta t} \cdot Vol = [Flux\ out - Flux\ \text{in}] \cdot \text{area}$$
$$= \left\{ \left[-D\frac{dn}{dz} + \frac{\partial}{\partial z}\left(-D\frac{dn}{dz}\right)\Delta z \right] - \left[-D\frac{dn}{dz} \right] \right\} \cdot S.$$

So for a small element we have: $\frac{\partial n}{\partial t} = -D\frac{d^2 n}{dz^2}$, which is the partial differential equation of diffusion, with the diffusion coefficient, D.

MATHEMATICAL FORMULATION OF EQUILIBRIUM KINETIC THEORY

Statistical information about a gas is embodied in the distribution function. In the following, there will first be a discussion of the properties of the distribution function in general, and then there will be a derivation of the distribution function for gases.

Distribution Function and Average Values

We are considering variables with a large number of values, a continuum of values. We will begin with an elementary discussion assuming discrete values of a variable.

The Calculation of Average Values of Property, i, *When that Property has Discrete (Integer) Values*

First, let i = value (amount) of property; n_i = number of particles that have i as a property i (this is referred to a distribution number), then $N = \sum_i n_i$ = the total number of particles.

The total amount of property i, is $\sum_i n_i \cdot i$ with $n(i)$ shown in Figure 2.10.

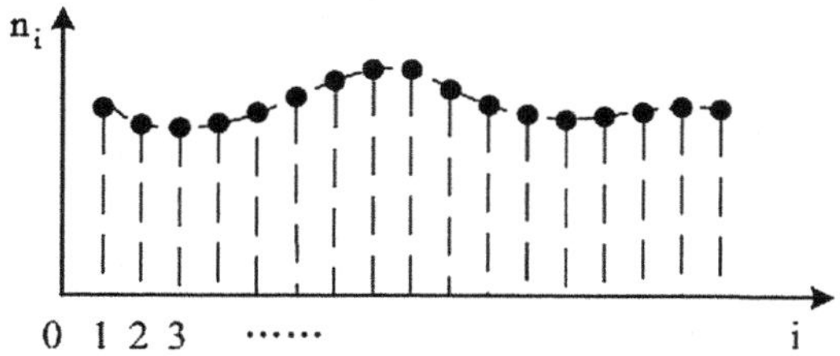

Figure 2.10 Distribution when property has discrete (integer) values.

The average value of property $= \frac{\text{Total amount of property}}{\text{Total number of particles}}$, or mathematically:

$$\bar{i} = \frac{\sum_i n_i \cdot i}{N} = \sum_i \frac{n_i}{N} \cdot i = \sum_i \eta_i \cdot i,$$

where $\eta_i = \frac{n_i}{N}$, the probability that a particle chosen at random will have the property, i.

The Calculation of Average Values of Property, i, When the Property has a Continuum of Values (As Opposed to Values at Discrete Integers)

In the diagram (Figure 2.11) below, the horizontal axis is 0 to arbitrary i, and vertical axis is n_i; a segment from i and $i + \Delta i$ of the horizontal axis is referred to as Δi. Now: $n_i\left(\lim_{\Delta i \to 0}\right) = 0$, so:

$$\frac{n_i}{N}\left(\lim_{\Delta i \to 0}\right) = 0, \quad \text{but} \quad \frac{n_i}{N}(\text{in } di) \neq 0.$$

We define the probability of finding a particle with value between i and $i + di$ as:

$$f(i)di = \frac{n_i \text{ (with value between } i \text{ and } i + di)}{N} = \frac{dn_i}{N},$$

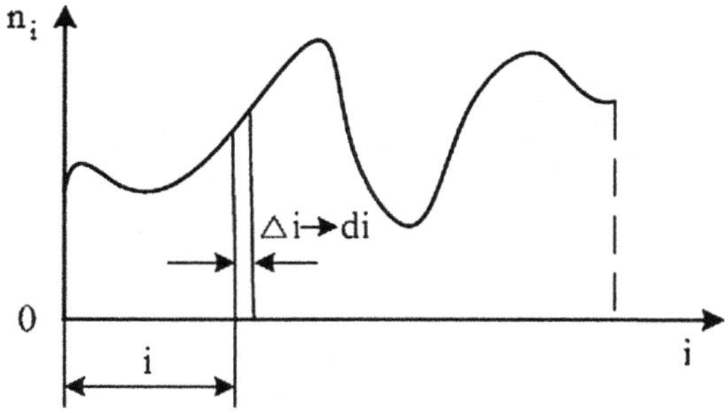

Figure 2.11 Distribution when property has a continuum of values.

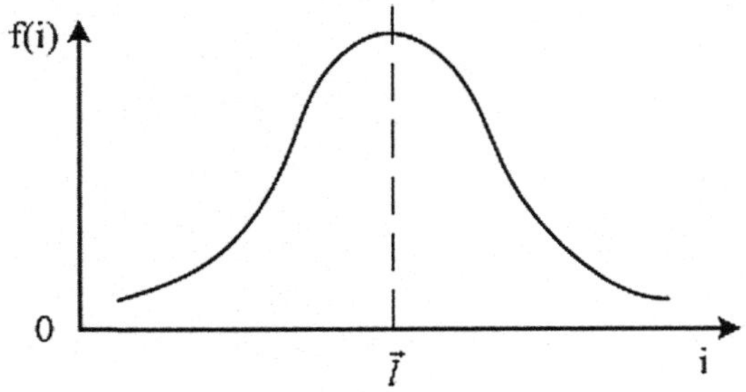

Figure 2.12 Normal distribution function.

where $f(i)$ is called the normalized distribution function. So the total number of particles is:

$$N = \int_{all\ i} dn_i, \quad \text{so} \quad \frac{N}{N} = \frac{1}{N}\int_{all\ i} dn_i = \int_{all\ i} \frac{dn_i}{N} = \int_{all\ i} f(i)di = 1.$$

Therefore, for any variable, Q', which is a function of i, the average value is equal to:

$$\overline{Q'} = \int_{all\ i} Q'(i)f(i)di.$$

As an example, Figure 2.12 shows the familiar normal distribution function, which is symmetrical about the average value, $\bar{i}$. It can be anticipated that the molecular speed distribution function will not be a normal distribution or symmetrical.

Determination of the Speed and Velocity Distribution Functions

Equilibrium Distribution of Molecular Kinetic States

Consider a volume V, with a total number of molecules N possessing a total system energy E. Molecules are identical except for individual energy content; energy level is represented by ε_i. Let:

N_1 *number of molecules have energy* ε_1

N_2 *number of molecules have energy* ε_2 then $E = N_1\varepsilon_1 + N_2\varepsilon_2 + \cdots = \sum_i N_i\varepsilon_i$

$\vdots$ $\vdots$ $N = N_1 + N_2 + \cdots = \sum_i N_i$

N_i *number of molecules have energy* ε_i

We now ask: What is the equilibrium distribution of energy for the molecules in the volume?

It depends on the constraints on the energy of each molecule.

Equilibrium $\underset{is\ equivalent\ to}{\Longleftrightarrow}$ most probable, so we must now consider probability.

Probability Considerations

The probability that a system will exist in a particular (energy) state is proportional to the number of distinguishable ways in which a state can be realized (attained) (Kreysig, 1980). From probability theory:

N molecules can be arranged *N*! different ways.

Example 1: Consider four molecules (A, B, C, D) capable of possessing one of four energies *and* each being different: $\varepsilon_1 \rightarrow \varepsilon_4$. Therefore, $N! = 24$ and there are 24 possible different ways to arrange these molecules. To illustrate (Table 2.1), set molecule D having energy ε_4; the table shows ε_4 always being connected to molecule D; ε_1 can be assigned to molecule A, molecule B, or molecule C; ε_2 can be assigned to molecule C, molecule A, or molecule B; and ε_3 can be assigned to molecule A, molecule B, or molecule C.

If we sum up all these results, we have 24 possible arrangements. But, again, in this case, we specified that the number of molecules with energy. $N_{\varepsilon_1} = N_1 = 1,\ N_2 = 1,\ N_3 = 1,\ N_4 = 1.$

Example 2: Consider four molecules (A, B, C, D) capable of possessing one of two possible energy states ε_1, ε_2. Specify that each will, in turn, possess ε_2, all others will be ε_1, then, in tabular form (Table 2.2).

Table 2.1 Possible Arrangements for D Having Energy ε_4; Other Particles, ε_1–ε_3

ε_1	ε_2	ε_3	ε_4			
				$\xrightarrow{yields}$	6 *arrangements holding*	D(ε_4)
A	B	C	D	and	6 *arrangements holding*	D(ε_3)
A	C	B	D	and	6 *arrangements holding*	D(ε_2).
B	A	C	D	and	6 *arrangements holding*	D(ε_1).
B	C	A	D			
C	A	B	D			
C	B	A	D			

Table 2.2 Possible Arrangements for One Molecule with ε_2, Others Have ε_1

ε_1	ε_2
ABC	D
ABD	C
ADC	B
DBC	A

There are $N! = 1 \times 2 \times 3 \times 4 = 24$ possible arrangements, but they are not different, as there are constraints. However, as $\begin{matrix} N_1 = 3 \\ N_2 = 1 \end{matrix}$, which is the same as $\begin{matrix} N_1 = 1 \\ N_2 = 3 \end{matrix}$, we really only have four arrangements.

The above two examples demonstrate the mechanics of a rule of probability. The number of possible arrangements of a system of N particles, with N_i particles possessing an energy ε_i ($i = 1,2,3,\ldots$) and accordingly $N_i!$ permutations in that energy is expressed as:

$$W = \frac{N!}{N_1!\ N_2!\ N_3! \cdots N_i!} = \frac{N!}{\Pi_i N_i!}.$$

In Example 1, $W_1 = \frac{4!}{1!\ 1!\ 1!\ 1!} = 24$; in Example 2, $W_2 = \frac{4!}{3!\ 1!} = 4$.

We want to reach a general statement of energy and probability. So, for a given state (energy arrangement, (k)) we can define the probability of finding particles in an energy arrangement as:

$$P_k = \frac{W_k}{\sum_k W_k}, \quad \text{and} \quad \sum_k P_k = 1.$$

Analysis of the Equilibrium State (To Determine the Equilibrium Distribution Function)

Thermodynamically, equilibrium is the state of maximum entropy (minimum energy). Statistically, equilibrium is the state of maximum probability or W_{max}. (Here, we follow the development of the equilibrium distribution provided by Moelwyn-Hughes (1961).)

Boltzmann concluded that as the two statements are true, maximum entropy and maximum probability of a given state must be related as $S = f(P) = f(W)$. Now, from the previous knowledge of thermodynamic systems, entropy is additive: ΔS(sublimation) $= \Delta S$(melting) $+ \Delta S$(evaporation). However, probabilities are multiplicative. Probability of a single event is $\sim W_1$, and the probability of a second, independent event is $\sim W_2$, and the probability ofboth events occurring is $W_{1,2} = W_1 \cdot W_2$. In general, $W = \frac{N!}{\Pi_i N_i!}$, where W is the number arrangements of N with N_i and ε_i energy; therefore, the maximum disorder state has maximum arrangements. As:

$$\text{Probability} = \frac{\text{the number with property}}{\text{the total number}} = \frac{\text{the number of arrangement}}{\text{the total number of arrangements}}.$$

Therefore: Probability of a given state $= \frac{W(\text{given state})}{\sum W_i(\text{pro. of all } W)}$. Accordingly, for two given (component) states of a gas molecular system: $S_1 = f(W_1)$, $S_2 = f(W_2)$; then for the total system $S = S_1 + S_2$ and $W = W_1\ W_2$.

Since S is related to W, the only way this is possible is through a logarithmic function: $S = \text{C} \ln W + \text{B}$, where C, B are constants (to be determined). Therefore, we have: $\frac{S-\text{B}}{\text{C}} = \ln W = \ln N! - \ln N_1! - \ln N_2! - \cdots = \ln N! - \sum_i \ln N_i!$. Now, when N is large, we can use the Stirling approximation to make this a simpler expression. First, $N! = (2\pi N)^{\frac{1}{2}}(N/e)^N$,

$$\text{or}: \quad \ln N! = \frac{1}{2}\ln\,(2\pi N) + N\ln N - N \approx N\ln N - N \quad \text{for} \quad \ln N \ll N.$$

$$\text{So,} \quad \frac{S-\text{B}}{\text{C}} = N\ln N - N - \sum_i \{N_i \ln N_i - N_i\} = N\ln N - \sum_i N_i \ln N_i.$$

In equilibrium, entropy is a maximum, but $S = f(N_i)$ where i can have many values. The condition of a maximum is that the total differential is zero, i.e., $\delta S = \sum_i dS_i = 0 = \sum_i \frac{dS}{dN_i} dNi$, and using the above expression for S, we can calculate:

$$\frac{dS}{dN_1} = -\text{C}(\ln N_1 + 1) \Rightarrow dS_1 = -\text{C}(\ln N_1 + 1)dN_1$$

$$\frac{dS}{dN_2} = -C(\ln N_2 + 1) \Rightarrow dS_2 = -C(\ln N_2 + 1)dN_2$$

Since C is not zero, then for maximum entropy:

$$\sum_i (\ln N_i + 1)dN_i = 0.$$

The constraints on this equation are the conservation of number of the molecules and the conservation of the total energy of the system, therefore:

$$\delta N = \sum dN_i = 0;$$

$$\delta E = \sum \epsilon_i dN_i = 0.$$

The solution can be developed using the method of Lagrange multipliers. The method states that with variables $x = 0,\ y = 0,\ z = 0$, then $x + \lambda y + \mu z = 0$, where λ and μ are undetermined multipliers. In this case, $x = \delta S$ (entropy is maximum) $= 0,\ y = \delta N = 0$, (number of particles is conserved), and $z = \delta E = 0$ (energy is conserved). So substituting into this, we have (with ϵ_i representing the general energy state):

$$\sum (dS_i + \lambda dN_i + \mu \epsilon_i dN_i) = 0; \quad \text{or}: \quad \sum (\ln N_i + 1 + \lambda + \mu \epsilon_i)dN_i = 0,$$

which must be true for all forms of dN_i (even $dN_i \neq 0$), so each must be zero independently, i.e.,

$$\ln N_i + 1 + \lambda + \mu \epsilon_i = 0.$$

Rearranging this, $N_i = Ke^{-\mu\epsilon_i}$, where $K = e^{-(\lambda+1)}$. Now since: $N = \sum N_i = K\sum e^{-\mu\epsilon_i}$, then $\frac{N_i}{N} = \frac{e^{-\mu\epsilon_i}}{\sum e^{-\mu\epsilon_i}} \equiv \frac{e^{-\mu\epsilon_i}}{Q}$, where Q is defined as the partition function. This expression is a probability, $\frac{N_i}{N}$, and the probability of finding a molecule in energy state ϵ_i is given by the above expression. However, the question remains, what is μ and what is Q? We now need to determine these terms.

We consider the general group of states i to consist of all the states having energy in the small range from ϵ to $+d\epsilon$, where ϵ is a continuous variable. If $\xi(\epsilon)$ is the distribution function for the molecular energy (not velocity), we can alternatively write for the group:

$$N_i = N\xi(\epsilon)d\epsilon.$$

Using the functional form determined above for a distribution function, we can write that $N\xi(\epsilon)d\epsilon = K'Ne^{-\mu\epsilon}$. Therefore we can define the energy distribution function as $\xi(\epsilon)d\epsilon = K'e^{-\mu\epsilon}$. To place this in convenient form, $K'' \equiv \frac{K'}{d\epsilon}$, so then $K'e^{-\mu\epsilon} = K''e^{-\mu\epsilon}d\epsilon$ and $\xi(\epsilon)d\epsilon = K''e^{-\mu\epsilon}d\epsilon$. In order to determine K'' and μ, we now turn to the form of energy which we are describing random translational energy. Then $\epsilon = m\frac{c^2}{2}$, $d\epsilon = mcdc$. So, $K''e^{-\mu(\frac{1}{2}mc^2)}mcdc \equiv K'''e^{-\mu(\frac{1}{2}mc^2)}dc$. Now we can write $f(c)dc = K^{iv}e^{-\mu(\frac{1}{2}mc^2)}dc$, as the correct functional form for the speed distribution function. This is the probability of finding the speed of a molecule between c and $c + dc$. However, we still need to find values for the coefficients μ, K^{iv}.

We proceed by turning to the velocity distribution function, i.e., in the Cartesian reference frame it is the probability of finding x speed between c_1 and $c_1 + dc_1$, y speed between c_2 and $c_2 + dc_2$ and z speed between c_3 and dc_3. The probability of the occurrence of all directional components is Prob. Vel. = (Prob. x speed)(Prob. y speed)(Prob. z speed). So, using the above functional form, we have:

$$f(c)dc_1dc_2dc_3 = Ae^{-\mu\left(\left(\frac{1}{2}m\right)\left(c_1^2+c_2^2+c_3^2\right)\right)}dc_1dc_2dc_3,$$

where $f(c)$ is the velocity distribution function, a probability per unit interval of velocity. To specify $f(c)$ we must determine A and μ. (Here we follow the formalism of Vincenti and Kruger (1975). We evaluate these by using two conditions. First, appealing to the basic property of any distribution function:

$$\int_{-\infty}^{+\infty} f(c_i)dV_c = 1,$$

where dV_c is an element of volume in c space. We can then express:

$$\iiint_{-\infty}^{+\infty} Ae^{-\mu\left(\left(\frac{1}{2}m\right)\left(c_1^2+c_2^2+c_3^2\right)\right)}dc_1dc_2dc_3 = 1, \text{ or:}$$

$$A\int_{-\infty}^{+\infty} e^{-\mu\left(\left(\frac{1}{2}m\right)c_1^2\right)}dc_1 \cdot \int_{-\infty}^{+\infty} e^{-\mu\left(\left(\frac{1}{2}m\right)c_2^2\right)}dc_2 \cdot \int_{-\infty}^{+\infty} e^{-\mu\left(\left(\frac{1}{2}m\right)c_3^2\right)}dc_3 = 1.$$

This integral can be evaluated in standard fashion using tables for this type of function and can be shown to be:

$$A\left(\frac{2\pi}{\mu m}\right)^{\frac{1}{2}}\cdot\left(\frac{2\pi}{\mu m}\right)^{\frac{1}{2}}\cdot\left(\frac{2\pi}{\mu m}\right)^{\frac{1}{2}}=1.$$

Therefore, $A\left(\frac{2\pi}{\mu m}\right)^{\frac{3}{2}}=1\rightarrow A=\left(\frac{\mu m}{2\pi}\right)^{\frac{3}{2}}$, and we have expressed A as a function of μ.

In order to determine μ, we utilize the additional fact that we know, $\overline{c_3^2}=\frac{\overline{c^2}}{3}=\frac{kT}{m}$. So now we can write a correct expression for $\overline{c_3^2}$ using the distribution function, as:

$$\overline{c_3^2}=\int_{-\infty}^{+\infty}c_3^2 f(c_i)\,dV_c$$

$$=\int_{-\infty}^{+\infty}c_3^2 A e^{-\mu\left(\frac{1}{2}m\right)c_3^2}dc_3\cdot\int_{-\infty}^{+\infty}e^{-\mu\left(\frac{1}{2}m\right)c_2^2}dc_2\cdot\int_{-\infty}^{+\infty}e^{-\mu\left(\frac{1}{2}m\right)c_1^2}dc_1.$$

This can be evaluated again in standard fashion, and we get:

$$\overline{c_3^2}=A\frac{1}{2}\left(\frac{8\pi}{\mu^3m^3}\right)^{\frac{1}{2}}\cdot\left(\frac{2\pi}{\mu m}\right)^{\frac{1}{2}}\cdot\left(\frac{2\pi}{\mu m}\right)^{\frac{1}{2}}=\frac{A}{\mu m}\cdot\left(\frac{2\pi}{\mu m}\right)^{\frac{3}{2}}=\frac{1}{\mu m},\quad\text{but,}\quad\overline{c_3^2}=\frac{kT}{m}$$

$\therefore \mu=\frac{1}{kT}$, and

$A=\left(\frac{m}{2\pi kT}\right)^{\frac{3}{2}}$, where k is Boltzmann's constant, and we have now evaluated all terms.

The equilibrium velocity distribution function (also referred to as the Maxwellian distribution function) can be written as:

$$f(c_i)=\left(\frac{m}{2\pi kT}\right)^{\frac{3}{2}}\cdot\exp\left[-\frac{m}{2kT}\left(c_1^2+c_2^2+c_3^2\right)\right].$$

Average Values of Speed and Velocities

We can now determine the speed distribution function. As above, the velocity distribution function can be written as $f(c_i)=\phi(c_1)\cdot\phi(c_2)\cdot\phi(c_3)$, where $\phi(c_1)$ is the directional velocity distribution function expressed as:

$$\phi(c_1)=\left(\frac{m}{2\pi kT}\right)^{\frac{1}{2}}\cdot\exp\left[-\frac{m}{2kT}\left(c_1^2\right)\right].$$

From this functional form, we can express the equation of state, pressure relationships, and transport properties in gases. Starting with the velocity distribution function, we can define and derive the particle motion kinetic properties. A diagram expressing the terminology of velocity space for a particle is shown in Figure 2.13.

With axes, c_1, c_2, c_3 as shown, and a vector $\overrightarrow{c}$ to an arbitrary point in velocity space, the magnitude of that vector is the speed $|c|$ at that point, and that speed c can be related through three different angles to the component speeds. So an element of velocity space is expressed:

$$dV_c = dc_1 dc_2 dc_3 = (cd\phi)(c \sin\phi\ d\theta)dc,$$
$$dV_c = c^2 \sin\phi\ d\phi d\theta dc.$$

Therefore, the probability of finding a molecule with the magnitude of velocity (that is, the speed) c to $c + dc$, the angle θ to $\theta + d\theta$, and the angle ϕ to $\phi + d\phi$, is:

$$f(c,\phi,\theta)dcd\phi d\theta = \left(\frac{m}{2\pi kT}\right)^{\frac{3}{2}} c^2 e^{-\left(\frac{m}{2kT}\right)c^2} \sin\phi\ d\phi d\theta dc.$$

We want to eliminate direction dependence, so define the speed distribution function, as:

$$\chi(c)dc = \int\limits_{\phi=0}^{\pi} \int\limits_{\theta=0}^{2\pi} f(c,\phi,\theta)dcd\phi d\theta$$
$$= \int\limits_{0}^{\pi} \sin\phi\ d\phi \int\limits_{0}^{2\pi} d\theta \left(\frac{m}{2\pi kT}\right)^{\frac{3}{2}} c^2 e^{-\left(\frac{m}{2kT}\right)c^2} dc,$$
$$\chi(c) = 4\pi \left(\frac{m}{2\pi kT}\right)^{\frac{3}{2}} c^2 e^{-\left(\frac{m}{2kT}\right)c^2}.$$

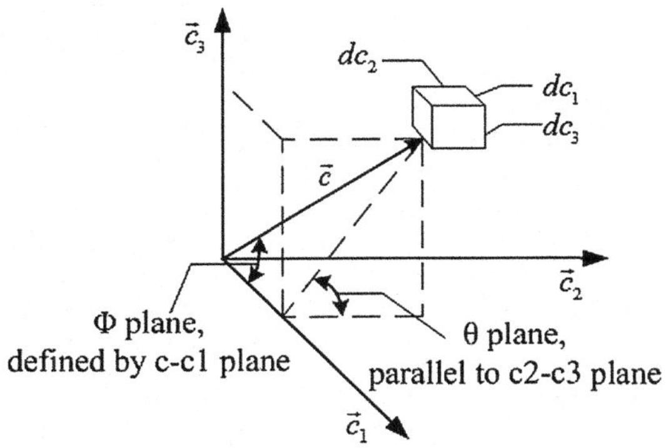

Figure 2.13 Diagram of velocity space.

Note that the speed distribution function is $\chi(c) = 4\pi c^2 f(c_i)$, where $f(c_i)$ is the velocity distribution function, and $4\pi c^2$ is the area of a sphere of radius c which incorporates the total solid angle. So we can write, $\chi(c) = \int_0^{4\pi} f(c_i)d\omega$, where ω is a solid angle.

We can now consider the calculation of some appropriate kinetic quantities. First, for average speed:

$$\bar{c} = \int_0^{\infty} c\chi(c)dc = \int_0^{\infty} 4\pi\left(\frac{m}{2\pi kT}\right)^{\frac{3}{2}} c^3 e^{-\left(\frac{m}{2kT}\right)c^2} dc = \left(\frac{8kT}{\pi m}\right)^{\frac{1}{2}}.$$

The mean square speed is $\overline{c^2} = \int_0^{\infty} c^2\chi(c)dc = \int_0^{\infty} 4\pi\left(\frac{m}{2\pi kT}\right)^{\frac{3}{2}} c^3 e^{-\left(\frac{m}{2kT}\right)c^2} dc = \left(\frac{3kT}{m}\right)$. This is in agreement with the relationship $\frac{1}{2}m\overline{c^2} \equiv \frac{3}{2}kT$, which was derived earlier. We can compare $\bar{c}$ and $(\overline{c^2})^{\frac{1}{2}}$ and see that $\bar{c} = \left(\frac{8}{3\pi}\right)^{\frac{1}{2}}(\overline{c^2})^{\frac{1}{2}} = 0.992(\overline{c^2})^{\frac{1}{2}}$. We can also define a most probable speed; it has a maximum where $\frac{d\chi(c)}{dc} = 0$, as:

$$\frac{d\chi(c)}{dc} = 4\pi\left(\frac{m}{2\pi kT}\right)^{\frac{3}{2}}\left\{c^2 \cdot e^{-\left(\frac{m}{2kT}\right)c^2} \cdot -\frac{m}{2kT}\cdot 2c + 2ce^{-\left(\frac{m}{2kT}\right)c^2}\right\} = 0,$$

$$\text{then} \quad -\frac{m}{2kT}c^2 + 1 = 0, \quad \text{so}: c_{mp} = \left(\frac{2kT}{m}\right)^{\frac{1}{2}} < \bar{c}.$$

With the mathematical form of the speed distribution function, we can present the behavior of particles in equilibrium in a graph of $f(c)$, the probability of finding a particle with speed between c and $c + dc$, as a function of c. In Figure 2.14 the average speed, rms speed and most probable speeds are shown. Note, most importantly, that this function is not symmetrical about $\bar{c}$. Note also that the directional component of speed, c_i, is symmetrical about 0, and that $\bar{c_i} = 0$.

We return now to the earlier proposition that the average speed in a given direction is equal to 1/4 of the average speed, as $\bar{w} = \frac{\bar{c}}{4}$. This is distinctly different from a simple arithmetic evaluation that proposes that if 1/3 of the particles are going in any given direction, their average speed would be 1/3 the average speed. To derive this important result which is

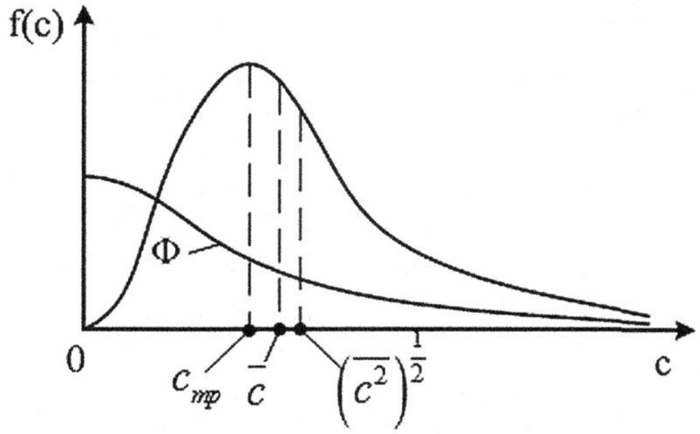

Figure 2.14 Distribution of average speed, rms speed and most probable speeds.

due to the fact that the equilibrium distribution is not symmetrical, we begin with the velocity distribution function in one direction as:

$$\phi(c_1) = \left(\frac{m}{2\pi kT}\right)^{\frac{1}{2}} \cdot \exp\left(-\frac{m}{2kT}c_1^2\right).$$

This is the probability of finding a molecule with velocity between c_1 and $c_1 + dc_1$. To define the average in a given direction, we take the positive direction limits as $0 \rightarrow \infty$, then:

$$\overline{c_1} = \int_0^\infty c_1 \left(\frac{m}{2\pi kT}\right)^{\frac{1}{2}} e^{-\left(\frac{m}{2kT}\right)c_1^2} dc = \left(\frac{kT}{2\pi m}\right)^{\frac{1}{2}} \quad \text{but with}$$

$$\bar{c} = \left(\frac{8kT}{\pi m}\right)^{\frac{1}{2}}, \quad \text{we have:} \quad \overline{c_1} = \frac{1}{4}\left(\frac{8kT}{\pi m}\right)^{\frac{1}{2}} = \frac{1}{4}\bar{c}.$$

In summary, we have considered several basic topics:

1. We have considered statistical calculations of the equilibrium state.
2. We have derived the velocity and speed distribution functions and calculated some values.
3. We have begun the process of connecting statistical to physical concepts of a gas by the relationship, $S = k \cdot \ln W$, but we have not yet determined the exact relationship between entropy and probability.

There are natural significant extensions and applications to these fundamentals, and they include:

1. How to extend the relationships to account for the relative velocities of molecules.
2. The calculation of the exact collision frequency and the equilibrium law of mass action.

3. The intermolecular forces and the derivation of an exact equation of state.
4. Revisiting statistical mechanics and including the above results into an exact calculation.

Extended consideration of these topics can be found in the literature (Chapman & Cowling, 1939 and Slater, 1963.)

REFERENCES

Chapman, S., Cowling, T.G., 1939. The Mathematical Theory of Non-Uniform Gases. Cambridge University, Cambridge.
Kreysig, E., 1980. Advanced Engineering Mathematics. Wiley, New York.
Moelwyn-Hughes, E.A., 1961. Physical Chemistry. Pergamon, New York.
Present, R.D., 1958. Kinetic Theory of Gases. McGraw-Hill, New York.
Slater, J.C., 1963. Introduction to Chemical Physics. McGraw-Hill, New York.
Vincenti, W.G., Kruger, C.H., 1975. Introduction to Physical Gas Dynamics. Krieger, Huntington, NY.

CHAPTER 3

Molecular Energy Distribution and Ionization in Gases

INTRODUCTION

The previous chapter developed the functional form for the equilibrium distribution of molecular speeds that occur relative to an average temperature for a gas. This behavior is for kinetic theory that assumed only random translational motion and the related kinetic energy. Again, the result demonstrated a nonsymmetric distribution about an average speed, which results in unique formulations of macroscopic properties. More importantly, the distribution function demonstrated that particles can have speeds and related energies that are significantly greater than the average values. As molecules can transfer energy in collisions and real gases have component internal energies, this sets the stage for an understanding of the behavior of real gases, which exhibit dissociation of molecules and ionization of atoms. This chapter will present a sequence of topics that develop a framework for the understanding of energies resident in molecules and atoms, the excitation of energy modes that lead to dissociation and ionization, and for hot gases that are in expansion processes, the aspects of the nonrecovery of internal energy modes that are known as frozen flow loss. The loss of that energy is a primary source of inefficiencies in plasma expansion processes.

MOLECULAR ENERGY

The following discussion and development of a molecular energy formalism has been well documented in the literature (Slater, 1939; Moelwyn-Hughes,1961; Clark & McChesney, 1964; Vincenti & Kruger, 1965).

Energy Distribution Function

The speed distribution function was derived in Chapter 2 to have the functional form:

$$\chi(c)dc = 4\pi\left(\frac{m}{2\pi kT}\right)^{\frac{3}{2}} c^2 e^{-\frac{m}{2kT}c^2}\, dc;$$

Introduction to Plasmas and Plasma Dynamics
ISBN 978-0-12-801661-9

however, we know that random thermal molecular energy can be expressed as:

$$\in = \frac{1}{2}mc^2 \text{ and } d\in = mcdc,$$

so:

$$\chi(c)dc = 4\pi\left(\frac{m}{2\pi kT}\right)^{3/2} c^2 e^{-\in/(kT)} \frac{d\in}{mc}$$

$$= 4\pi\left(\frac{m}{2\pi kT}\right)^{3/2} \frac{(2\in)^{1/2}}{m^{3/2}} e^{-\in/(kT)} d\in .$$

We can write an energy distribution function, $f(\in)$, as $f(\in)d\in = \frac{2\in^{1/2}}{\pi^{1/2}(kT)^{3/2}}e^{-\in/(kT)}d\in$, which is the probability of finding a molecule with energy between $\in$ and $\in + d\in$.

This formalism has been introduced (Chapter 2); then: $f(\in)d\in = \frac{dN}{N} = \text{Constant} \cdot e^{-\in/(kT)}d\in$.

So: $N_i = \text{Constant} \cdot e^{-\in_i/(kT)}$ with $N = \sum_i N_i = \text{Constant} \cdot \sum_i e^{-\in_i/(kT)}$

and:

$$\frac{N_i}{N} = \frac{e^{-\in_i/(kT)}}{\sum_i e^{-\in_i/(kT)}} \equiv \frac{e^{-\in_i/(kT)}}{Q}$$

where Q is the partition function.

Applying the energy distribution function, we can calculate the average random thermal energy, as:

$$\overline{\in} = \int_0^\infty \in f(\in)d\in = \int_0^\infty \frac{2\in^{3/2}}{\pi^{1/2}(kT)^{3/2}} e^{-\in/(kT)} d\in = \frac{3}{2}kT.$$

So it can be seen that the calculated random thermal energy component is consistent with basic formulations.

Molecular Energy Calculations

For the energy of the molecular system, we know that the total energy is:

$$E = \sum_i N_i \in_i = \sum_i \frac{N}{Q} e^{-\in_i/(kT)} \in_i = \frac{N}{Q}\sum_i \in_i e^{-\in_i/(kT)},$$

but for molecules with independent component energies:

$$Q = \sum_i e^{-\in_i/(kT)} = e^{-\in_1/(kT)} + e^{-\in_2/(kT)} + \cdots,$$

and:

$$\left(\frac{\partial Q}{\partial t}\right)_V = \frac{\epsilon_1}{kT^2}e^{-\epsilon_1/(kT)} + \frac{\epsilon_2}{kT^2}e^{-\epsilon_2/(kT)} + \cdots = \frac{1}{kT^2}\sum_i \epsilon_i e^{-\epsilon_i/(kT)}.$$

So:

$$kT^2\left(\frac{\partial Q}{\partial t}\right)_V = \sum_i \epsilon_i e^{-\epsilon_i/(kT)},$$

and:

$$kT^2\,\frac{1}{Q}\left(\frac{\partial Q}{\partial T}\right)_V = kT^2\left(\frac{\partial \ln Q}{\partial T}\right)_V = \frac{\sum_i \epsilon_i e^{-\epsilon_i/(kT)}}{Q}.$$

So finally:

$$E = NkT^2\left(\frac{\partial \ln Q}{\partial T}\right)_V,$$

where: N is the total number of particles, T is the temperature, and Q is the energy partition function. Functionally, this has allowed the expression $E = f(T, N, Q)$. From thermodynamics, it is possible to derive similar expressions for entropy (S) and other properties such as Helmholtz free energy ($F = E - TS$) and chemical potential $\left(\mu = -T\left(\frac{\partial S}{\partial N}\right)E, V\right)$.

Partition Function and Energy Evaluation

Energy Evaluation

Since we know that all thermodynamic quantities (F, S, E, p, μ) can be expressed in terms of the partition function, N and T, it is important to consider the evaluation of the partition function and its form for various components of the energy of the molecule. (The question to answer is: what is Q?)

Suppose that a molecule can have several *independent* types of energy denoted by $\epsilon', \epsilon'', \epsilon'''$, etc. Then (Vincenti and Kruger, 1965, Chapter IV) the total energy of the molecule is given by:

$$\epsilon = \epsilon' + \epsilon'' + \epsilon'''.$$

For gases with weakly interacting particles we can identify and separate energy components as:

$$\epsilon = \epsilon_{tr} + \epsilon_{rot} + \epsilon_{vib} + \epsilon_{el}.$$

Now since we have defined:

$$Q = \sum_i e^{-\in_i/(kT)} \quad \text{where } i \Rightarrow tr, rot, vib, el,$$

and:

$$\begin{aligned} e^{-\in/(kT)} &= e^{-(\in_{tr}+\in_{rot}+\in_{vib}+\in_{el})/(kT)} \\ &= e^{-\in_{tr}/(kT)} \cdot e^{-\in_{rot}/(kT)} \cdot e^{-\in_{vib}//(kT)} \cdot e^{-\in_{el}//(kT)}, \end{aligned}$$

then:

$$\begin{aligned} Q &= \sum_{i,tr} e^{-(\in_{tr})_i/(kT)} \cdot \sum_{i,rot} e^{-(\in_{rot})_i/(kT)} \cdot \sum_{i,vib} e^{-(\in_{vib})_i/(kT)} \cdot \sum_{i,el} e^{-(\in_{el})_i/(kT)} \\ &= Q_{tr} \cdot Q_{rot} \cdot Q_{vib} \cdot Q_{el}. \end{aligned}$$

Furthermore:

$$Q \equiv Q_{tr} \cdot Q_{int} \quad \text{where } Q_{int} = Q_{rot} \cdot Q_{vib} \cdot Q_{el}.$$

We have been considering energy per molecule, $\in$; we can extend this to energy per unit mass, as:

$$e\left(\frac{\text{energy}}{\text{mass}}\right) = \in\left(\frac{\text{energy}}{\text{molecule}}\right) \cdot \frac{1}{m}\left(\frac{\text{molecule}}{\text{mass}}\right) = \frac{E}{N} \cdot \frac{1}{m}.$$

Directly:

$$e = \frac{E}{N \cdot m} = \frac{1}{N \cdot m}\left[NkT^2\left(\frac{\partial \ln Q}{\partial T}\right)_V\right],$$

$$e = RT^2 \frac{\partial(\ln Q)}{\partial T},$$

and:

$$e = RT^2 \frac{\partial}{\partial T} \ln Q_{tr} + RT^2 \frac{\partial}{\partial T} \ln Q_{int},$$

$$e = e_{tr} + e_{int},$$

and by extension:

$$C_V = \left(\frac{\partial e}{\partial T}\right)_V = \frac{\partial}{\partial T}(e_{tr} + e_{int}) = C_{V_{tr}} + C_{V_{int}}.$$

In other words, specific heat can be envisioned as made up of *linear* components.

Partition Function Analytic Forms

We will here briefly summarize the results of quantum mechanical analysis to identify component partition functions and related quantities (Vincenti and Kruger, 1965, Chapter IV).

Translational Component

From quantum mechanical analysis assuming *weak* interactions:

$$Q_{tr} = V\left(\frac{2\pi mkT}{h^2}\right)^{3/2} \quad \text{where } h = \text{Planck's constant } \left(6.6 \times 10^{-27}\ \text{erg}\cdot\text{s}\right).$$

Electronic Component

For a monatomic gas ϵ_{rot}, $\epsilon_{vib} \rightarrow 0$, so the combination of translation and electronic components will fully describe the gas. When describing electronic energies, it must be recognized that the form will be dictated by internal atomic properties, i.e., electron shell structure implies electrons with the same value of energy level or degeneracy, g, so we write:

$Q_{el} = \sum_l g_l e^{-\epsilon_l/(kT)} = g_0 e^{-\epsilon_0/(kT)} + g_1 e^{-\epsilon_1/(kT)} + g_2 e^{-\epsilon_2/(kT)} + \cdots$ with $\epsilon_0 \equiv 0$. We define $\frac{\epsilon}{kT}\left(\text{i.e.,} \frac{\text{energy}}{\text{Random th.en.}}\right) \equiv \frac{\theta}{T}$ where $\theta = \frac{\epsilon}{k}$ and is called the characteristic temperature (a constant).

Then:

$$Q_{el} = g_0 + g_1 e^{-\theta_1/T} + g_2 e^{-\theta_2/T} + \cdots,$$

but if $\theta \gg T, ge^{-\theta/T} \rightarrow 0$, then we approximate for an atom with one "soft" mode,

$$Q_{el} = g_0 + g_1 e^{-\theta_1/T},$$

and we can write for the component molecules in air, for example,

$$Q_{el}(O_2) = 3 + 2e^{-\frac{11390}{T}}; \quad Q_{el}(N_2) = 1; \quad Q_{el}(NO) = 2 + 2e^{-\frac{174}{T}}.$$

For atomic oxygen:

$$Q_{el}(O) = 5 + 3e^{-228/T} + e^{-336/T} \approx 5 + 4e^{-270/T},$$

and for atomic nitrogen:

$$Q_{el}(N) = 4.$$

Note that:

$$Q_{el} = \text{constant} \Rightarrow e_{el} = 0, \quad C_{V_{el}} = 0$$

Rotational Component (Molecules)

From quantum mechanics it is found that degenerate rotation levels can exist at low temperatures, i.e., $\in_l = \frac{h^2}{8\pi^2 I} l(l+1)$, where I = moment of inertia and $l = 0, 1, 2, 3, \ldots$. The partition function for rotation can be written as

$$Q_{rot} \rightarrow \int_0^\infty (2l+1) e^{-l(l+1)\theta_{rot}/T} dl = \int_0^\infty e^{-z\theta_{rot}/T} dz,$$

where θ_{rot} (for $O_2 = 2.1$ K, $N_2 = 2.9$ K, $NO = 2.5$ K) is the characteristic temperature for rotation, and for $\frac{\theta_{rot}}{T} \ll 1$, $Q_{rot} = \frac{T}{\theta_{rot}}$, and with $e = RT^2 \frac{\partial(\ln Q)}{\partial T}$, $e_{rot} = RT$ and $C_{V_{rot}} = R$.

Vibrational Component (Molecules)

We will make the approximation that molecular vibration can be represented as a harmonic oscillator of frequency, υ (again following Vincenti and Kruger, 1965). From quantum mechanics, the permissible energy states are given by:

$$\in_i = \left(i + \frac{1}{2}\right) h\upsilon, \quad i = 0, 1, 2, \ldots$$

For:

$$i = 0, \in_0 = \frac{1}{2} h\upsilon, \text{ zero point energy (constant), so we take, } \in_0 = 0,$$

and:

$$\in_i = ih\upsilon \quad \text{with} \quad i = 0, 1, 2, 3, \ldots$$

So,

$$Q_{vib} = \sum_{i,vib} e^{-(\in_{vib})_i/(kT)} \approx \frac{1}{1 - e^{-\theta_{vib}/T}} \quad \text{where } \theta_{vib} = h\upsilon/k$$

Directly,

$$e_{vib} = \frac{R\theta_{vib}}{e^{\theta_{vib}/T} - 1}, \quad C_{V_{vib}} = R\left(\frac{\theta_{vib}}{T}\right)^2 \frac{e^{\theta_{vib}/T}}{\left(e^{\theta_{vib}/T} - 1\right)^2} = R\left\{\frac{\theta_{vib}/(2T)}{\sinh[\theta_{vib}/(2T)]}\right\}^2.$$

So,
as $T \to 0$, $e^{\theta_{vib}/T} \to \infty$ and e_{vib}, $C_{V_{vib}} \to 0$;

as $T \to \infty$, $e^{\theta_{vib}/T} = 1 + \frac{\theta_{vib}}{T} + \left(\frac{\theta_{vib}}{T}\right)^2 \frac{1}{2!} + \cdots \approx 1 + \frac{\theta_{vib}}{T}$;

and $e_{vib} = \frac{R\theta_{vib}}{1+\frac{\theta_{vib}}{T}-1} = RT, \quad C_{V_{vib}} = R\left(\frac{\theta_{vib}}{T}\right)^2 \frac{1+\frac{\theta_{vib}}{T}}{\left(1+\frac{\theta_{vib}}{T}-1\right)^2} \approx R.$

For the components of air, the following values are indicative of vibration excitation:

$$\theta_{vib}(O_2) = 2270 \text{ K};$$
$$\theta_{vib}(N_2) = 3390 \text{ K};$$
$$\theta_{vib}(NO) = 2740 \text{ K}.$$

So, it can be seen that e and C_V can vary significantly at temperatures encountered in high-temperature gas dynamics and plasma dynamics.

In summary, for *diatomic* gases, including translation, rotation, and vibration:

$$C_V = C_{V_{tr}} + C_{V_{rot}} + C_{V_{vib}} \left(C_{V_{el}} \approx 0\right),$$

$$C_V = \frac{3}{2}R + R + R\left\{\frac{\theta_{vib}/(2T)}{\sinh[\theta_{vib}/(2T)]}\right\}^2,$$

$$= \mathrm{R}\left\{\frac{5}{2} + \left\{\frac{\theta_{vib}/(2T)}{\sinh[\theta_{vib}/(2T)]}\right\}^2\right\} = f(T).$$

Partition Function and Dissociation Energy

At elevated temperatures, excited vibration leads to the breakup of molecules by dissociation. As part of building a conceptual and computational model of gases at high temperatures, dissociation is accommodated as follows.

We consider a gas mixture with reaction:

$$A + B \rightleftharpoons AB.$$

Then the energy of the system is expressed as the sum of component energies for each of the species,

$$E^A = \sum_i N_i^A \epsilon_\iota^A;$$

$$E^B = \sum_i N_i^B \epsilon_\iota^B;$$

$$E^{AB} = \sum_i N_i^{AB}\left(\epsilon_\iota^{AB} - D\right);$$

and:

$$E = \sum_i N_i^A \in_i^A + \sum_i N_i^B \in_i^B + \sum_i N_i^{AB} \left(\in_i^{AB} - D\right).$$

Energy is referenced relative to atoms at zero temperature. Therefore, $\in_i^{AB}$ includes disassociation energy when at rest.

At equilibrium we know that:

$$N_i^A = \frac{N^A e^{-\in_i^A/(kT)}}{Q^A};$$

$$N_i^B = \frac{N^B e^{-\in_i^B/(kT)}}{Q^B};$$

$$N_i^{AB} = \frac{N^{AB} e^{-\in_i^{AB}/(kT)}}{Q^{AB}}.$$

So we can write that:

$$\frac{N^{AB}}{N^A \cdot N^B} = \frac{Q^{AB} N_i^{AB}}{e^{-\in_i^{AB}/(kT)}} \cdot \frac{e^{-\in_i^A/(kT)}}{Q^A N_i^A} \cdot \frac{e^{-\in_i^B/(kT)}}{Q^B N_i^B},$$

$$= \frac{Q^{AB}}{Q^A Q^B} \cdot \underbrace{\frac{N_i^{AB}}{e^{-\in_i^{AB}/(kT)}} \cdot \frac{e^{-\in_i^A/(kT)}}{N_i^A} \cdot \frac{e^{-\in_i^B/(kT)}}{N_i^B}}_{e^{D/(kT)}},$$

$$\frac{N^{AB}}{N^A \cdot N^B} = \left(\frac{Q^{AB}}{Q^A Q^B} e^{D/(kT)} \right),$$

and to complete the equations for the solution of N^A, N^B, and N^{AB}, we can also write:

$$N^A + N^{AB} = N^A,$$
$$N^B + N^{AB} = N^B.$$

Equilibrium Composition of High-Temperature Air

We will now consider the general problem of determining the equilibrium composition of air at high temperatures. At elevated temperatures with significant energy input, vibration will be excited, electronic modes will be excited, and disassociation will occur. Temporarily, we will not consider ionization reactions and ionization equilibrium.

We will present here an approximate model for a symmetrical diatomic gas (Lighthill, 1951; Vincenti and Kruger, 1965) to illustrate gas behavior; this is useful conceptually and convenient for analysis. Let us consider a disassociation reaction where:

$$A + A \rightleftharpoons A_2, (A_2 \Leftrightarrow A^{aa}),$$

and if:

$$N^a = \text{number of } A \text{ atoms},$$

then:

$$N^A = N^a + 2N^{aa}, \quad \text{the total number of } A \text{ atoms in the mixture.}$$

Furthermore, let us define $\alpha \equiv \frac{N^a}{N^A}$, the degree of disassociation, with
$\alpha = 0$, no dissociation
$\alpha = 1$, fully dissociation

Then, since the total mass of gas is mN^A, where $m\left(\frac{\text{mass}}{\text{atom}}\right)$, then $\alpha = \frac{\text{mass of dissociation atoms}}{\text{total mass of gas}}$.

Directly,

$$N^a = \alpha N^A;$$

$$N^{aa} = \frac{N^A - N^a}{2} = \frac{N^A - \alpha N^A}{2} = \frac{1-\alpha}{2} N^A.$$

Accordingly, the relationship of the numbers of species particles is given by:

$$\frac{N^{aa}}{N^a N^a} = \frac{Q^{aa}}{Q^a Q^a} e^{D/kT}, \quad \text{where } D = \text{energy of dissociation},$$

and substituting

$$\frac{1-\alpha}{2} N^A \cdot \frac{1}{(\alpha N^A)^2} = \frac{Q^{aa}}{(Q^a)^2} e^{D/kT},$$

$$\frac{1-\alpha}{\alpha^2} = 2N^A \frac{Q^{aa}}{(Q^a)^2} e^{D/kT} = 2N^A \frac{Q^{aa}}{(Q^a)^2} e^{\theta_D/T}$$

$$\text{where } \theta_D = D/k, \quad N^A = \frac{\text{mass}}{m/\text{atom}} = \frac{\rho V}{m},$$

and:

$$\frac{\alpha^2}{1-\alpha} = \frac{m}{2\rho V} \cdot \frac{(Q^a)^2}{Q^{aa}} e^{-\theta_D/T}, \quad \text{with} \quad \begin{matrix} \theta_D(O_2) = 59{,}500 \text{ K}; \\ \theta_D(N_2) = 113{,}000 \text{ K}. \end{matrix}$$

Since $Q = V \cdot f(T)$ $\therefore$ $\alpha \sim \frac{f(T)}{\rho}$. Since $\alpha \sim f(T)$ how are p, ρ, T related?

For a gas mixture, $p = \sum_i p_i = \sum_i n_i k T_i$, or: $pV = (N^a + N^{aa})kT$, assuming temperature equilibrium, so $pV = \left(\alpha N^A + \frac{1-\alpha}{2} N^A\right)kT = (1+\alpha)\frac{N^A}{2} kT$.

Now, since $R_{A_2} = \frac{k}{2m}$, $\therefore$ $\frac{p}{\rho} = (1+\alpha)R_{A_2}T$, where R_{A_2} is the gas constant for A_2.

For undissociated gas,

$$\frac{p}{\rho} = Z(\rho, T) R_{A_2} T,$$

where Z = the compressibility factor $= 1 + \alpha$ where $\begin{matrix} \alpha = 0 \rightarrow \text{minimum} \\ \alpha = 1 \rightarrow \text{maximum} \end{matrix}$.

We now consider energy per unit mass of gas mixture in terms of the degree of dissociation.

First, let us note from above, that:

$$E = \sum_i N_i^a \in_i^a + \sum_i N_i^{aa} \left(\in_i^{aa} - D\right);$$

$$E = \sum_i \in_i^a N^a \frac{e^{-\in_i^a/(kT)}}{Q^a} + \sum_i \left(\in_i^{aa} - D\right) N^{aa} \frac{e^{-\in_i^{aa}/(kT)}}{Q^{aa}},$$

but:

$$\frac{\partial}{\partial T} \ln Q = \frac{\partial}{\partial T} \ln \sum_i e^{-\in_i/(kT)} = \frac{\frac{1}{kT^2} \sum_i e^{-\in_i/(kT)}}{Q},$$

so we can write:

$$E = kT^2 \left(N^a \frac{\partial}{\partial T} \ln Q^a + N^{aa} \frac{\partial}{\partial T} \ln Q^{aa} \right) - N^{aa} D,$$

and since:

$$e = \frac{E}{\rho V}, \quad R_{A_2} = \frac{k}{2m}, \quad N^A = \frac{\rho V}{m}$$

then,

$$e = \frac{kT^2}{\rho V}\left(\alpha N^A \frac{\partial}{\partial T}\ln Q^a + \frac{1-\alpha}{2} N^A \frac{\partial}{\partial T}\ln Q^{aa}\right) - \frac{1-\alpha}{2}\frac{N^A D}{\rho V},$$

$$= R_{A_2} T^2\left[2\alpha\frac{\partial}{\partial T}\ln Q^a + (1-\alpha)\frac{\partial}{\partial T}\ln Q^{aa}\right] - (1-\alpha)R_{A_2}\theta_D.$$

Since:

$$Q = V \cdot f(T) \text{ and } \frac{\partial}{\partial T}\ln Q = f(T), \text{ then } \alpha = \alpha(\rho, T) \Rightarrow e = e(\rho, T).$$

Recall that for an ideal gas, $e = e(T)$.
Now, with:

$$R_{A_2} = \frac{k}{2m} = \frac{1}{2}\left(\frac{k}{m}\right) = \frac{1}{2}R_A,$$

then:

$$e = \underbrace{\alpha R_A T^2 \frac{\partial}{\partial T}\ln Q^a}_{e_A} + (1-\alpha)\underbrace{R_{A_2} T^2 \frac{\partial}{\partial T}\ln Q^{aa}}_{e_{A_2}} - (1-\alpha)R_{A_2}\theta_D;$$

$$e = \alpha e_A + (1-\alpha)e_{A_2} - (1-\alpha)R_{A_2}\theta_D,$$

or:

Energy = Energy of atomic species + Energy of molecular species
× (relative to molecular O) – Energy of dissociation.

We still have $\alpha = f(Q)$, $\frac{p}{\rho} = f(\alpha)$, $e = f(Q)$. But what are the Q's?

The Ideal Dissociating Gas

Since air is composed mostly of diatomic gases, oxygen and nitrogen, the specific form of the partition functions will allow simplifications. This model was originally developed by Lighthill (1951) as an "ideal dissociating gas" and described in detail by Vincenti and Kruger (1965) whose derivation we follow here.

For the atomic species:

$$Q^a = Q^a_{tr} \cdot Q^a_{el};$$

For the molecular species:

$$Q^{aa} = Q^{aa}_{tr} \cdot Q^{aa}_{rot} \cdot Q^{aa}_{vib} \cdot Q^{aa}_{el} \cdot$$

Now, for translation:

$$Q^{a}_{tr} = V\left(\frac{2\pi m_A kT}{h^2}\right)^{3/2}, \quad Q^{aa}_{tr} = V\left(\frac{2\pi m_{A_2} kT}{h^2}\right)^{3/2};$$

for rotation:

$$Q^{aa}_{tr} = \frac{1}{2}\left(\frac{T}{\theta_{rot}}\right);$$

and for vibration:

$$Q^{aa}_{vib} = \frac{1}{1 - e^{-\theta_{vib}/T}}.$$

Then, as:

$$\frac{\alpha^2}{1-\alpha} = \frac{m}{2\rho V} \cdot \frac{\left(Q^A\right)^2}{Q^{AA}} e^{-\theta_D/T},$$

get:

$$\frac{\alpha^2}{1-\alpha} = \frac{m}{2\rho V} \cdot \frac{V\left(\frac{2\pi m_A kT}{h^2}\right)^{3/2}}{2^{3/2}} \cdot \frac{\left(Q^a_{el}\right)^2}{Q^{aa}_{el}} \cdot \frac{1}{\frac{1}{2}\left(\frac{T}{\theta_{rot}}\right)} \cdot \frac{1 - e^{-\theta_{vib}/T}}{1} \cdot e^{-\theta_D/T},$$

$$= \frac{e^{-\theta_D/T}}{\rho}\left[m\left(\frac{\pi mk}{h^2}\right)^{3/2} \theta_{rot}\sqrt{T}\left(1 - e^{-\theta_{vib}/T}\right)\frac{\left(Q^a_{el}\right)^2}{Q^{aa}_{el}}\right],$$

$$\equiv \frac{e^{\frac{-\theta_D}{T}}}{\rho}[\rho_D],$$

where ρ_D is defined as the characteristic density for dissociation and has values of:

$$\rho_D = 145 \rightarrow 123, \quad \text{for} \quad O_2, 1 \rightarrow 7000\text{ K};$$

$$\rho_D = 113 \rightarrow 118, \quad \text{for} \quad N_2, 1 \rightarrow 7000\text{ K}.$$

It is useful here to identify the meaning and results of the approximations proposed. Taking ρ_D = constant (average value)$\Rightarrow f(Q)$ = constant, specifically:

$$\rho_D = \frac{m}{2V} \cdot \frac{(Q^a)^2}{Q^{aa}} = \text{constant},$$

implies that

$$\rho_D = \frac{m}{2V} \cdot \frac{(Q^a)^2}{Q^{aa}} = \text{constant},$$

then $\ln(Q^a)^2 = \ln \text{constant} + \ln Q^{aa}$, and with $Q^a = Q^a_{tr} \cdot Q^a_{el} \approx Q^a_{tr} = \text{constant} \cdot VT^{3/2}$, $\therefore \frac{\partial}{\partial t} \ln Q^a = \frac{3}{2} \cdot \frac{1}{T}$. Finally: $e = 3R_{A_2}T - (1-\alpha)R_{A_2}\theta_D$.

For the atoms, neglecting electronic energy:

$$e_A = (e_A)_{tr} = \frac{3}{2}R_A T = 3R_{A_2}T.$$

For the molecules:

$$e_{A_2} = (e_{A_2})_{tr} + (e_{A_2})_{rot} + (e_{A_2})_{vib},$$

or:

$$3R_{A_2}T = \frac{3}{2}R_{A_2}T + R_{A_2}T + (e_{A_2})_{vib}; \quad \therefore (e_{A_2})_{vib} = \frac{1}{2}R_{A_2}T.$$

In other words, the ideal dissociating gas model introduces a half-excited vibration as an approximation to achieve analytic simplification.

Summary of Properties of High-Temperature Gases

We now present here, as examples, some results for the properties of air (which is primarily N_2 and O_2) over a range of increasing temperatures and for a range of pressures. As temperature increases and pressure decreases, the gases become further dissociated and ionized, with the resulting changes of composition, thermodynamic properties, and transport properties. The general effects are shown in Figure 3.1.

Specific changes in composition with temperature for two different pressures are shown in Figure 3.2; the effects of temperature on specific heat for three pressures are shown in Figure 3.3.

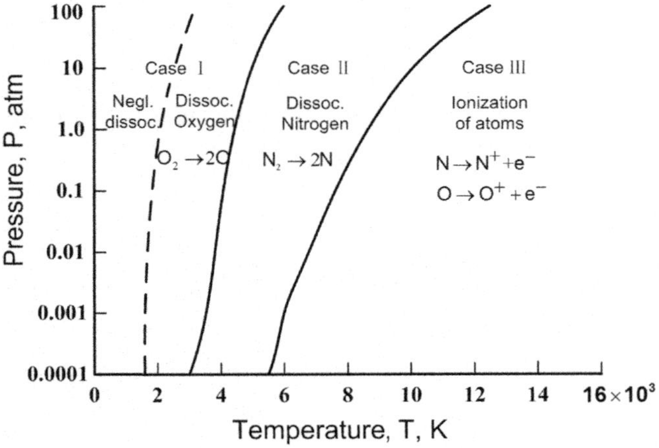

Figure 3.1 Domains of pressure and temperature for the chemical reactions of air. *From Hansen (1958) with permission.*

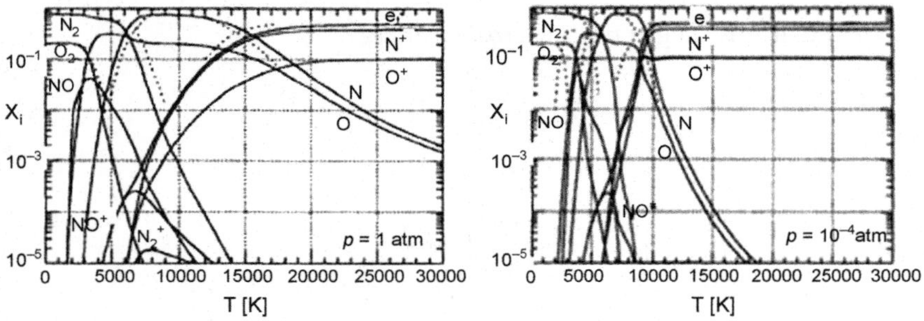

Figure 3.2 Equilibrium composition (mole fractions) of dissociated and ionized air, p = 1 atm and 10^{-4} atm. The dotted lines are mole fractions of the species O_2, N_2, O, N, and e^- from calculations of Gupta et al. (1991, NASA RP-1260) at temperature regions where dissociation and ionization are important. *From Selle et al. (1998) with permission.*

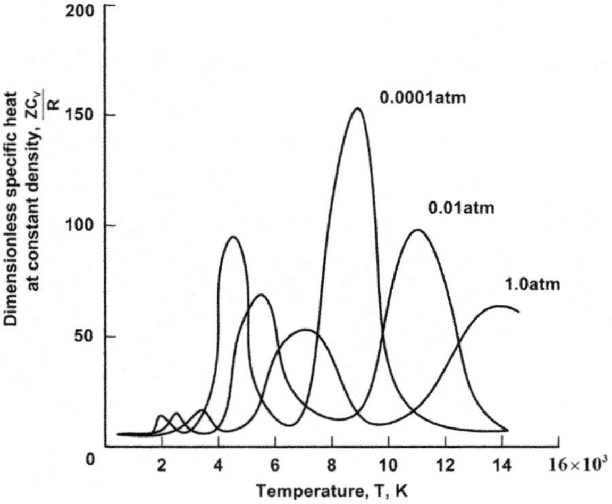

Figure 3.3 Specific heat of air at constant density as a function of temperature. *From Hansen, 1958 with permission.*

In considering the behavior of gases at high temperatures, it is important to emphasize the following points:

$$e, h = f(p, T), \quad \text{not only } f(T) \text{ as at low temperatures;}$$

$$C_V, C_p = f(T), \quad \text{not constant as for low temperatures;}$$

$$\gamma = f(T), \quad \text{not constant as for low temperatures.}$$

"Frozen Flow"—Definitions and Effects on Properties

We have been considering equilibrium gas mixtures (different species, energy components, and chemical reactions) where reaction rates are infinitely fast. In reality, reaction rates can be on the same time scale as flow events or even slower. Such events result in situations where energies that have been deposited in atomic and molecular excitation are not able to return to the random, kinetic energy component. Such energies are lost to the process or "frozen," and this occurrence is known as frozen flow, i.e., reaction rate $\ll$ flow rate. Frozen flow is the source of significant inefficiencies in high temperature and plasma flows.

We will consider here the basics of finite rate (time related) problems where at a point in the flow field the composition becomes fixed at some value. We will consider the effects that this has on specific heats, ratio of specific heats (gamma), and other properties. It can be specified that translation and rotation will always be in equilibrium. However, vibrational energy and dissociation–recombination chemical reactions can "freeze" and affect plasma properties.

Effect of Reaction Rate on Specific Heats

We consider here dissociation–recombination freezing (Vincenti and Kruger, 1965, VI.5). For a reacting gas mixture, the thermal equation of state can be written as $\frac{p}{\rho} = \eta \overline{R} T$, where $\overline{R}$ is the universal gas constant and η is the number of moles per unit mass of gas. However, also, for a reacting gas $\frac{p}{\rho} = ZR_0 T$, where $Z = 1 + \alpha$ and R_0 is the low-temperature gas constant. With $ZR_0 = \eta \overline{\mathrm{R}}$ and $R_0 = \eta_0 \overline{\mathrm{R}}$, $\mathrm{Z} = \frac{\eta}{\eta_0} = \frac{(H)}{(H_0)}$, where H is the number of moles of mixture at high temperature and H_0 is the number of moles of mixture at low temperature (a constant).

Now, frozen flow $\Rightarrow \eta$ (moles of mix in active flow) $=$ constant, $\therefore Z =$ constant, and we can write $\frac{p}{\rho} = ZR_0 T = \eta \overline{\mathrm{R}} T \equiv R_f T$, where the gas constant $R_f = \eta \overline{\mathrm{R}} = ZR_0 =$ constant.

R_f is the gas constant for "frozen flow" affected by dissociation but not by vibration freezing.

Let us now consider the effects of freezing on specific heats for a gas consisting only of monatomic (m) and diatomic (*d*) species. Then we can write:

$$C_{V_f}\left(\frac{\text{constant}}{\text{mass mix}}\right) = \eta_m\left(\frac{\text{moles}}{\text{mass mix}}\right)\overline{C_{V_m}}\left(\frac{\text{constant}}{\text{moles}}\right) + \eta_d\overline{C_{V_d}}.$$

But for monatomic species (*tr only*)

$$\overline{C_{V_m}} = \frac{3}{2}\overline{\mathrm{R}},$$

and for diatomic species (*tr, rot, vib*)

$$\overline{C_{V_d}} = A\overline{\mathrm{R}} \text{ where: } \begin{cases} A = \dfrac{5}{2} & \mathit{No\ VIB.(FROZEN)} \\ A = \dfrac{7}{2} & \mathit{VIB.EQUIL.} \end{cases}$$

Then:

$$C_{V_f} = \eta_m\,\frac{3}{2}\overline{\mathrm{R}} + \eta_d\,A\overline{\mathrm{R}} = \overline{\mathrm{R}}\left(\frac{3}{2}\eta_m + A\eta_d\right).$$

But if the gas is all diatomic at low temperature, $\eta_0 = \eta_{d_0}$, and the moles of diatomic particles dissociated at high temperature, $\eta_0 - \eta_d$, we get $\eta_m = 2(\eta_0 - \eta_d)$. However, also, $\eta_m = \eta - \eta_d$.

Then:

$$\eta_d = 2\eta_0 - \eta \rightarrow \eta_0(2 - Z) \quad \text{and} \quad \eta_m = 2(\eta - \eta_0) \rightarrow 2\eta_0(Z - 1).$$

So:

$$C_{V_f} = R_0\left[(2A - 3) + (3 - A)Z\right],$$

where values of A relate to frozen or equilibrium vibration and Z values relate to dissociation.

Also:

$C_{p_f} = C_{V_f} + R_f = C_{V_f} + ZR_0$, and values of $\gamma = C_p/C_V$, change accordingly.

Example 1: Composition and vibration frozen at the low-temperature state (e.g., through a strong normal shock wave).

$$\text{Low temperature} \Rightarrow \mathrm{A} = \frac{5}{2}, \quad Z \approx 1.0, \quad \text{so } C_{V_f}$$
$$= R_0\left[(5-3) + \left(3 - \frac{5}{2}\right)\cdot 1\right] = \frac{5}{2}R_0.$$

Then:

$$C_{p_f} = \frac{7}{2}R_0 \quad \text{and} \quad \gamma = \frac{C_p}{C_V} = \frac{7}{5} = 1.4.$$

Example 2: Composition and vibration frozen at high-temperature state (e.g., flow in nozzle with high expansion ratio).

$$\text{High temperature} \Rightarrow \mathrm{A} = \frac{7}{2}, \quad Z \approx 2.0, \quad \text{so } C_{V_f}$$
$$= R_0\left[(7-3) + \left(3 - \frac{7}{2}\right)\cdot 2\right] = 3R_0$$

Then:

$$C_{p_f} = 5R_0 \quad \text{and} \quad \gamma = \frac{C_p}{C_V} = \frac{5}{3}.$$

With $R_f = ZR_0 = 2R_0$, the sound speed is $c_f^2 = \gamma RT = 3.33R_0T$ for high-temperature frozen flow versus $1.4R_0T$ for low-temperature gas. As compressible flow effects are related to Mach number, $M_f = \frac{v}{c} \rightarrow \frac{v}{c_f}$, frozen flow can result in flow fields (in an expansion, for example) that are modified significantly by these reduced Mach number effects.

Chemical Kinetics—Law of Mass Action

We have discussed particle kinetics in the case where no changes in molecules occur during the collision process, i.e., the gas is chemically inert. However, suppose that we now consider chemical reactions. Consider a typical reaction:

$$2HI \rightleftharpoons H_2 + I_2,$$

i.e., 2 mol of HI decompose into 1 mol of H_2 and 1 mol of I_2, or 1 mol of H_2 and 1 mol of I_2 combine to 2 mol of HI.

Why would this reaction move forward or backward? What energies are involved?

Chemical equilibrium implies a balance between forward and backward reaction rates, and:

$$\text{Reaction} \Rightarrow \text{collision of particles!}$$

Accordingly, the rate of reaction can be expressed as:

$$\left(\frac{\text{reactions}}{\text{time}-\text{vol}}\right) = \left(\frac{\Delta\text{molec}}{\text{time}-\text{vol}}\right) = \frac{\partial n_1}{\partial t} = -k^* n_1^{S_1} n_1^{S_2},$$

where $S = \sum_i S_i$ is called the order of the reaction and $k^*(T)$ is called the reaction rate constant.

From binary collision theory:

$$Z_{12}\left(\frac{\text{collision}}{\text{time}-\text{vol}}\right) = n_1 n_2 \sigma_{12}^2 \left(\frac{8\pi\pi kT}{\mu}\right)^{1/2}.$$

So, by analogy for $H_2 + I_2 \rightarrow 2HI$: $\frac{\text{collision}}{\text{time}-\text{vol}} = n_{H_2} n_{I_2} k_B(T)$ $(B \Rightarrow$ backward).

Similarly, $HI + HI \rightarrow H_2 + I_2$: $\frac{\text{collision}}{\text{time}-\text{vol}} = n_{HI} n_{HI} k_F(T)$ $(F \Rightarrow$ forward).

However, for *equilibrium* the number of decomposing collisions equals the number of formative collisions, so:

$$n_{HI}^2 \cdot k_F(T) = n_{H_2} \cdot n_{I_2} k_B(T),$$

or:

$$\frac{n_{H_2} \cdot n_{I_2}}{n_{HI}^2} = \frac{k_F(T)}{k_B(T)}; \text{ this is the } \textit{Law of Mass Action}.$$

We can also define the equilibrium constant as $\mathcal{K}_c(T) = \frac{k_F(T)}{k_B(T)}$, based on moles, or we can define $\mathcal{K}_p(T)$ based on species pressure.

IONIZATION IN GASES

Introduction

The process of ionization involves the creation of electrons that are free from the bonds of atomic structure. In gases the creation of free electrons can occur in a number of ways, including collisions and radiation. The process of ionization is oftentimes balanced by the process of deionization.

The effectiveness of ionization processes depends very much on the gas pressure. Pressure is a variable that indicates the random thermal energy of particle motion. Pressure is a product of particles per unit volume and energy per particle. Gases at low pressure with higher energy per particle ionize more easily.

The understanding of ionization depends very much on an understanding of the quantum model atom. In this model electrons reside in shells with various energies. The electrons in the outermost shells are bound to the nucleus with lower energies, so it is easier to ionize electrons in the outermost shells. The ionization process involves delivery of energy to an electron that is greater than the energy holding the electron in its atomic shell. This energy can be delivered by collision with other electrons, with ions, with neutral atoms or molecules, or with radiation. Generally though, as collision processes dominate, the ionization process can be analytically described by collision formulations. One of the more interesting questions in understanding ionization is how ionization takes place: how to define the probability of occurrence in one step, that is, bound to free, and that of ionization in multisteps, that is, how the electron gains energy in increments before a final step in which the electron gains a cumulative energy that makes it free of the atomic or molecular bond. The discussions to follow are basic and are intended to provide a framework for understanding the detailed physical processes that are component in the composite of events in any particular plasma.

The details of the creation and dynamics of ionization depend in many ways upon how energy is provided to the gas (Brown, 1959). In most cases the energy is provided in a way that allows electrons to respond by absorption, and then with subsequent equilibration with other species. This is typical in devices that create potential differences to drive current flow. So most general discussions of ionization occur within the context of gas discharge devices.

There are ionizing events, however, such as with collisional shock waves, in which all the species of the gas are suddenly exposed to a common heating mechanism. Plasma shock waves can have much more complicated heating processes, and this will be discussed later. Likewise, some plasma devices require heating after formation of plasmas, and specific heating techniques that can heat electrons or ions and can be carried out for plasma within certain density and temperature ranges. Finally, we define the physics of different mechanisms for radiation loss, as this is always an important factor in plasma energy balance.

The calculation of degree of ionization and properties of species undergoing ionization can be difficult in general. The details of the ionization process for a given gas, for a given range of properties for participating particles, remain very much a subject of research. Perhaps because of the complexities of the physics, most works of applied research on plasmas do not dwell on the specific details of particle excitation that defines gas behavior. Rather, the assumption of equilibrium ionization is usually made, and the related Saha equation for species densities is applied. We review the derivation of that relationship in the final section of this chapter, and we recognize its widespread application in plasma analysis.

Atomic Structure and Electron Arrangements

To explain experimental observations, Bohr proposed a model for the hydrogen atom that defines a dense core with an electron moving in a circular orbit (Richtmeyer et al., 1955). This dense core has the mass of the H atom, and it has a positive charge equal to that of the rotating electron. Quantum mechanics was necessary to explain the behavior of this system, which did not radiate energy from the rotating electron. Bohr proposed that angular momentum, and therefore the radius a, must exist at only certain multiples and that electrons do not radiate at those radii. The moment of momentum can be expressed as $m_e v_e a = \frac{nh}{2\pi}$, where n is the integer quantum number and h is the Planck's constant.

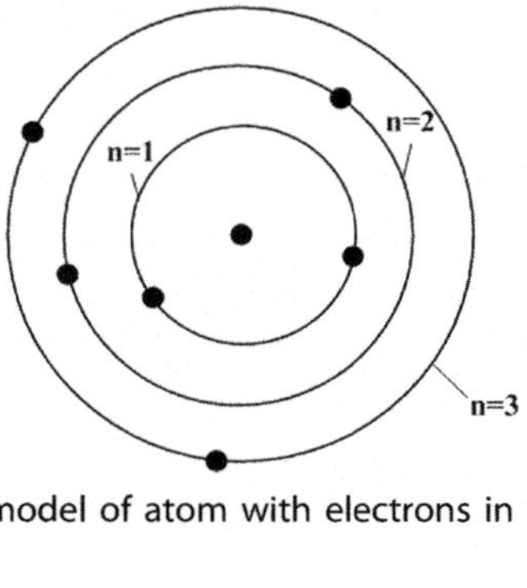

Figure 3.4 Bohr's model of atom with electrons in integer values of *n*.

Table 3.1 Shell Distribution of Electrons (Richtmeyer et al., 1955)

Shell symbol	n	l	Designation of electrons in shell	Number of electrons in complete subshell	Elements in which subshell is completed	Configuration of completed shell
K	1	0	1s	2	He	$1s^2$
L	2	0	2*s*	2	Be	$2s^2 2p^6$
	2	1	2*p*	6	Ne	
M	3	0	3*s*	2	Mg	$3s^2 3p^6 3d^{10}$
	3	1	3*p*	6	A	
	3	2	3*d*	10	Cu	
M	4	0	4*s*	2	Zn	$4s^2 4p^6 4d^{10} 4f^{14}$
	4	1	4*p*	6	Kr	
	4	2	4*d*	10	Pd	
	4	3	4*f*	14	Lu	
O	5	0	5*s*	2	Cd	
	5	1	5*p*	6	Xe	
	5	2	5*d*	10	Au	
P	6	0	6*s*	2	Ba	
	6	1	6*p*	10	Rn	
Q	7	0	7*s*	2	Ra	

Most importantly, radiation would occur when an electron moved from one orbit to another for $n = 1, 2, 3, \ldots$ For more complex atoms, the nucleus becomes a multiple of the hydrogen nucleus and the number of electrons in shells becomes equal to the electric charge of the nucleus.

Accordingly, the orbital angular momentum will change, having the designation l. The simple Bohr model is shown in Figure 3.4, and the arrangement of shells and angular momentum for characteristic atoms is shown in Table 3.1 (Richtmeyer et al., 1955).

Ionization Potentials for Different Gases

Compatible with the radius and angular momentum of the electrons, each electron possesses energy for a specific quantum level. Changes in shell position are consistent with absorption and emission of energy increments. For different atoms, the energies compatible with ionization from different shells are shown in Table 3.2 ($1\ \text{eV} = 1.6 \times 10^{-19}\ \text{J}$). The periodic table is presented (Table 3.3) to emphasize the order and regularity of the values.

Table 3.2 Representative Ionization Potentials in Electronvolts

Element	Atomic number	Stage of ionization					
		I	II	III	IV	V	VI
Aluminum	13	5.984	18.823	28.44	119.96	153.77	190.42
Argon	18	15.755	27.62	40.90	59.79	75.0	91.3
Calcium	20	6.111	11.87	51.21	67	84.39	
Carbon	6	11.264	24.376	47.864	64.476	391.986	480.84
Cesium	55	3.893	25.1				
Helium	2	24.580	54.400				
Hydrogen	1	13.505					
Iron	24	7.90	16.18	30.64			
Krypton	34	13.99	24.56	36.9			
Magnesium	12	7.644	15.03	80.12	109.29	141.23	186.86
Mercury	80	10.44	18.8				
Niobium	41	6.77	14				
Nitrogen	7	14.54	29.605	47.426	77.450	97.963	551.92
Oxygen	8	13.614	35.146	54.934	77.394	113.873	138.08
Platinum	78	8.9	18.5				
Potassium	19	4.339	31.81	46	60.90		90.7
Silicon	14	8.149	16.34	33.46	45.13	166.73	205.11
Silver	47	7.574	21.48				
Sodium	11	5.138	47.29	71.65	98.88	138.60	172.36
Strontium	38	5.692	11.027		57		
Titanium	81	6.83	13.63	28.14	43.24	99.8	120
Tungsten	74	7.94					
Xenon	54	12.13	21.2				

Table 3.3 Periodic Table of Elements

Period	Series	GROUP I	II	III	IV	V	VI	VII	VIII	O
1	1	1 H 1.0080								2 He 4.003
2	2	3 Li 6.940	4 Be 9.013	5 B 10.82	6 C 12.010	7 N 14.008	8 0 16.000	9 F 19.00		10 Ne 20.183
3	3	11 Na 22.997	12Mg 24.32	13Al 26.97	14Si 28.06	15 P 30.98	16 S 32.066	17Cl 35.457		18 A 39.944
4	4	19 K 39.096	20Ca 40.08	21Se 45.10	22Ti 47.90	23 V 50.95	24Cr 52.01	25Mn 54.93	26Fe 27Co 28Ni 55.85 58.94 58.69	
	5	29Cu 63.54	30Zn 65.38	31Ga 69.72	32Ge 72.60	33As 74.91	34Se 78.96	35Br 79.916		36 Kr 83.7
5	6	37Rb 85.48	38Sr 87.63	39 Y 88.92	40Zr 91.22	41Nb 92.91	42Mo 95.95	43Te	44Ru 45Rh 46Pd 101.7 102.91 106.7	
	7	47Ag 107.88	48Cd 112.41	49In 114.76	50Sn 118.70	51Sb 121.76	52Te 127.61	53 I 126.92		54 Xe 131.3
6	8	55Cs 132.91	56Ba 137.36	57-71 Rare earths[a]	72Hf 178.6	73Ta 180.88	74W 183.92	75Re 186.31	76Os 77Ir 78Pt 190.2 193.1 195.23	
6	8	55Cs 132.91	56Ba 137.36	57-71 Rare earths[a]	72Hf 178.6	73Ta 180.88	74W 183.92	75Re 186.31	76Os 77Ir 78Pt 190.2 193.1 195.23	
	9	79Au 197.2	80Hg 200.61	81Ti 204.39	82Pb 207.21	83Bi 209.00	84Po 210	85At		86 Rn 222
7	10	87Fr	88Ra 226.05	89 Actinide Series[b]						

[a]Rare earths:

57La 132.92	58Ce 140.13	59Pr 140.92	60Nd 144.27	61Pm	62Sm 150.43	63Eu 152.0	64Gd 156.9	65Tb 159.2	66Dy 162.46
67Ho 164.94	68Er 167.2	69Tm 169.4	70Yb 173.04	71Lu 174.99					

[b]Actinide series:

89Ac 227	90Th 232.12	91Pa 231	92 U 238.07	93Np	94Pu	95Am	96Cm	97Rk	98Cf

Atomic weights as adopted by the International Union of Chemistry.

Ionization Processes

Ionization by Electron Collision

Electron collisions with neutral particles are elastic at low energies—so kinetic energy is preserved in collision. The scattering of electrons during the collisions is of much interest and is extensively documented for different atoms. Excitation collisions can occur when the energy of the electrons exceeds that of the excitation potential of the atom. As momentum must be conserved, the change of electron momentum must be balanced by that of the atom. The angular distribution of scattered electrons is similar to that for elastic collisions. The bound electrons in the atom then exist in an excited state. It is possible for more than one electron in an atom to exist in an excited state and be potential emitters of radiation.

Ionization by single-electron impact is possible when the electron energy exceeds the ionization potential, eV_i. This probability can be approximately expressed as ionization efficiency (Francis, 1960) $S_e = \frac{dN}{dx} = ap(V_e - V_i)$, where S_e is the number (N) of ion pairs produced by one electron with energy $\varepsilon = eV_e$ moving 1 cm, a is a constant, and p is the gas pressure. This behavior is shown for various gases at a pressure $p = 1$ mm Hg in Figure 3.5. With increasing electron energy the ionization efficiency increases to a maximum and then decreases.

The ionization of excited atoms begins at much lower energies, and the ionization energy can be many times larger than those for electron collisions with unexcited atoms (Burgess et al., 1977). The binding energy of outer electrons with quantum numbers n, l can be written for hydrogen as (Book, 1990, p.52):

$$E_{\infty}^{z}(n, l) = -\frac{Z^2 E_{\infty}^{H}}{(n - \Delta l)^2},$$

where $E_{\infty}^{H} = 13.6$ eV is the hydrogen ionization energy, $\Delta l = 0.75 l^{-5}$, and Z is the charge state of the atom ($Z = 0$ for the neutral atom).

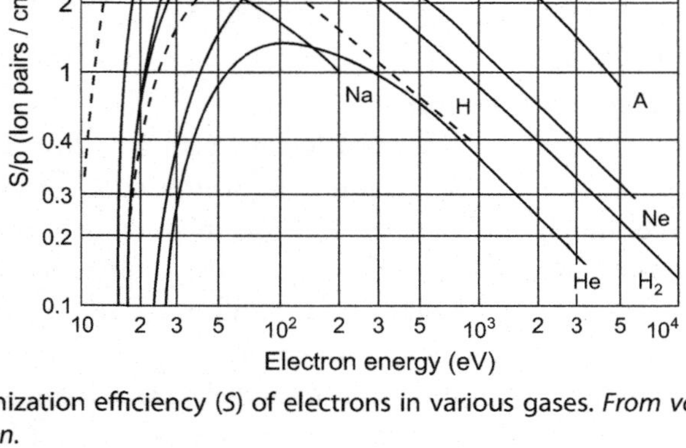

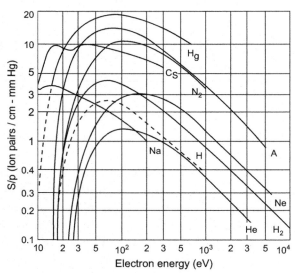

Figure 3.5 Ionization efficiency (*S*) of electrons in various gases. *From von Engle, 1956 with permission.*

The classical ionization cross-section for an atomic shell, j, is (Book, 1990, p. 54):

$$\sigma_i = 6 \times 10^{-14} b_j g_j(x)/v_j \ (\text{cm}^2),$$

where b_j is the number of shell electrons; v_j is the binding energy of the ejected electrons; $x = \in/v_j$, where $\in$ is the incident electron energy; and g is a universal function with minimum value, $g_{\text{min}} = 0.2$ at $x = 0.4$.

The ionization rate from ground state averaged over a Maxwellian electron distribution for $0.2 \leq T_e/E_\infty^Z \leq 100$ (Book, 1990, p.54) is:

$$S(z) = 10^{-5} \frac{\left(T_e/E_\infty^Z\right)^{1/2}}{\left(E_\infty^Z\right)^{3/2}\left(6.0 + T_e/E_\infty^Z\right)} e^{-E_\infty^Z/T_e} \ (\text{cm}^3/\text{s}),$$

where E_∞^Z is the ionization energy of Z charge state and T_e is the electron temperature, both in electronvolts.

Ionization by Heavy Particle Collisions

The collision of two atoms has a low probability of resulting in electrons being ejected, resulting in ionization. However, charge transfer collisions that result in electron transfer from a neutral atom to an impinging ion can be an important process. Also, in some gases in some energy ranges, atoms and ions can be effective in producing ionization, particularly for heavier particles. However, the energies where this effect becomes significant are generally more than twice the value of eV_i.

Photoionization

Ionization by photon impingement occurs when $h\nu > eV_i$. Conservation of energy dictates that an electron will be ejected with an energy equal to $(h\nu - eV_i)$. The wavelength of radiation necessary is $\lambda_i \leq 12{,}400/V_i$ (eV). The wavelength range of 200–1000 $\times$ 10^3nm will be effective for ionization. Photoabsorption cross-sections for various gases are shown in Figure 3.6 (Francis, 1960).

The photoionization cross-section for ions of level n, l is (Book, 1990, p. 34):

$$\sigma_{ph}(n, l) = 1.64 \times 10^{-16} \frac{Z^5}{n^3 K^{7+2l}} \ (\text{cm}^2),$$

where K is the wave number in Rydbergs (1.0974×10^5/cm).

The development and availability of lasers with a wide range of wavelength output and power has added considerable flexibility to the use

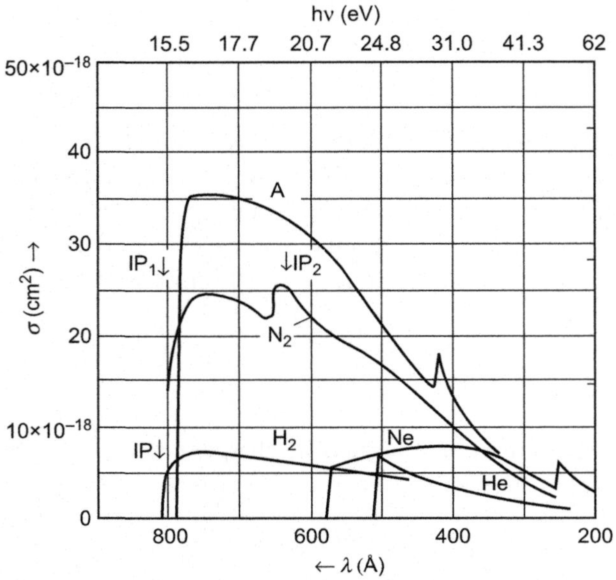

Figure 3.6 Photoabsorption cross-sections in various gases. *From Francis, 1960 with permission.*

of radiation for the initiation of ionization and the absorption of radiated energy. Early studies (Tozer, 1965) described and analyzed laser ionization processes. A general functional form for the effects of absorption and the number of electrons produced during the pulse time takes the form $N_e \sim e^{n_\nu \sigma t}$, where n_ν is the photon flux ($1/\text{cm}^2$ s), σ is the cross-section for excitation of the atom by N_ν photons, and t is the pulse time. However, the complexity of the atomic processes oftentimes results in anomalous laser-induced ionization rates (Bettis, 2009). Generally, once electrons are available, the effects of electric fields, as in laser beams, result in the significant growth of the electron population.

Ionization (Electron) Loss Mechanisms

Recombination

As plasmas are generally electrically neutral with $n_i = n_e$, reactions that recombine electrons and ions are active and in balance. Their rate can be expressed as:

$$\frac{dn_{+,-}}{dt} = \alpha_r n_{+,-}^2 \quad \text{where } \alpha_r \text{ is the recombination coefficient.}$$

Most plasmas have different types (+, −) of ions, i.e., different species of ions as well as multiple charges on ions, so recombination can be a complicated process. Ion–ion (+, −) recombination usually results in an electron in an excited state, and there is subsequent emission of radiation. This recombination coefficient is usually several orders of magnitude higher than that for electron–ion recombination as the ions are moving at lower speeds. Electrons can also combine with molecules, which then results in subsequent dissociation.

A relationship for electron–ion radiative recombination rate has been given (Book, 1990, p.54 cites Seaton) as (1 eV $< T_e/Z^2 <$ 15 eV):

$$\alpha_r(z) = 2.7 \times 10^{-13} Z^2 T_e^{-1/2}\ (\text{cm}^2/\text{s}),$$

where Z is the ion charge and T_e is in electronvolts.

Electron Diffusion

There are a number of factors that result in loss of electrons and so reduce the degree of ionization. A basic property of electrons is that with energy equilibration among species, their small mass means that electrons possess high random speed. This property results in enhanced electron transport, and so diffusion loss processes must always be considered. Diffusion is related to the gradient in population, and in the absence of magnetic fields, it can be expressed as:

$$(\text{Particles/time-area})\ \Gamma_e = -D_e \nabla n_e \quad \text{with} \quad D_e = \frac{1}{2}\bar{c}_e \lambda_e.$$

We can also define $\Gamma_e \equiv n_e v_{e,\text{drift}}$, so an average drift velocity can be evaluated and is an appropriate descriptor for diffusive movement of the electrons.

Radiation (Energy) Loss from Plasmas

While plasmas are created, generally, by energy input from external sources and exist at high temperatures, it is natural that plasma can lose energy in a number of ways, including radiation. At elevated temperatures with excited atomic and molecular states, radiation from electrons, atoms, and ions will occur and will be a significant factor in the energy balance. Some important diagnostics utilize this characteristic in a useful application.

The optical thickness (τ) of a plasma is an important indicator of radiation properties and is related to the coefficient of absorption per unit length, α_w, in the plasma; it is given by $\tau = \int_0^s \alpha_w dS$, and plasma is defined as optically thick when $\tau \ll 1$. The radiation from the surface of an (optically thick) blackbody can be expressed as $W = 1.03 \times 10^5 T^4$ (Watt/cm^2). Bremsstrahlung from hydrogen-like plasmas can be expressed as (Book, 1990, p. 56):

$$P_{Br} = 1.69 \times 10^{-32} N_e T_e^{1/2} \sum [Z^2 N(Z)] \left(\frac{\text{Watt}}{\text{cm}^2}\right),$$

and the bremsstrahlung optical depth can be expressed as:

$$\tau = 5.0 \times 10^{-38} N_e N_i Z^2 \bar{g} \Delta s T^{-7/2},$$

where $\bar{g}$ is the Gaunt factor and N_e, and N_i are the number densities. Recombination (free–bound) radiation can be expressed as:

$$P_{Br} = 1.69 \times 10^{-32} N_e T_e^{1/2} \sum \left[Z^2 N(Z) \frac{E_\infty^{Z-1}}{T_e} \right]$$

An additional important radiation component from line (bound–bound) transitions can be calculated for different species. Also, the effects on radiation related to particles in magnetic fields can be substantial.

Equilibrium Ionization in Gases—The Saha Equation

We consider here an analysis of equilibrium based on chemical reaction kinetics as was developed above. An alternate approach is based on thermodynamic equilibrium and statistical mechanics. We examine atomic gases; the extension of this analysis to include other reactions is relatively straightforward (Cambel, 1963).

In equilibrium, every atomic excitation is exactly balanced by the reverse process (the law of detailed balancing). For a monatomic gas, ionization and recombination are represented as:

$$A + \epsilon_i \rightleftharpoons A^+ + e,$$

where ϵ_i is the ionization energy. We can describe this reaction by:

$$\frac{N_+ N_e}{N_A} = K(T), \quad \text{the Equilibrium Constant.}$$

This is really the law of mass action. To solve for n_+, n_e, n_A, we take $n_+ = n_e$, and we have $n_+ + n_A \equiv n_0$, the total number density of heavy particles, which is known. To determine the equilibrium constant, we have from atomic physics, in equilibrium (Vincenti and Kruger, 1965):

$$\frac{N_i}{N} = \frac{g_i e^{-\frac{u_i}{kT}}}{Z} \rightarrow \frac{e^{-\frac{u_k}{kT}}}{Z},$$

for each species. It can be shown:

$$\frac{N_+ N_e}{N_A} = \frac{Z_+ Z_e}{Z_A},$$

where Z_x is the total partition function for the species, x. All terms must be based on the same energy reference level. Then:

$$\frac{n_+ n_e}{n_A} V = \frac{Z_+ Z_e}{Z_A}, \quad \text{with } Z_A(\text{total}) = Z_A^{tr} Z_A^{int},$$

where *int* refers to the electron excitation to a higher quantum number state. As above,for the atoms:

$$Z_A(\text{total}) = \frac{V(2\pi m_A kT)^{3/2}}{h^3} \cdot \sum_{j,int} g_j e^{-\frac{\epsilon_j}{kT}}.$$

For the ions, however, care must be taken for the energy level. This is best envisioned with a Grotrian diagram (Figure 3.7). Then for the ions:

$$Z_+ = Z_+^{tr} \cdot Z_+^{int} = \frac{V(2\pi m_+ kT)^{3/2}}{h^3} \cdot \sum_j g_j^+ e^{-\frac{\tilde{\epsilon}_j^+}{kT}}$$

$$= \frac{V(2\pi m_+ kT)^{3/2}}{h^3} \cdot e^{-\frac{\epsilon_i}{kT}} \sum_j g_j^+ e^{-\frac{\epsilon_j^+}{kT}}.$$

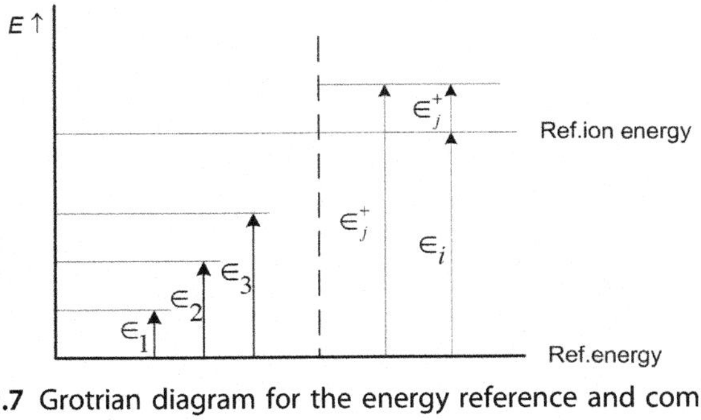

Figure 3.7 Grotrian diagram for the energy reference and components.

For the electrons:

$$Z_e = Z_e^{tr} \cdot Z_e^{int} = \frac{V(2\pi m_e kT)^{3/2}}{h^3} \cdot Z_e^{int}, \quad \text{with } Z_e^{int} = g_j^- = z.$$

Due to the fact that electrons have spins in opposite direction, there is degeneracy; these levels can be split by magnetic field. Then:

$$\frac{n_+ n_e}{n_A} = \frac{1}{V}\frac{Z_+ Z_e}{Z_A} = 2\frac{(2\pi m_e kT)^{3/2}}{h^3}\frac{\sum_j g_j^+ e^{-\frac{\epsilon_j^+}{kT}}}{\sum_j g_j e^{-\frac{\epsilon_j}{kT}}} e^{-\frac{\epsilon_L}{kT}}.$$

Now:

1. $\alpha = \frac{n_+}{n_A + n_+} = \frac{n_+}{n_0} = \frac{n_e}{n_0}$,
2. $n_+ = n_e$,
3. $p = (n_+ + n_e + n_A)kT = n_0(1+\alpha)kT$.

On substituting:

$$\frac{\alpha n_0 \cdot \alpha n_0}{(1-\alpha)n_0} = 2\frac{(2\pi m_e kT)^{3/2}}{h^3}\frac{Z_+^{\text{int}}}{Z_A^{\text{int}}} e^{-\frac{\epsilon_L}{kT}}.$$

But with:

$$p = (1+\alpha)n_0 kT,$$

$$\frac{\alpha^2}{1-\alpha^2} = 2\frac{(2\pi m_e)^{3/2}(kT)^{5/2}}{ph^3}\frac{Z_+^{\text{int}}}{Z_A^{\text{int}}} e^{-\frac{\epsilon_L}{kT}}.$$

And we have $\alpha = \alpha(p, T, \text{atomic properties})$. So, given gas pressure, temperature, and the specification of equilibrium we can calculate the degree of ionization. This is known as the Saha equation for equilibrium ionization.

We can develop some simpler, approximate dependencies (Cambel et al., 1963). For low temperatures, $(Z_+/Z_A)^{int}$ is approximately constant and can be given a value. Also, $kT \ll \epsilon_i$; $\alpha \to 0$, so:

$$\frac{\alpha^2}{1-\alpha^2} \to \alpha^2 = \text{constant} \cdot p^{-1} T^{5/2} e^{-\frac{\epsilon_L}{kT}}. \quad \text{For:} \quad \alpha \ll 1 \Rightarrow p \approx n_0 kT,$$

$$\therefore \alpha = \text{constant} \cdot n_0^{-1/2} k^{-1/2} T^{3/4} e^{-\frac{\epsilon_L}{2kT}} \quad \text{and} \quad n_+ \approx \alpha n_0 \sim n_0^{1/2}.$$

The standard approximation, $\frac{2Z_1^{elec}}{Z_0^{elec}} = 1$, is not accurate in all cases. Improved approximations have been derived for some gases of interest (Cambel et al., 1963). For argon, we can write:

$$\frac{2Z_1^{elec}}{Z_0^{elec}} = \frac{(2)\left[4 + 2e^{-2062/T}\right]}{1}.$$

For singly ionized nitrogen (5000–30,000 K), we can write:

$$\frac{n_+ \times n_-}{n_o} = 35.3\left(\frac{2\pi m_e k}{h^2}\right)^{3/2} T^{5/4} e^{-\frac{\epsilon_L}{\kappa T}}.$$

REFERENCES

Bettis, J.R., 2009. Anomalous laser-induced ionization rates of molecules and rare-gas atoms. Phys. Rev. A 80, 063420.

Book, D.L., 1990. NRL Plasma Formulary. Naval Research Laboratory, Washington, DC.

Burgess, A., Summers, H.P., Cochrane, D.M., McWhirter, R.W.P., 1977. Cross-sections for ionization of positive ions by electron impact. Mon. Not. R. Astro. Soc. 179, 275–292.

Brown, S.C., 1959. Basic Data of Plasma Physics. Wiley & MIT, New York.

Cambel, A.B., 1963. Plasma Physics and MagnetoFluidmechanics. McGraw-Hill, New York.

Cambel, A.B., Duclos, D.P., Anderson, T.P., 1963. Real Gases. Academic, New York.

Clarke, J.F., McChesney, M., 1964. The Dynamics of Real Gases. Butterworths, London.

von Engle, A., 1956. In: Flugge, S. (Ed.), Handbuch der Physik, vol. 21. Springer, Heidelberg.

Francis, G., 1960. Ionization Phenomena in Gases. Butterworths, London.

Gupta, R.N., et al., 1991. RP-1260 NASA Langley Research Center, Hampton, VA.

Hansen, C.F., 1958. Approximations for the Thermodynamic and Transport Properties of High-temperature Air. NACA, Washington, DC. TN 4150.

Lighthill, M.J., 1951. Dynamics of dissociating gases, Part 1. J. Fluid Mech. 2 (1).

Moelwyn-Hughes, E.A., 1961. Physical Chemistry. Pergamon, New York.

Richtmeyer, F.K., Kennard, E.H., Lauritsen, T., 1955. Introduction to Modern Physics, fifth ed. McGraw-Hill, New York.

Slater, J.C., 1939. Introduction to Chemical Physics. McGraw-Hill, New York.

Selle, S., Riedel, U., Warnatz, J., 1998. Reaction rates and transport coefficients of ionized species in high temperature air. In: Proc. 3rd Eur. Symp. On Aerothermo-dynamics of Space Vehicles. ESA, Noorwijk, Neth.

Tozer, S.A., 1965. Theory of ionization of gases by laser beams. Phys. Rev. Lett. 137, A1665.

Vincenti, W.G., Kruger, C.H., 1965. Introduction to Physical Gas Dynamics. Wiley, New York.

CHAPTER 4

Electromagnetics

INTRODUCTION

There are a large number of widely used textbooks on Electricity and Magnetism that can be reviewed for a detailed and comprehensive treatment of the subject in this chapter. We will be concerned here with the following basic problems: (1) How charged particles generate electric and magnetic fields; (2) How charged particles interact with each other; and (3) How charged particles interact with applied electric and magnetic fields. The presentation here utilizes the material covered in the works of Stratton (1941) and Jackson (1962) as fundamental.

There are two sets of widely used units: Gaussian and rationalized meter, kilogram, second (MKS). The basic units for each system are as follows: Gaussian—q(statcoul), i(statamp), B(Gauss); Rationalized MKS—q(Coulomb), i(Amperes), B(Weber/m^2). Maxwell's equations have six constants. In Gaussian units, three constants have unity value, but the equations include c, the speed of light. In MKS units, two constants are unity, and the equations units are physically meaningful; these will be preferred in this work and used whenever convenient. A table with conversion between these sets of units is provided in Appendix A (Jackson, 1962).

We accept the fact that atomic-scale particles can possess electrical charges, and those charges can be positive (proton) or negative (electron); charges will be multiples of the electron unit charge (1.6×10^{-19} Coulomb). The electron mass, $m_e = 9.1 \times 10^{-31}$ kg, is much smaller than the proton unit mass, $m_p = 1.67 \times 10^{-27}$ kg.

ELECTRIC CHARGES AND ELECTRIC FIELDS—ELECTROSTATICS

Consider two particles with charges q_1, q_2 and separated in vacuum by a distance r.

Then the force acting on charge q_2 due to the presence of q_1 is:

$$\vec{F_2} = \frac{q_1 q_2}{4\pi\varepsilon_0 r^2}\vec{r}.$$

Introduction to Plasmas and Plasma Dynamics
ISBN 978-0-12-801661-9

This is Coulomb's law, where $\varepsilon_0 = 8.854 \times 10^{-12}$ farad/m (permittivity of free space) and $\overrightarrow{r}$ is a unit vector.

It can be seen that like charges repel and unlike charges attract; the "Stuff" between particles can affect the force on the particles. There is no direct connection—this is action at a distance—yet forces are exerted on the particles. Action at a distance is referred to as due to a "field." As we are dealing with plasmas, we are interested in generating fields that make particles do what we want.

The electric field due to an object (particle) with charge (q_1) will act on a second, small "test" charge, q, as:

$$\overrightarrow{E_{q_1}} = \frac{\overrightarrow{F_q}}{q} = \frac{1}{q}\cdot\frac{qq_1}{4\pi\varepsilon_0 r^2}\overrightarrow{r} = \frac{q_1}{4\pi\varepsilon_0 r^2}\overrightarrow{r} = \overrightarrow{E}(r),$$

so:

$$\overrightarrow{F_q} = q\overrightarrow{E}.$$

Now, Force/particle $= q\overrightarrow{E}$, but in plasma we have a large number of (mixed) particles in a volume. For a volume (of plasma) filled with a large number of charged particles and with the volume (V) having a net charge, q_1, we can write:

$$q_1 = \int_V \rho_e dV,$$

where the charge density, $\rho_e(\frac{\text{charge}}{\text{vol}}) = e(n^+ - n^-)$.

Then:

$$\frac{\text{Force}}{\text{vol}} = \frac{\text{Force}}{\text{particle}}\cdot\frac{\text{particle}}{\text{vol}} = qE\cdot\frac{\rho_e}{e} = \rho_e E.$$

Now interpreting q_1 in Coulomb's law for a particle to predict the field generated from the group of particles:

$$\overrightarrow{E} = \frac{1}{4\pi\varepsilon_0 r^2}\overrightarrow{r}\int_V \rho_e dV,$$

and evaluating both sides over the surface area (normal, $\overrightarrow{n}$) of the volume, we can write:

$$\int_A \overrightarrow{E}\cdot\overrightarrow{n}dA = \int_A \frac{\overrightarrow{r}\cdot\overrightarrow{n}}{4\pi\varepsilon_0 r^2}\left(\int_V \rho_e dV\right)dA.$$

Applying the divergence theorem on the left-hand side (LHS), and integrating the right-hand side (RHS), we get:

$$\int_V \nabla \cdot \vec{E}\, dV = \frac{1}{\varepsilon_0} \int_V \rho_e dV, \quad \text{the integral form, Gauss law, or}$$

$$\nabla \cdot \vec{E} = \frac{\rho_e}{\varepsilon_0}, \quad \text{the differential form, Gauss law.}$$

From this equation we can see:

1. If $n^+ = n^-$, $\rho_e = 0$, then $\nabla \cdot \vec{E} = 0$ i.e., electric field is conserved ($E_{\text{in}} = E_{\text{out}}$).
2. In a medium other than a vacuum:

$\nabla \cdot \varepsilon \vec{E} = \rho_e$, and $\vec{D}$ (electric displacement vector) $\equiv \varepsilon \vec{E}$, so, $\nabla \cdot \vec{D} = \rho_e$.

This accounts for polarization of a medium. But what does this mean physically?

The mathematical statements can be interpreted physically as follows:

1. $\nabla \cdot \vec{E} = \frac{\rho_e}{\varepsilon}$, If $n^- \neq n^+$, electric field can be created or destroyed in the volume of plasma;
2. $\nabla \cdot \vec{E} = 0$, If $n^- = n^+$, electric field cannot be created or destroyed in the volume of plasma.

With:

$$\vec{E} = \frac{q_1}{4\pi\varepsilon_0 r^2}\vec{r}, \text{ then } \nabla \times \vec{E} = 0 \Rightarrow \vec{E} = -\nabla\phi \text{ where, } \phi = \frac{q_1}{4\pi\varepsilon_0 r}.$$

This defines how to "make" an electric field, i.e., with a gradient in ϕ. (We still have not defined how physically we will make a field in the laboratory. That is clarified in simple terms below.)

The behavior of particles in a field involves motion of particles, and motion under forces involves work on or by the particles, i.e., energy transfer. So we now look at the question of work required to move a charge in a field:

$$W = F \cdot \text{distance} - \text{work done on a particle.}$$

So, $dW = -q\vec{E} \cdot d\vec{s}$, the work done on a particle over distance $d\vec{s}$ against the force of the field, and:

$$\frac{dW}{q} = -\vec{E} \cdot d\vec{s} = \frac{q_1 \vec{r}}{4\pi\varepsilon_0 r_2^2} \cdot (\vec{r_2}) - \frac{q_1 \vec{r}}{4\pi\varepsilon_0 r_1^2} \cdot (\vec{r_1}) = \nabla\phi \cdot d\vec{s} = \frac{d\phi}{ds} \cdot d\vec{s} = d\phi,$$

$$\therefore \frac{W}{q} = \int_1^2 \frac{dW}{q} = \int_1^2 d\phi = \phi_2 - \phi_1,$$

where the units of ϕ are Volts.

So, $E \cdot ds = \Delta V$ and $E = \frac{\Delta V}{ds}$ with units of $\frac{\text{Volts}}{\text{m}}$, and $\frac{\text{Work}}{\text{charge}} = \frac{\text{n} \cdot \text{m}}{\text{coulomb}} = \text{Volt}$.

Therefore, a potential (voltage) difference, ΔV, applied across a gap, ds, will create an electric field in the gap region.

ELECTRIC CURRENTS AND MAGNETIC FIELDS—MAGNETOSTATICS

In the above discussion of charged particles, we observed that static charges are the source for existence of electric fields. Experimentally, it has also been observed that moving charges (current) are the source for existence of magnetic fields. First, we define:

$$I(\text{current}) = \frac{dq}{dt} = n_e\left(\frac{\text{electrons}}{\text{vol}}\right) u \left(\frac{\text{dist.}}{\text{time}}\right) A_{cs}(\text{dist.}^2) q_e \left(\frac{\text{Coulomb}}{\text{electron}}\right),$$

where dq is the flow (with velocity, u) of charge through a control surface, $u \perp A$.

But we define:

$$\frac{I}{A_{cs}} = J\left(\text{current density}, \frac{\text{amp}}{\text{m}^2}\right) = n_e u q_e = \rho_e u.$$

The discovery that the origin of magnetic interactions is due to currents was made by Ampere, and for currents in a vacuum environment we can write:

$$d\overrightarrow{F_{21}} = \frac{\mu_0}{4\pi} I_1 I_2 \frac{d\overrightarrow{l_2} \times (d\overrightarrow{l_1} \times \overrightarrow{r})}{r_{12}^2},$$

where $\mu_0 = 4\pi \times 10^{-7}\left(\frac{\text{Henry}}{\text{m}}\right)$, permittivity of free space.

It can be seen that Ampere's law for force generation from currents (moving charges) is completely analogous to Coulomb's law for force generation from (static) charges.

From a more basic point of view, let us examine the field generated ($d\overrightarrow{B}$) at an arbitrary point (1) at $\overrightarrow{r}$ due to a source element length of conductor dl_1 carrying I_1 at the origin; this can be written as:

$$d\overrightarrow{B}(r) = \frac{\mu_0 I_1}{4\pi}\frac{d\overrightarrow{l_1} \times \overrightarrow{r}}{r^2}, \quad \text{this is the Biot-Savart Law.}$$

Then for another current $(I_2, d\overrightarrow{l_2})$ at a point (2) we can write:

$$d\overrightarrow{F_{21}} = I_2 d\overrightarrow{l_2} \times d\overrightarrow{B}.$$

Generally:

$$Id\overrightarrow{l} = (\overrightarrow{J}A_{cs})\cdot d\overrightarrow{l} = \overrightarrow{J}dV \rightarrow \frac{d\overrightarrow{F_{21}}}{dV} = \overrightarrow{J} \times d\overrightarrow{B}.$$

Further:

$$B(r) = \frac{\mu_0}{4\pi}\int \frac{\overrightarrow{J} \times \overrightarrow{r}}{r^2} dV,$$

from which we can get:

$$\nabla\cdot\overrightarrow{B} = 0,$$

and:

$$\nabla \times \overrightarrow{B} = \mu_0\overrightarrow{J} \quad \text{or} \quad \nabla \times \overrightarrow{H} = \overrightarrow{J}$$

where

$$\overrightarrow{B} = \mu_0\overrightarrow{H}.$$

CONSERVATION OF CHARGE

We will consider the flow of charged particles, rates of change, and the condition, $\frac{\partial}{\partial t} \neq 0$. Consider a volume V with charge density, $\rho_e = e(n^+ - n^-)$. Then the total number of (net) charges in the control volume is, $\int_V \rho_e dV$. Let us also consider charge flow in and out:

$$\frac{\text{charge particles (in-out)}}{\text{time}} = \frac{\text{Change of net particles in } V}{\text{time}}.$$

Using:

$$\overrightarrow{I} = n_e \cdot \overrightarrow{u} \cdot A_{cs} \cdot q_e,$$

or:

$$\overrightarrow{J} = \frac{\overrightarrow{I}}{A_{cs}} = n_e \cdot \overrightarrow{u} \cdot q_e,$$

and assuming a gain of charge (in > out):

$$-\int_A \overrightarrow{J} \cdot \overrightarrow{n}\, dA = \frac{d}{dt} \int_V \rho_e dV,$$

or by Gauss' theorem:

$$-\int_V \nabla \cdot \overrightarrow{J}\, dV = \int_V \frac{\partial \rho_e}{\partial t} dV.$$

So:

$$\int_V \left(\nabla \cdot \overrightarrow{J} + \frac{\partial \rho_e}{\partial t} \right) dV = 0,$$

this is the conservation of charge in a volume.

At a point, we can write the differential expression:

$$\nabla \cdot \overrightarrow{J} + \frac{\partial \rho_e}{\partial t} = 0 \text{ and if } \rho_e = 0, \text{ then } \nabla \cdot \overrightarrow{J} = 0.$$

This is similar in meaning to the statement in fluid mechanics, $\Delta \cdot \overrightarrow{V} = 0$, which is the mathematical statement of conservation of mass: the amount of fluid flow "in" equals the fluid flow "out" of a volume.

But, we note:

$\nabla \cdot \overrightarrow{D} = \rho_e,$ $\overrightarrow{D}$ is the electric displacement vector;

$\nabla \cdot \overrightarrow{J} + \frac{\partial}{\partial t}(\nabla \cdot \overrightarrow{D}) = 0,$ $\dot{\overrightarrow{D}}$ is the rate of change of $\overrightarrow{D}$;

$\nabla \cdot \left(\overrightarrow{J} + \frac{\partial \overrightarrow{D}}{\partial t} \right) = 0,$ $\dot{\overrightarrow{D}}$ is called the displacement current.

Physically, a changing electric field is related to an equivalent current flow (which must take place if charge is transported to create a field).

This term is critical in Maxwell's equations, and its existence was originally postulated on mathematical, not physical, grounds. So, we have made the recognition that $\dot{D}$, a time-changing electric field term, is related to a current. But currents are related to magnetic fields. So, how are electric fields related to magnetic fields? A basic experiment provides the answer.

FARADAY'S LAW

Experimentally, Faraday noticed that when a pulse of current is passed through a metal coil, there will be a momentary flow of current in a second metal coil that is nearby. Accordingly, the second metal loop, or test loop, is located in a changing magnetic field created by the original current and experiences an electric field along its length which is proportional to the time rate of change of flux of $\overrightarrow{B}$ through the loop, as:

$$V(\text{Volts}) = -\frac{\partial}{\partial t}\phi(\text{flux}),$$

or:

$$\oint \overrightarrow{E}\cdot d\overrightarrow{l} = -\frac{\partial}{\partial t}\int_A \overrightarrow{B}\cdot\overrightarrow{n}\,dA.$$

Stokes Law is applied to get:

$$\int_A \nabla\times\overrightarrow{E}\cdot\overrightarrow{n}\,dA = -\frac{\partial}{\partial t}\int_A \overrightarrow{B}\cdot\overrightarrow{n}\,dA,$$

or:

$$\nabla\times\overrightarrow{E} = -\frac{\partial\overrightarrow{B}}{\partial t}.$$

(Note: In electrostatics, $\nabla\times\overrightarrow{E} = 0$.)

Further:

$$\nabla\cdot\left(\nabla\times\overrightarrow{E}\right) = \nabla\cdot\left(-\frac{\partial\overrightarrow{B}}{\partial t}\right),$$

so:

$$\frac{\partial}{\partial t}\left(\nabla\cdot\overrightarrow{B}\right) = 0,$$

and:

$$\nabla \cdot \overrightarrow{B} = \text{constant} = 0.$$

So, physically, by comparison with mass and charge conservation, we can say that there is no source of magnetic fields similar to a point charge. More importantly, recognizing that $\dot{\overrightarrow{B}}$ implies the existence of $\overrightarrow{E}$, we ask the inverse: can $\dot{\overrightarrow{E}}$ imply the existence of $\overrightarrow{B}$?

AMPERE'S LAW

It has been noted that $\nabla \times \overrightarrow{B} = \mu_0 \overrightarrow{J}$; this is Ampere's law for magnetostatics, and $\nabla \cdot \nabla \times \overrightarrow{B} \equiv 0$.

Recall the conservation of charge: $\nabla \cdot \overrightarrow{J} + \frac{\partial \rho_e}{\partial t} = 0$, or, $\nabla \cdot \overrightarrow{J} = -\nabla \cdot (\frac{\partial \overrightarrow{D}}{\partial t})$. Therefore, for changing fields we must take, $\overrightarrow{J} \rightarrow \overrightarrow{J} + \frac{\partial \overrightarrow{D}}{\partial t}$. This is Maxwell's triumph! His study of the equations of electromagnetics led him to postulate that this transformation was necessary. This had far-reaching physical and mathematical consequences by allowing the prediction of electromagnetic wave motion. So, Ampere's law becomes:

$$\nabla \times \frac{\overrightarrow{B}}{\mu_0} = \overrightarrow{J} + \frac{\partial \overrightarrow{D}}{\partial t},$$

and $\nabla \cdot (LHS) = 0 = \nabla \cdot (RHS)$ for any case. Note that, indeed, $\dot{\overrightarrow{D}} = \varepsilon \dot{\overrightarrow{E}} \Rightarrow \overrightarrow{B}$.

MAXWELL'S EQUATIONS

We can now summarize the complete set of Maxwell's equations in two forms:

	Differential form;	Integral form;
(3 eq.)	$\nabla \times \overrightarrow{E} = -\frac{\partial \overrightarrow{B}}{\partial t}$	$\oint_c \overrightarrow{E} \cdot d\overrightarrow{l} = -\int_A \dot{\overrightarrow{B}} \cdot \overrightarrow{n} dA$
(3 eq.)	$\nabla \times \overrightarrow{H} = \overrightarrow{J} + \frac{\partial \overrightarrow{D}}{\partial t}$	$\oint_c \overrightarrow{H} \cdot d\overrightarrow{l} = I + \int_A \dot{\overrightarrow{D}} \cdot \overrightarrow{n} dA$
(1 eq.)	$\nabla \cdot \overrightarrow{D} = \rho_e$	$\int_A \overrightarrow{D} \cdot \overrightarrow{n} dA = Q(\text{net charge})$
	$\nabla \cdot \overrightarrow{B} = 0$	$\int_A \overrightarrow{B} \cdot \overrightarrow{n} dA = 0$

In this a set of seven equations; there are 10 unknowns. In order to reach a solution, they are supplemented by two constitutive relations, $\overrightarrow{D} = \varepsilon \overrightarrow{E}$ and $\overrightarrow{B} = \mu \overrightarrow{H}$, and one functional relationship, $\overrightarrow{J} = f(\overrightarrow{E}, \overrightarrow{B})$, which is to be determined.

FORCES AND CURRENTS DUE TO APPLIED FIELDS

Forces

We will first consider the relationship for forces on a particle of a single species:

$$\overrightarrow{F_{EM}} = \overrightarrow{F_E} + \overrightarrow{F_M} = q\overrightarrow{E} + q\left(\overrightarrow{V} \times \overrightarrow{B}\right) = q\left(\overrightarrow{E} + \overrightarrow{V} \times \overrightarrow{B}\right),$$

where $\overrightarrow{V}$ is the velocity of a particle. For a distribution of particles:

$$\frac{F_{EM}}{Vol} = \frac{q}{V}\left(\frac{\text{charge}}{\text{vol}}\right)\left(\overrightarrow{E} + \overrightarrow{V} \times \overrightarrow{B}\right),$$

$$\begin{aligned} f_{EM} &= \rho_e \overrightarrow{E} + \rho_e\left(\overrightarrow{V} \times \overrightarrow{B}\right) = \rho_e \overrightarrow{E} + en^{+}\overrightarrow{V_{+}} \times \overrightarrow{B} - en^{-}\overrightarrow{V_{-}} \times \overrightarrow{B} \\ &= \rho_e \overrightarrow{E} + \overrightarrow{J_{+}} \times \overrightarrow{B} + \overrightarrow{J_{-}} \times \overrightarrow{B}, \end{aligned}$$

and then for a plasma of electrons and ions:

$$f_{EM} = \rho_e \overrightarrow{E} + \overrightarrow{J} \times \overrightarrow{B},$$

where, $\overrightarrow{J} = \overrightarrow{J_{+}} + \overrightarrow{J_{-}} = \sum_i n_i q_i \overrightarrow{V}_i$, in the normal sense (positively charged particles carrying the current). Clearly, $\overrightarrow{J} \sim \overrightarrow{E}, \overrightarrow{B}$ and that important relationship will now be developed.

Current Conduction—Electrical Conductivity and Ohm's Law

Recall from simple circuit theory that Ohm's law is written as $V\,(\text{or } \Delta V) = I \cdot R$, where I is current and R is resistance. We can also write, $I\,(\text{Amperes}) = \Delta V/R$ with units (Volt/Ohm), then:

$$JA_{cs} = \frac{\Delta V/x}{R/x} = \frac{E}{R/x}, \quad \therefore J = \frac{E}{R \cdot A_{cs}/x} \rightarrow \frac{E}{\rho(\text{Ohm} \cdot \text{m})},$$

where ρ (*Ohm* · *m*) is resistivity or we can write, $J = \sigma \cdot E$, where σ is conductivity (1/Ohm-m) and E has units (Volt/m). So, J has units of (Volt/Ohm-m^2) or Ampere/m^2, current per unit area through which the current is passing; this is referred to as current density. Note that we have considered the medium as a "wire" in the circuit sense, and we have not included the effects of magnetic field.

When we have a moving medium (velocity, $\vec{V}$), Ohm's law transforms (is invariant), so, $\vec{J'} = \sigma\vec{E'}$, where the prime refers to the rest frame of the medium, i.e., moving with velocity, $\vec{V}$. Specifically, for a deformable flowing conductor, it can be shown from Faraday's law (Jackson, 1962, pp. 170–173) that the electric field in the moving frame, $\vec{E'}$, is expressible as $\vec{E'} = \vec{E} + \vec{V} \times \vec{B}$.

Note that for different species with different flow velocities, the flow or measurable velocity is the average for each species: the velocity is averaged for the mixture flow. Then for a flowing medium (velocity, $\vec{V}$), we have: $\vec{J} = \vec{J'} + \rho_e\vec{V} = \sigma\vec{E'} + \rho_e\vec{V}$, and:

$$\vec{J} = \sigma\left(\vec{E} + \vec{V} \times \vec{B}\right) + \rho_e\vec{V} = \vec{J}_{\text{cond}} + \vec{J}_{\text{conv}},$$

where $\vec{V}$ is the average for all species $\left(\rho_e\vec{V} = en^+\vec{V_+} - en^-\vec{V_e}\right)$, and if $n^- = n^+$,

$$\vec{J} = \sigma\left(\vec{E} + \vec{V} \times \vec{B}\right).$$

This is an expression of Ohm's law for a plasma, but it neglects the perturbation velocity due to Hall effect on the particles of different charge. What we are really representing here is the equation of motion for particles of different mass and charge, and the only exact way of getting the correct relationship is by writing an exact equation of motion for the particles. The appropriate terms will be included in later considerations.

Evaluation of Electrical Conductivity

Let us first consider the simplified case where $\vec{B} = 0$, and let us consider the average "drift" of a charged particle species (electrons) as being the current conduction. We consider electrons to be accelerated by a field and decelerated by collisions with heavy atoms and ions, as:

$$F = \frac{\Delta \text{ momentum}}{\text{time}},$$

or:

$$eE = \left(\frac{\Delta\ \text{momentum}}{\text{collision}}\right)\left(\frac{\text{collision}}{\text{time}}\right) = m_e u_e \cdot \nu_{ei} = 2m_e u_d \cdot \nu_{ei},$$

$$\therefore\ u_d = \frac{eE}{2m_e\nu_{ei}}.$$

Since:

$$\vec{J} = nq\vec{u}_d \rightarrow n_e e\vec{u}_d,$$

then:

$$J = n_e e\left(\frac{eE}{2m_e\nu_{ei}}\right) = \frac{n_e e^2}{2m_e\nu_{ei}}\cdot E \equiv \sigma E,$$

$$\therefore\ \sigma = \frac{n_e e^2}{2m_e\nu_{ei}} = f(n_e, \nu_{ei}).$$

This is Lorentz conductivity, the model presumes that electrons drift in static ions. Using this concept, we can then interpret interactions to express conductivity, as:

$$\nu_{ei}\left(\frac{\text{collision}}{\text{time}}\right) = n\left(\frac{\text{particles}}{\text{vol}}\right)u_{avg}\left(\frac{\text{dist.}}{\text{time}}\right)Q_{ei}(\text{cross-section area}).$$

If we account for the several types of collisions:

$$\nu_{ci} = u_{avg}(el)\sum_j n_j Q_{ij} = \left(\frac{8}{\pi}\right)^{1/2}\left(\frac{kT_e}{m_e}\right)^{1/2}\sum_j n_j Q_{ij},$$

$$\therefore\ \sigma = f\left(\frac{n_e}{n_j}, T_e^{-1/2}, Q_{ij}^{-1}\right).$$

It is important to note that the model assumes that the particle velocity is dominated by random thermal velocity, i.e.,

$$u_{avg}(el) \approx u_{th},\ \frac{1}{2}m_e\overline{u_{th}^2} = \frac{3}{2}kT_e, \quad \text{and} \quad \overline{u_e} = \left(\frac{8}{\pi}\right)^{1/2}\left(\frac{kT_e}{m_e}\right)^{1/2}.$$

The collision cross-section term must be considered in detail in order to model the plasma behavior. What we have done here is to develop a simple conduction model with no magnetic field effects, no ion current, and without several effects due to real particle motion (Sutton and Sherman, 1965).

One of the most widely used formulas for calculating conductivity is based on the Lorentz model. If we include factors for Coulomb collisions, for a fully ionized plasma we have (Spitzer, 1962):

$$Q_{ei} = \frac{\pi^3 \ln \Lambda}{16}\left(\frac{e^2 Z_i}{4\pi\varepsilon_0}\right)^2 \frac{1}{kT^2} \rightarrow \text{which leads to, } \sigma_{Spitzer} = \frac{(1.508 \times 10^{-2})T^{3/2}}{\ln \Lambda},$$

where $\ln \Lambda$ is the Coulomb logarithm, which can be given numerical values based on plasma properties. We will return later to the question of an exact calculation of conductivity.

Now, we will examine some aspects of the physical behavior of plasmas under the influence of applied fields; we will examine gas discharges and field sensitive diagnostics.

Plasma Dielectric Properties

In the development of physical and mathematical models for a plasma, we take the perspective (Jackson, 1962; Chen, 1984) that all charge in the medium is considered "free" (as opposed to bound, as in a solid dielectric). Accordingly, we utilize the electromagnetic equations with:

$\mu \approx \mu_0 = 4\pi \times 10^{-7}$ (Henry/m) and $\varepsilon \approx \varepsilon_0 = 8.85 \times 10^{-12}$ (Farad/m), with B, magnetic induction (Gauss) and E, electric field (V/m) considered basic.

PLASMA BEHAVIOR IN GAS DISCHARGES

Introduction

The ambient state of gases (at standard temperature and pressure (STP)) typically has very few charged particles, and so the gas does not easily conduct electrical current. Air at STP is an insulator that can hold off potential differences of 30 kV/cm before the gas will break down and allow passage of electric current between electrodes in an electric circuit. The voltage across an electrode gap at which electron current will initiate is "breakdown" or spark potential and it depends on the gas species and pressure; this occurrence is related to the parameter, p-d, the product of pressure and electrode gap separation. Accordingly, it is evident that it is easier to create breakdown and initiate current conduction at lower pressures. However, it is possible to create discharges at higher pressures with small gaps at moderate voltages and larger gaps at higher voltages. The behavior of plasmas formed in the laboratory in response to external circuit characteristics and properties is the general subject of gas discharge plasmas (Cobine, 1958), gaseous electronics (Hirsh and Oskam, 1978), plasma technology (Capitelli and Gorse, 1992), and plasma chemistry (Fridman, 2008).

While the study of gas discharges forms a specialist area of science and engineering, the study of plasmas in general can be enhanced by understanding the developments in gaseous conductors and gas discharge plasmas. In particular, a brief treatise published by S.C. Brown, "*A Short History of Gaseous Electronics*," (in Hirsh and Oskam, 1978) can provide invaluable context for this field of study. Based upon a strong foundation of more than 100 years of science and technology, there has been the development of a number of techniques for generating gas discharges and the technology of devices to utilize their characteristics. General reviews of the subject of gas discharges and their current applications have been presented (Conrads and Schmidt, 2000; Bogaerts et al., 2002); a general discussion of some of the basic ideas will be reviewed here.

The behavior of gas discharges has distinct differences for direct current (DC) and alternating current (AC) (high frequency) source circuits, and these will be briefly reviewed. Along with the relationship of gas discharge phenomena to pressure and current level, there is a difference in plasma behaviors to temperature ranges that relate to levels of atomic excitation states. Low-temperature behavior occurs at $T < 1$ eV, $\sim 10{,}000$ K, and high-temperature plasmas occur at $T > 1$ eV, $\sim 10{,}000$ K; furthermore, due to more rapid energy gain or loss by the lower mass electrons, there can be situations where $T_e \lesseqgtr T_i$; this will be discussed later.

Discharge Formation

With a gas-filled tube fitted with electrodes connected into a (DC) circuit with voltage source and (ballast) resistor in series, the type of discharge that establishes in a stable condition depends upon the magnitudes of both voltage and resistance. The electrode voltage-discharge current behavior can be seen in Figure 4.1, where secondary lines are drawn to define properties for parametric values, $R_1 \gg R_2 > R_3$.

The figure shows points of stable discharge for given voltages that are compatible with different values of circuit resistance. With high resistance (R_1) and low voltage ($V < V_s$), the currents are low ($<10^{-6}$ A), with little

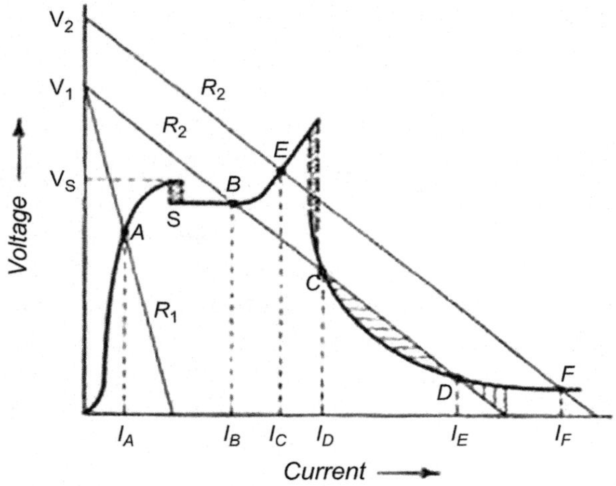

Figure 4.1 Current–voltage behavior of a gas discharge for various values of ballast resistance. *From Cobine (1958), with permission.*

luminosity: these are called Townsend or "dark" discharges. At the (spark) voltage (V_s) the currents are larger (10^{-4} A) and electrons gain enough energy to stimulate visible light of a form referred to as a "glow" discharge. Note that the discharge behavior near V_s depends critically on the value of R; with values of R_2 and R_3 there are two stable points: one with a glow discharge and one with a much higher current "arc" discharge. The arc discharge is high power and has a characteristic small-diameter turbulent high-intensity light. Both the glow and arc discharges evidence sharp voltage drops near the electrodes and a central region with approximately linear variation along the gap.

Glow discharges can be operated over a wide range of pressures, including atmospheric pressures, which can be useful in some technological applications (Conrads and Schmidt, 2000; Bogaerts et al., 2002). However, as the scaling follows the *p-d* function, the size of the discharge is reduced. There are some advantages to operation of glow discharges in the pulsed mode; with pulse durations of 10–100 μs, the plasma discharge avoids the effects of high-energy ion sputtering on electrodes.

Ionization Growth in Electric Fields

Once incipient electrons are formed, as electron collision is the most effective ionization mechanism, the transfer of energy from electric fields by acceleration of particles allows for rapid growth of the electron population

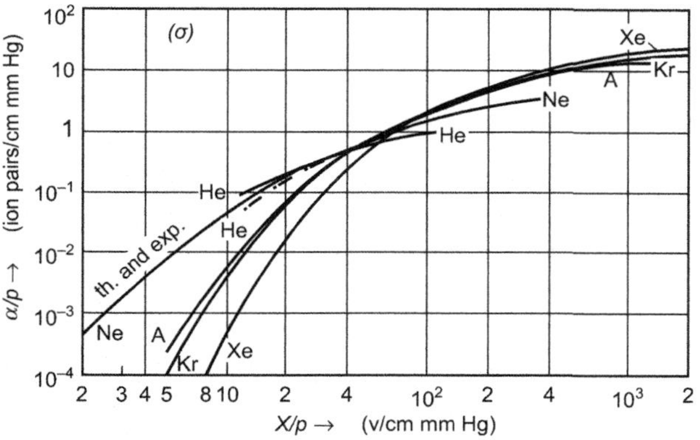

Figure 4.2 Ionization coefficient (α/p) as a function of field per unit pressure (X/p). *From Francis (1960), with permission.*

(Francis, 1960). If a flux of electrons is accelerated by an electric field, E, the growth in flux can be expressed as $n(x) = e^{+\alpha x}$, where α is Townsend's coefficient. The Townsend (growth) coefficient is a function of field, species, and pressure, $\frac{\alpha}{p} = f\left(\frac{E}{p}\right)$. The variation is shown for some gases in Figure 4.2; the data were derived from von Engle, 1956.

Ionization in High-Frequency Gas Discharges

The previous discussion of ionization and discharge behavior related to the application of steady DC electric fields. The generation of ionization with the application of periodic electromagnetic fields in discharge tube geometries has proved to be unique and useful in a number of technologies (Cobine, 1958; Hirsh and Oskam, 1978; Conrads and Schmidt, 2000; Bogaerts et al., 2002; Francis, 1960). With alternating electric fields, the electrons are not unidirectional in their motion toward electrodes, but they reverse direction in response to the fluctuating electric field. The discharge does not need electrodes to be sustained, so the vessel may be an insulator, and the size of the vessel becomes important in setting up an equilibrium. The frequency of the discharge affects the mode of energy delivery to electrons. The basic parameter for a given species of gas is pressure, which fixes the electron mean free path, λ_e. High-frequency (>1 MHz) AC discharges are different from DC or low-frequency discharges in that the electrons reverse direction before coming in contact with a boundary.

Radio frequency gas discharges operate at frequencies of 1–100 MHz and corresponding wavelengths of about 100–1 m. While the wavelength dimension is larger than typical discharge devices, the accelerated particles must gain sufficient energy to produce secondary electrons and so create and sustain the discharge. Microwave frequency discharges operate with frequencies on the order of 3 GHz and have compatible wavelengths of 10 cm, closer to the order of the discharge vessel dimensions.

At medium or high pressures, we have $\lambda_e < d$ (*length*), r (*radius*) of the vessel and ν (*electric field frequency*) $\gg \nu_c$ (*collision frequency*). In this case, the electrons make many oscillations between collisions with molecules, and the physical situation can be described as that the electrons move as a cloud. The periodic discharge applies an electric field, $E = E_0 \sin \omega t = E_0 \sin 2\pi\nu t = E_0 \sin \frac{2\pi}{\lambda} ct$, and this can add energy to the electrons; when the electrons collide with molecules, they change phase and so have a net energy addition. In high-frequency discharges, there can be no energy addition without collisions. The electrons gain energy from the field and lose it in collision.

Analysis of this process can be carried out in terms of a "frictional factor." Based on a specific model (Francis, 1960) identifying energy gain and loss of the electron cloud, the rate at which electrons gain energy (i.e., power) per unit volume from an applied field, $E_0 e^{iwt}$, was calculated to be:

$$\overline{P} = \frac{n^- e^2 E_0^2}{2m_e}\left(\frac{\nu_c}{\nu_c^2 + \omega^2}\right).$$

For any given discharge tube, diffusion of electrons is important, and at very high frequencies the tube can act as a waveguide/resonator system.

In the high-frequency discharge, the two primary processes are electron collision with molecules and ion and electron diffusion to the walls. Generally, for this type of discharge, $p > 10^{-2}$ mmHg and $\omega > 100$ Mc/s. Breakdown begins when an electron creates ionization during the time it takes it to diffuse to the walls. A discharge can occur when the electron density rises rapidly, as (Francis, 1960):

$$D_e\left(\frac{\pi}{d}\right)^2 = \nu_i,$$

where D_e is the electron diffusion coefficient, d is the tube length and ν_i is the number of ionizing collisions per second. Using Townsend's ionization coefficient, $\alpha = \nu_{drift}/\nu_i$:

$$(pd^2) = \frac{\pi^2 kT_e}{e\left(\frac{E}{p}\right)\left(\frac{\alpha}{p}\right)} \quad \text{for} \quad \nu_{c,inelastic} > \omega > \nu_{c,elastic}.$$

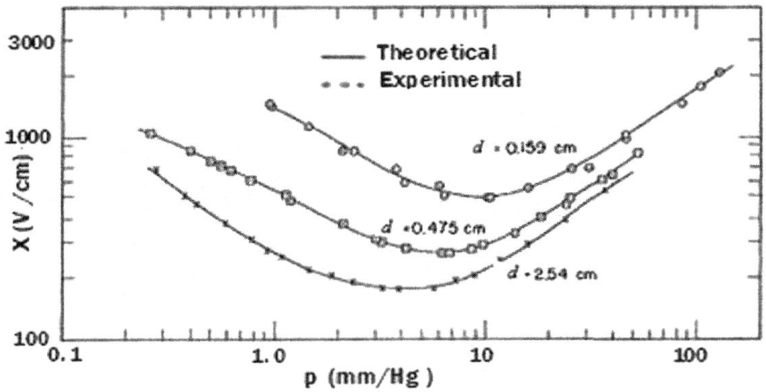

Figure 4.3 Breakdown electric field (X) in H_2 at 3 MHz, variable dimensions. *From Francis (1960), with permission.*

Table 4.1 Gas Discharge Plasma Sources

Device	Pressure (mbar)	n_e (cm^{-3})	T_e (eV)
DC glow	10^{-3} to 100	10^{11}	1–10
DC arc	1–1000	10^{11} to 10^{13}	0.1
AC capacitive coupled	10^{-3} to 10^{-1}	10^{11}	1–10
AC inductive coupled	10^{-3} to 10	10^{12}	1
Helicon	10^{-4} to 10^{-2}	10^{13}	1
AC microwave			
(Closed chamber)	10^3	10^{12}	3
(Open system)	100	10^{11}	2

Typical breakdown curves are shown for hydrogen in Figure 4.3; the data were derived from Brown, 1959. Note that wall effects and magnetic fields can exert considerable influence on the breakdown and ionization.

A useful comparison of the parameters of typical gas discharge plasma devices has been presented by Conrads and Schmidt (2000), and a summary is presented in Table 4.1.

ILLUSTRATIVE APPLICATIONS OF MAXWELL'S EQUATIONS

As was pointed out in the introduction, the intent of the material presented is not simply to provide a catalog of equations that may be applied to numerous problems that the student or researcher may encounter in practice, but rather to recognize that understanding the application of the equations defining plasma behavior is the best way to understand the equations. This is so particularly with respect to Maxwell's equations, as in many works, the equations are presented in differential form, with the indirect implication that the solution of all problems simply requires an understanding of the solution of a set of differential equations with boundary conditions. The following examples are of problems involved with real devices, and they are presented in the context of being typical applications.

Closed-Loop Magnetic Probes

A magnetic probe is a physical device that is inserted into a plasma or region of magnetic flux in order to determine the strength of the magnetic field. There are several types of probes that can be used to measure magnetic field, but the simplest type is the twisted pair with open loop area, and it will be examined here (Huddlestone and Leonard, 1965); it will then be applied to measure magnetic fields in pulsed electromagnetic discharges.

The basic equation used is Faraday's law:

$$\nabla \times \vec{E} = -\frac{\partial \vec{B}}{\partial t},$$

or integrating over the loop area:

$$\int_A \left(\nabla \times \vec{E}\right) \cdot d\vec{A} = -\int_A \frac{\partial \vec{B}}{\partial t} \cdot d\vec{A},$$

then:

$$\oint_C \vec{E} \cdot d\vec{l} = -\int_A \frac{\partial \vec{B}}{\partial t} \cdot d\vec{A}.$$

So, the voltage induced in the loop, which would be measured across the twisted pair is:

$$V(\text{Volts, emf}) = -\frac{\partial}{\partial t}\int_A \vec{B} \cdot d\vec{A} = -\frac{\partial}{\partial t}\left(\vec{B} \cdot \vec{A}\right) = A \cdot \frac{\partial B}{\partial t}.$$

But what is this result physically? Directly, $V \sim \dot{B}$, the time rate of change of the magnetic field in the loop. So, as we are physically measuring voltage across the twisted-pair leads, we can use this as input to other

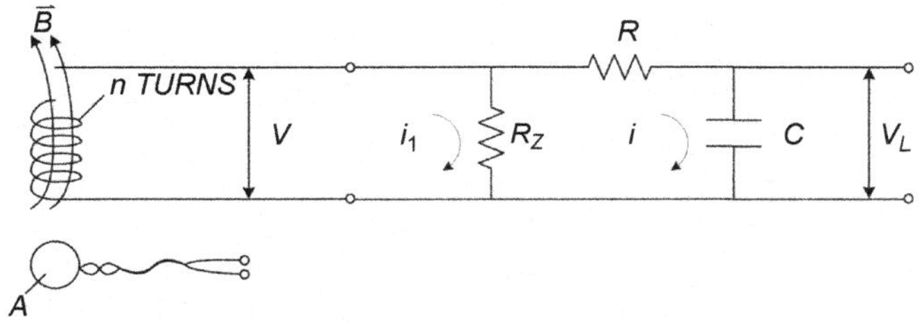

Figure 4.4 Magnetic probe (twisted loop) with passive integration circuit.

electrical circuit elements to carry out an electrical signal integration so that the output is proportional to the strength of the magnetic field, B. A schematic is shown in Figure 4.4.

Then:

$$V_L = \frac{Q(t)}{C} = \frac{\int_0^t i dt}{C} = \frac{\int_0^t \left(\frac{V}{R}\right) e^{-t/(R+R_Z)C} dt}{C},$$

and if $(R + R_z)\, C \gg t$, then:

$$V_{out} = \frac{1}{RC} \int_0^t V_L dt = \frac{1}{RC} \int_0^t \left(A \cdot \frac{\partial B}{\partial t}\right) dt = \frac{nA}{RC} \Delta B.$$

Magnetic Compression (Z-Pinch) and Magnetic Probe Response

The dynamic (linear) Z-pinch (Jackson, 1962) is a device that creates an electrical discharge in gases enclosed in a right-cylindrical chamber with electrodes at the ends, as shown in Figure 4.5. The discharge is pulsed, being supplied by the current from high-voltage capacitors. The circuit is designed to create an initial current sheet discharge at the largest radius of

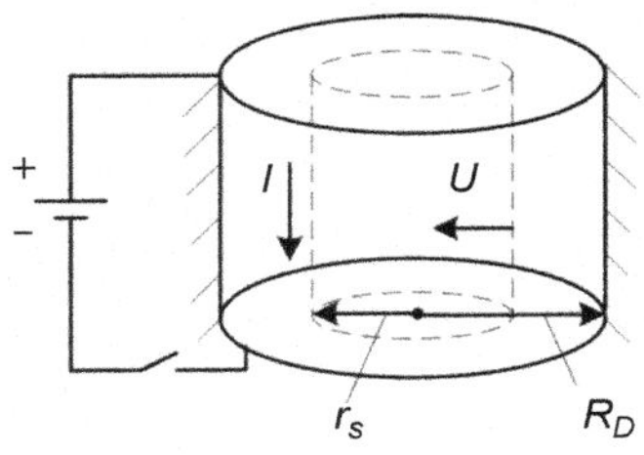

Figure 4.5 Linear Z-pinch schematic.

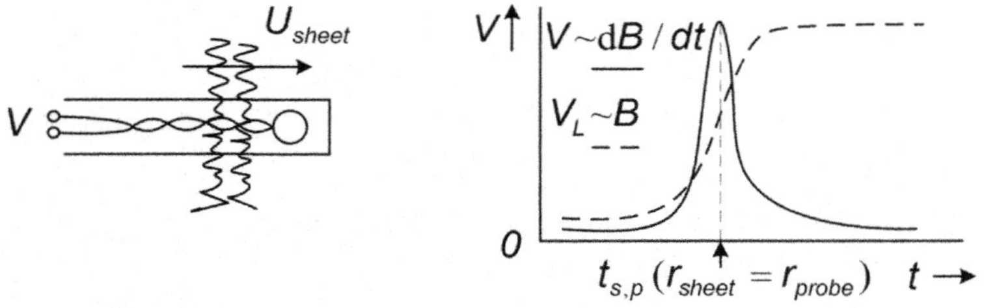

Figure 4.6 Magnetic probe geometry and typical response to the collapsing current sheet.

the chamber; at radii inside the sheet the magnetic field is zero and outside the sheet the azimuthal magnetic field satisfies Maxwell's equations. The probe geometry and typical response to the collapsing current sheet is shown in Figure 4.6.

The basis of the measurement is Ampere's law:

$$\nabla \times \frac{\vec{B}}{\mu_0} = \vec{J} + \frac{\partial \vec{D}}{\partial t}.$$

For axial current in a cylindrical system:

$$\mu_0 J_z = \frac{1}{r}\frac{\partial (B_\theta r)}{\partial r} = \frac{1}{r}\left(r\frac{\partial B_\theta}{\partial r} + B_\theta \right) = \frac{\partial B_\theta}{\partial r} + \frac{B_\theta}{r}.$$

Experimentally, the calibrated output shows B_θ as a function of time, t, at a given radius of the probe (Figure 4.6). The sketch shows output from a probe at fixed radius as the collapsing current sheet passes by, with $\dot{B}$ and B_θ based on electrical integration. With data at different radii, we can construct a graph of B_θ as a function of radius at a given time.

REFERENCES

Bogaerts, A., et al., 2002. Gas discharge plasmas and their applications. Spectrochim. Acta, Part B 57, 609–658.

Brown, S.C., 1959. Basic Data of Plasma Physics. MIT and Wiley, New York.

Capitelli, M., Gorse, C., 1992. Plasma Technology: Fundamentals and Applications. Plenum, New York.

Chen, F.F., 1984. Introduction to Plasma Physics and Controlled Fusion, second ed. Plenum, New York.

Cobine, J.D., 1958. Gaseous Conductors – Theory and Engineering Applications. Dover, New York.

Conrads, H., Schmidt, M., 2000. Plasma generation and plasma sources. Plasma Sources Sci. Technol. 9, 441–454.

von Engle, A., 1956. In: Flugge, S. (Ed.), Handbuch der Physik, vol. 22. Springer, Heidelberg.

Francis, G., 1960. Ionization Phenomena in Gases. Butterworths, London.

Fridman, A., 2008. Plasma Chemistry. Cambridge, New York.

Hirsh, M.N., Oskam, H.J., 1978. Gaseous electronics. In: Electrical Discharges, vol. 1. Academic, New York.

Huddlestone, R.H., Leonard, S.L. (Eds.), 1965. Plasma Diagnostic Techniques. Academic, New York.

Jackson, J.D., 1962. Classical Electrodynamics. Wiley, New York.

Spitzer, L., 1962. Physics of Fully Ionized Gases, second ed. Interscience, New York.

Stratton, J.A., 1941. Electromagnetic Theory. McGraw-Hill, New York.

Sutton, G.W., Sherman, A., 1965. Engineering Magnetohydrodynamics. McGraw-Hill, New York.

CHAPTER 5

Plasma Parameters and Regimes of Interaction

INTRODUCTION

We began by examining the plasma medium in terms of the characteristics of ionization. In order to further understand some of the more complex behaviors, we now define some of the important parameters of the plasma state and define some of the regimes of interaction. We know that charged particles will interact among themselves, with fields and with boundaries. We will classify these interactions by defining properties of description.

EXTERNAL PARAMETERS

Geometric size: L, size of container, probe, etc.
Frequency of forcing function: ω, frequency of fields
For example, an electromagnetic wave: $\overrightarrow{E} = \overrightarrow{E}_0 e^{i(kx-\omega t)}$ where ω is the wave frequency, $k = \omega / v$ is the wave number, and v is phase velocity; $\omega = v/\lambda$ and λ is wavelength.

PARTICLE (COLLISION) PARAMETERS

Collision mean free path: λ, average distance between collisions, where $\lambda = 1/n\sigma_c$ and σ_c is the collision cross section.
Collision frequency: ν_c (collision/time) $= v$(distance/time)$/\lambda$(distance/collision).
Collision time: τ_c (time/collision) $= 1/\nu_c$

SHEATH FORMATION AND EFFECTS

Because of the internal (random kinetic) energy of plasma particles, the plasma has an inherent ability to separate charges. This separation occurs at boundaries and is referred to as a sheath. There can also be a separation of charges within the plasma due to electromagnetic wave motion.

Introduction to Plasmas and Plasma Dynamics
ISBN 978-0-12-801661-9

The physical equations of the sheath mirror the balance of kinetic energy and potential energy of the field due to charge separation. We will consider aspects of this balance now.

We have noted that the following is a good approximation:

$$\rho_e = e(n_i - n_e) \approx 0.$$

The basis for this approximation is as follows: $\nabla \cdot \vec{E} = \frac{\rho_e}{\varepsilon_0}$, or in radial coordinates, $\frac{dE}{dr} = \frac{e\Delta n_e}{\varepsilon_0}$, where Δn_e is an excess of electrons. For example, take: $n_e \approx 10^{16}/\text{m}^3$, which would be a diffuse plasma, and let us take $\Delta n_e \approx 0.01 n_e \approx 10^{14}/\text{m}^3$. Over distance, $\Delta r = 1$ cm, then $\Delta E = \Delta r \cdot \varepsilon_0^{-1} \cdot e\Delta n_e \approx 10^{-2} \cdot 10^{11} \cdot 10^{-19} \cdot 10^{14} = 10^2 \text{V/cm}$, and generally, if:

$$\Delta E_{\text{experiment}} \ll \Delta E, \quad \therefore \rho_e \approx 0.$$

In order to quantify the possible inherent separation of charges in plasma, let us calculate the distance across which the particle potential energy would equal the random kinetic energy. If we begin with an electrically neutral plasma of electrons and (positive) ions, and then assuming the ions stay fixed, we collect electrons and attach them to the boundary of a sphere of radius, R. We collect all of the electrons in the sphere, and assume that all the ions are static at their initial positions, and so an electric field is created. We then calculate the work done (W) in establishing the electric field, as:

$$dW = d(F \cdot \text{distance}) = -q\vec{E} \cdot d\vec{s} = q_e E dr,$$

and substituting:

$$dW = \left(\frac{4}{3}\pi r^3 n_e e\right) \frac{\frac{4}{3}\pi r^3 n_e e}{4\pi\varepsilon_0 r^2} dr,$$

so:

$$W = \frac{4}{45} \frac{\pi n_e^2 e^2}{\varepsilon_0} R^5 = \Delta PE.$$

That work is equal to the change in potential energy due to the particle arrangement. Now we equate the potential energy in the field to the initial kinetic energy in the electrons in the sphere, as:

$$\Delta PE = \Delta KE,$$

then:

$$\frac{4}{45}\frac{\pi n_e^2 e^2}{\varepsilon_0}R^5 = \frac{3}{2}kT\cdot\frac{4}{3}\pi n_e R^3,$$

so:

$$\therefore\ R^2 = \frac{45}{2}\frac{\varepsilon_0 kT}{n_e e^2} = 22.5\left(\frac{\varepsilon_0 kT}{n_e e^2}\right), \quad \text{and} \quad R \approx 5\left(\frac{\varepsilon_0 kT}{n_e e^2}\right)^{1/2}.$$

This result establishes the functional form and order of magnitude of possible charge separation in a plasma. We define a related characteristic of the separation length as:

$$\lambda_D(\text{Debye length}) = \left(\frac{\varepsilon_0 kT}{n_e e^2}\right)^{1/2}.$$

This is the characteristic length in a plasma over which deviations in charge neutrality can naturally occur. The approach taken above is a simplistic electrostatic approach. To further clarify the significance of this variable, we develop a more tractable electric field solution.

Assume that electrons are in thermal equilibrium at temperature, T, at a potential, ϕ, created by separation from the ions. From Maxwell–Boltzmann statistics for a group of particles in a potential field:

$$n_e(r) = \overline{n}_0 \exp\left(\frac{\phi(r)}{kT}\right).$$

Let us take the perturbation potential due to separation as $e\phi \ll kT$, then:

$$n_e(r) \approx \overline{n}_0\left(1 + \frac{e\phi}{kT}\right).$$

But ϕ must satisfy Poisson's equation:

$$\nabla^2\phi = -\frac{\rho_e}{\varepsilon_0} = -\frac{1}{\varepsilon_0}e(n_i - n_e),$$

and:

$$\nabla^2\phi = -\frac{1}{\varepsilon_0}e\left[\overline{n}_0 - \overline{n}_0\left(1 + \frac{e\phi}{kT}\right)\right] = \frac{e^2\phi\overline{n}_0}{\varepsilon_0 kT} = \frac{\phi}{\lambda_D^2}.$$

In a spherical system, $\frac{1}{r^2}\frac{d}{dr}\left(r^{2}\frac{d\phi}{dr}\right) = \frac{\phi}{\lambda_D^2}$, with $\phi = 0$ at ∞, $\phi \sim q/r$ and at small r, we get a solution, $\phi(r) = \frac{e}{4\pi\varepsilon_0 r}\exp\left(-\frac{r}{\lambda_D}\right)$.

For $r > \lambda_D$, there is shielding of the potential field, called the Coulomb shielded potential.

The shielding distance is λ_D, i.e., a charged particle will not feel (be shielded from) the field when $r > \lambda_D$.

Physically, from kinetic theory, the number of collisions per unit time per unit area of a surface is $\frac{1}{4}n\bar{c}$, and $\bar{c} = \left(\frac{8}{\pi}\right)^{1/2}\left(\frac{kT}{m}\right)^{1/2}$, so particles with smaller mass will have more collisions per unit time with a surface, and electrons will then have much more residence time at, or near, a surface. So, excess electrons in a region means that a potential is created, and this results in a layer or sheath near a surface. The thickness of this layer is a Debye length, as defined above.

PLASMA OSCILLATIONS AND PLASMA FREQUENCY

Let us first consider the dynamics of an electron-ion plasma system where particles separate and can move in only one direction, and where (as above with the charge separation on a spherical surface) electrons have been perturbed by some force to create an electric field, $\vec{E}$. With $\nabla\cdot\vec{E} = \frac{\rho_e}{\varepsilon_0}$, then $\int_S \vec{E}\cdot\vec{n}\,dA = \int_V \frac{\rho_e}{\varepsilon_0}dV$, and $E = \frac{\sigma_S}{\varepsilon_0}$, where the surface charge density is $\sigma_S = n_e q s$. Then the restoring force on the electrons is $f_R = -Ee = \frac{n_e e^2 x}{\varepsilon_0} \sim k_s x$, where k_s is an effective spring constant. As in any oscillating system, the frequency of oscillation is:

$$\omega^2 = \frac{k}{m_e} = \frac{ne^2 x}{\varepsilon_0 m_e} = \omega_p,$$

which in this case is defined as the plasma frequency, a characteristic of the plasma. But what does this mean?

We now take the standard mathematical approach (Chen, 1983) and consider the solution of the equations involved. The governing equations are:

Mass:

$$\frac{\partial n_e}{\partial t} + \nabla\cdot(n_e\vec{v}_e) = 0,$$

Momentum:

$$m_e \frac{\partial \vec{v}_e}{\partial t} = -e\vec{E},$$

Gauss:

$$\nabla \cdot \vec{E} = \frac{\rho_e}{\varepsilon_0} = \frac{e}{\varepsilon_0}(n_+ - n_e),$$

where $\vec{v}_e$ is electron velocity. Taking small perturbations, $n_e = n_0 + n'(r, t)$, and v_e small, we substitute in mass conservation:

$$\frac{\partial(n_0 + n')}{\partial t} + n_e \nabla \cdot \vec{v}_e + v_e \nabla n_e = 0,$$

so:

$$\frac{\partial n'}{\partial t} + n_0 \nabla \cdot \vec{v}_e + v_e \nabla n_0 = 0,$$

$$\therefore \quad \frac{\partial n'}{\partial t} + n_0 \nabla \cdot \vec{v}_e = 0.$$

With $m_e \frac{\partial \vec{v}_e}{\partial t} = -e\vec{E}$, and $\nabla \cdot \vec{E} = \frac{e}{\varepsilon_0}(n_+ - n_0 - n') = -\frac{e}{\varepsilon_0} n'$, combine, to get:

$$\frac{\partial^2 n'}{\partial t^2} + \frac{e^2 n_0}{\varepsilon_0 m_e} n' = 0.$$

Take $n' = \tilde{n}' e^{i\omega t}$ then $\frac{\partial n'}{\partial t} = \tilde{n}' i\omega e^{i\omega t}$, and $\frac{\partial^2 n'}{\partial t^2} = -\tilde{n}' \omega^2 e^{i\omega t} = -\omega^2 n'$. So:

$$\omega^2 = \frac{e^2 n_0}{\varepsilon_0 m_e} \equiv \omega_p^2, \quad \text{the plasma frequency.}$$

MAGNETIC FIELD RELATED PARAMETERS

Larmor Radius

Consider a charged particle moving with velocity, $\vec{u}$, $\perp$ to a magnetic field, $\vec{B}$ (assume no collisions). The force on the particle is $F = q(\vec{u} \times \vec{B})$; therefore we have a change in direction and centrifugal force.

The particle will move in an orbit of radius, r, where:

$$\text{Magnetic force} = \text{Centrifugal force, or } qu_{\perp}B = \frac{mu_{\perp}^2}{r}, \text{ and } r_{\perp} = \frac{mu_{\perp}}{qB}.$$

(Large $B \Rightarrow$ small r; large $m \Rightarrow$ large r; $q \Rightarrow$ opposite direction for opposite charges.)

Cyclotron Frequency

For the motion of a charged particle in a magnetic field, as above, we can define the angular velocity, as $\omega = \frac{u_{\perp}}{r} = \frac{qB}{m}$, cyclotron (Larmor) frequency.

Hall Parameter

The frequency of particle collisions was defined as ν_c (collisions/time), and also:

$\tau_c \equiv 1/\nu_c$, (time/collision). Then with respect to the rotational motion, we define:

$$\omega\tau_c(\text{revolutions/collision}), \quad \text{the Hall parameter.}$$

Classification of Some Regimes of Interaction

1.	$\lambda \gg L$ Reduced collisions	$L \gg \lambda_D$ Strong plasma	$\lambda \gg \lambda_D$ Collisionless sheath
2.	$\lambda \ll L$ Continuum	$L \ll \lambda_D$ Weak plasma	$\lambda \ll \lambda_D$ Collisional sheath
3.	$\lambda \ll L$ $\lambda \ll L$ Continuum	$\lambda \ll \lambda_D$ $\lambda \ll \lambda_D$ Weak plasma	$L \gg \lambda_D$, solvable $L < \lambda_D$, difficult solution. Fluid-Plasma coupling.

ELECTROSTATIC PARTICLE COLLECTION IN (LANGMUIR) PROBES

We had noted above that surfaces in contact with plasma will evidence sheath layers near the surface in which the local charge equilibrium ($n_e = n_i$) is disturbed. In order to understand this behavior more fully, we examine here an important application which involves the insertion of electrodes into the plasma in order to diagnose the properties of a plasma by applying voltages and collecting currents. This usefulness of data from this collection process depends upon plasma properties, and accurate diagnosis requires correct analysis of sheath behavior. We will consider here in detail one limiting condition: mean free paths much larger than the sheath size.

The electrostatic sheath forms because of the difference in the flux of positive (ions) and negative (electrons) particles. The sheath has a thickness on the order of the Debye length. In typical plasmas this is thin compared to body dimensions, but it is finite and in some cases it can be thick compared to body size, as is possible on space probes in ionosphere plasmas. This layer size must be accounted for in the application of any device, but it can be useful in the measurement of plasma properties where physical probes can be inserted into the plasma.

In a plasma volume with $n_i \approx n_e$, the plasma will exist at some value of potential, V_p. We can represent the variation of potential and density in the sketch, Figure 5.1.

At any point in the plasma we must satisfy Coulomb's law; so: $\nabla\cdot\vec{E} = \frac{(n_i - n_e)e}{\varepsilon_0}$. But,

$$\vec{E} = -\nabla V \quad \text{so,} \quad \nabla^2 V = \frac{-(n_i - n_e)e}{\varepsilon_0} \text{(Poisson's equation)}.$$

Considering a one-dimensional geometry for simplicity, if a probe with plane surface at potential, V_0, is introduced at $x = 0$, with an undisturbed density, n_0, and if we assume that n_i is constant near the surface, then $n_e = n_0 e^{-\frac{eV}{kT_e}}$, and $V = V_0 e^{-x/h}$. So the potential will decay to $1/e$ of its value at $h \approx \lambda_D$. This, again, is the Debye length and is given by:

$$h(m) = \left(\frac{\varepsilon_0 k T_e}{n_e e^2}\right)^{\frac{1}{2}} \approx 69.1\sqrt{\frac{T_e}{n_e}}\left(T, K;\ n,\ m^{-3}\right).$$

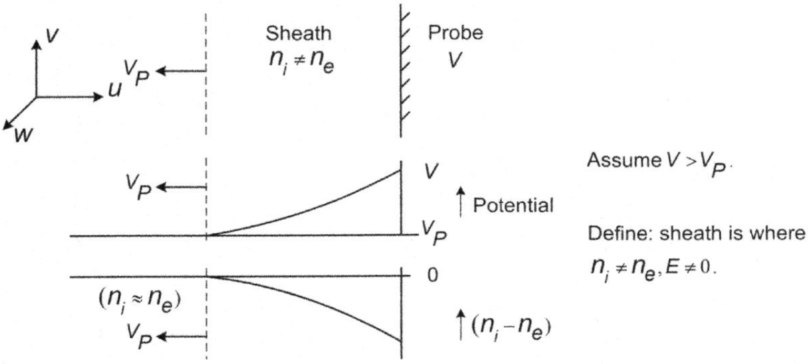

Figure 5.1 Schematic of densities and potential in sheath region.

Accordingly, as the probe potential has little affect on the plasma outside the sheath, we can assume that the normal particle flux of electrons and ions are presented to the sheath surface by the plasma, and then the electrons and ions participate in the sheath processes.

In order to quantify the physical behavior in the sheath, we assume that the particles possess a Maxwellian speed (u, v, w) distribution, as: $f(u,v,w) = \left(\frac{m}{2\pi kT}\right)^{3/2} e^{-\frac{m}{2kT}(u^2+v^2+w^2)}$, with flux: $Flux = nuf(u)du$, where:

$$f(u) = \int_{-\infty}^{+\infty}\int_{-\infty}^{+\infty} f(u,v,w)dvdw = \left(\frac{m}{2\pi kT}\right)^{1/2} e^{-\frac{m}{2kT}u^2}.$$

We note that there are different regimes for particle behavior, and this is best demonstrated by examining the variation of current collected by a (cylindrical) probe surface as the voltage of the probe cylinder is varied from strongly negative to strongly positive (Figure 5.2). We also sketch the current and density behaviors (Figure 5.3) with respect to the probe surface.

Relative to the probe behavior, we can define the following:

$V = V_p$—Plasma potential is the probe potential which is equal to that of the plasma and where there is no sheath, and both the ion and electron thermal flux reach the probe.

$V = V_F$—Floating potential is the potential where no current is allowed to flow to the external circuit, and the potential is due to the charge separation set up by the difference of electron and ion thermal speeds.

$\hat{J}_e$—Electron saturation current is where all the electron flux that moves to the sheath will be collected at the probe. (NOTE: In this section: current is upper case J; current density is lower case j.)

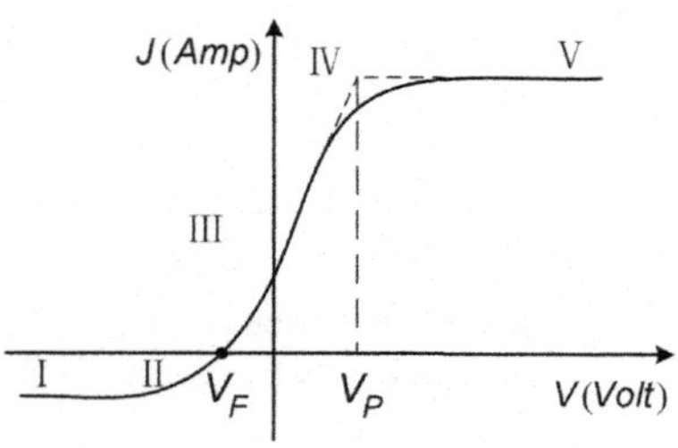

Figure 5.2 Current collected as a function of probe voltage.

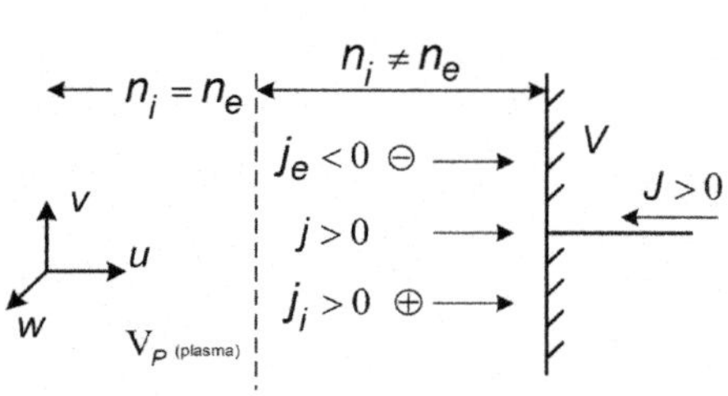

Figure 5.3 Particle currents due to probe voltage bias.

$\hat{j}_i$—Ion saturation current: is, generally, where the limiting ion flux entering the sheath is collected, however, collection extends beyond the sheath region and depends much on regime and probe geometry.

Collisionless Sheath Calculations

In this case we consider: $\lambda_D \ll d$ (probe size); $\lambda_D \ll \lambda_i$, λ_e and $\lambda > d$. There are specific regions of the particle collection that are unique and can be identified, as follows (Note: J (upper case) is current).

The saturation current for the electrons $(\hat{j}_e,\ V \gg V_p)$ will include all the electrons that move into the sheath from a Maxwellian distribution at the edge of the sheath, as:

$$\hat{j}_e = -n_e e \int_0^\infty u f(u) du = -n_e e \left(\frac{kT_e}{2\pi m_e} \right)^{1/2} \approx -0.4 n_e e \sqrt{\frac{kT_e}{m_e}}.$$

The electron current with adverse potential gradient $(V < V_p)$; in this case the electrons must possess more than enough kinetic energy to overcome the potential energy rise to reach the probe, so:

$$\frac{1}{2} m_e u^2 \geq e|V - V_p|, \quad \text{where} \quad u \gg \left(\frac{2e|V - V_p|}{m_e} \right)^{1/2} \equiv u_1,$$

so:

$$j_e = -n_e e \int_{u_1}^\infty u f(u) du = -n_e e \left(\frac{kT_e}{2\pi m_e} \right)^{1/2} e^{-\frac{m_e u_1^2}{2kT_e}} = -n_e e \left(\frac{kT_e}{2\pi m_e} \right)^{1/2} e^{-\frac{e|V - V_p|}{kT_e}}$$
$$= \hat{j}_e e^{-\frac{e|V - V_p|}{kT_e}}.$$

The ion current collection before saturation $(j_i < \hat{j}_i,\ V \approx V_p)$:

$$j_i = n_i e v_i = n_i e \left(\frac{kT_i}{2\pi m_i} \right)^{1/2}.$$

The saturation ion current $(\hat{j}_i,\ V \ll V_p)$:

When the probe is biased highly negative to collect ion saturation current, the potential energy of the ions is not small compared to the thermal energy. We assume that there is a presheath region (Figure 5.4) that

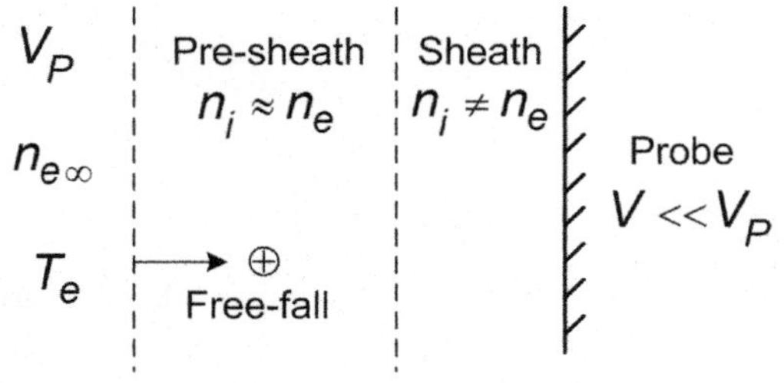

Figure 5.4 Schematic of presheath region with ion saturation bias.

forms outside the sheath; however, the electrons remain in a Maxwellian distribution in much of the sheath region. The presheath has $n_i \approx n_e$, while in the sheath $n_i \neq n_e$.

In the presheath we take the ions to be in free-fall behavior in ΔV, as follows:

$$n_i = \frac{j_i}{ev_i} \quad \text{with} \quad v_i = \sqrt{-\frac{2e\Delta V}{m_i}},$$

then:

$$n_i = \frac{j_i}{e\sqrt{-\frac{2e(\Delta V)}{m_i}}} \equiv n_{e,\infty} e^{-\frac{e(V-V_P)}{kT_e}}, \quad \text{as } n_i = n_e.$$

This functional form allows that j_i has a maximum as a function of ΔV, which would define saturation current, as $\frac{dj_i}{dV} = 0 \Rightarrow \left(\frac{e\Delta V}{kT_e}\right)_{\max} = -\frac{1}{2}$, so that, $\hat{j}_i = 0.607 en_{e,\infty}\sqrt{\frac{kT_e}{m_i}}$. Note that the saturation ion current is proportional to T_e, not T_i.

We can also examine the behavior of the sheath thickness with high negative bias. We analyze assuming only ions are present and electrons have been repelled. We take $j_i =$ constant with increasing voltage, so:

$$en_i = j_i \Big/ \sqrt{-\frac{2eV}{m_i}} \quad \text{with} \quad V \equiv \Delta V = V - V_p.$$

Now substituting into Poisson's equation:

$$\frac{d^2V}{dx^2} = -4\pi e(n_i - n_e) = -4\pi en_i = 4\pi j_i\sqrt{\frac{m_i}{2e|V|}},$$

and so, integrating, get $\left(\frac{dV}{dx}\right)^2 = 16\pi j_i\sqrt{\frac{m_i}{2e}}\sqrt{|V|} + \text{constant}$. We assume that the field is small near the edge of the presheath, so that constant ≈ 0, and with integration:

$$\frac{4}{3}V^{\frac{3}{4}} = Ax + c' \Rightarrow A = \left(16\pi j_i\sqrt{\frac{m_i}{2e}}\right)^{\frac{1}{2}}.$$

At $x = 0$, $|V| = \frac{kT_e}{2e} = |\Delta V_t| \equiv |V_t|$, so that, $\frac{4}{3}[(V)^{3/4} - (V_t)^{3/4}] = Ax$, and then the sheath thickness at ion saturation voltages is:

$$h_s = \frac{4}{3}\frac{(V)^{3/4} - (V_t)^{3/4}}{A}.$$

The thickness of the sheath can be seen to increase with voltage near saturation. This is a form of the Childs-Langmuir law. For a plane electrode, this will not directly affect current collected, but for finite shapes such as cylindrical probe elements, a sheath size increasing with voltage will affect the current collected by increasing the collection area.

With the above understanding of physical behavior and using the functional relationships, the plasma temperature and number density can be determined. Recall that for $V < V_p$ the electron current density collected is:

$$j_e = -n_e e\left(\frac{kT_e}{2\pi m_e}\right)^{\frac{1}{2}} e^{-\frac{e|V-V_p|}{kT_e}} = j - \hat{j}_i, \quad \text{(a measureable)}.$$

Accordingly,

$$\ln|j_e| = \text{constant} - \frac{e|V - V_p|}{kT_e}, \quad \text{and} \quad \frac{d(\ln|j_e|)}{d|V - V_p|} = -\frac{e}{kT_e}.$$

So, on a graph of $\ln j_e$ versus $V - V_p$, then the slope of the curve would allow the calculation of T_e. With $\hat{J}_e$ and T_e, then n_e can be determined, and with T_e and $\hat{J}_i$, then n_i can be determined.

Where electrodes for the electrical discharge of a plasma device can be in contact with the probe electrodes through the plasma, collection of electron current by a positively biased probe can be influenced, and so the data can be inaccurate. To reduce such effects, a configuration that has proven highly effective is that of a double probe (Huddlestone and Leonard, 1965), where two cylindrical electrodes are biased relative to each other, and in the limit, relative to that bias, either could collect ion saturation current.

The analysis presented above provides the basic framework of particle collection by biased probes. Collection by probes with other shapes and in other plasma regimes has been documented. Other effects on the current collected by probes immersed in plasmas and flowing plasmas becomes complex, but different effects have been evaluated and reported in the literature (Chung et al., 1975). Effects due to collisions and magnetic fields have been treated, and relatively straightforward calculation procedures have evolved. The effects of flow become complex, but in different regimes the difference of currents collected with flow aligned parallel and perpendicular to the probe axis can be used as an added part of the diagnosis.

REFERENCES

Chen, F.F., 1983. Introduction to Plasma Physics and Controlled Fusion, second ed. Plenum, New York.

Chung, P.M., Talbot, L., Touryan, K.J., 1975. Electric Probes in Stationary and Flowing Plasmas: Theory and Application. Springer, Berlin.

Huddlestone, R.H., Leonard, S.L. (Eds.), 1965. Plasma Diagnostic Techniques. Academic, New York.

CHAPTER 6

Particle Orbit Theory

INTRODUCTION

One of the more important regimes of plasma behavior is that which is termed collisionless; more specifically, it is defined as a region where mean free path, $\lambda \gg L$. There are some occurrences of low-density plasma where this condition is met exactly, such as in space plasmas, ion thrusters, and magnetic fusion mirror machine concepts. However, the real significance of understanding behavior of this type of plasma is that many plasma regimes exhibit some of these characteristics, but to a lesser extent. For example, it is possible that electrons will exhibit collisionless behavior while ions are collision dominated. Plasmas with relaxed effects of collisions demonstrate distinctly different effects of fields than those that exist for continuum plasmas, and these effects do provide insight into the unique behavior of charged particles in all plasmas, particularly with respect to transport phenomena. Understanding of collisionless behavior is fundamental for the study of more complex particle interactions (Chen, 1984; Boyd and Sanderson, 1969), particularly where there is energy transfer between particles and fields.

We will make some simplifying assumptions in this chapter. We will consider (1) collisions negligible, (2) nonrelativistic particle velocities ($v \ll c$), and (3) radiation effects negligible.

There is a basic direction reference to that of the magnetic field, i.e., $v_{\|}$ implies that the velocity is parallel to the magnetic field, while $v_{\perp}$ is perpendicular to the magnetic field. We have noted that:

$$\vec{F}_{EM} = q\left(\vec{E} + \vec{v} \times \vec{B}\right) \quad \text{with} \quad \vec{F}_E \parallel \vec{E}, \vec{F}_B \perp \vec{v}, \vec{B};$$

if:

$$\vec{v} \parallel \vec{B} \quad \therefore \text{no field interaction}\left(v_{\|}\right); \qquad \vec{v} \perp \vec{B} \quad \therefore \text{field interaction } \left(v_{\perp}\right).$$

However, $\vec{F}_B \perp \vec{v} \Rightarrow \vec{F}_B \perp$ direction of motion, so with $W = F \cdot \text{dist.}$, there is no work done, and energy is conserved.

Basically, we are considering the equation of motion for a particle as:

$$\vec{F} = m\ddot{\vec{r}};$$

Introduction to Plasmas and Plasma Dynamics
ISBN 978-0-12-801661-9

$$q\left[\vec{E}(\vec{r},t) + \dot{\vec{r}} \times \vec{B}(\vec{r},t)\right] = m\ddot{\vec{r}};$$

$$q\left(\vec{E} + \vec{v} \times \vec{B}\right) = m\dot{\vec{v}};$$

and:

$$q\left(\vec{E}_{\parallel} + \vec{E}_{\perp} + \vec{v}_{\perp} \times \vec{B} + \vec{\tilde{v}}_{\perp} \times \vec{B}\right) = m\dot{\vec{v}},$$

where:

$$\vec{v} = \vec{v}_{\perp} + \vec{\tilde{v}}_{\perp} + \vec{v}_{\parallel} \text{ and we define } \vec{\tilde{v}}_{\perp} \text{ as } \perp \text{ to } \vec{E} \text{ and } \vec{B}.$$

We will be considering the solution of such equations of motion for charged particles (Spitzer, 1962; Sutton and Sherman, 1965).

We assume that $\vec{E}$, $\vec{B}$ are given, but for the general problem we know that they must be consistent with Maxwell's equations.

CHARGED PARTICLE MOTION IN CONSTANT, UNIFORM MAGNETIC ($\vec{B}$) FIELD

We take:

$$\vec{E} = 0;\ \vec{B} = \vec{B}_z = B\vec{k};\ \vec{r} = \vec{i}x + \vec{j}y + \vec{k}z.$$

Then:

$$m\ddot{\vec{r}} = q\left(\dot{\vec{r}} \times \vec{B}\right).$$

1. Consider particles with $\dot{\vec{r}} = v_{\parallel} = v_z$.
   The equation of motion is then $m\ddot{\vec{r}} = 0$; integrate to get $\dot{\vec{r}} = v_{\parallel} = v_z =$ constant.
   So the velocity along the magnetic field direction is not affected.
2. To clarify general behavior, with the equation of motion, take: $\dot{\vec{r}} \cdot (LHS = RHS)$.

We get $m\ddot{\vec{r}} \cdot \dot{\vec{r}} = 0$; integrate, then $\frac{1}{2}m\dot{\vec{r}}^2 =$ constant $= W$ (kinetic energy), so:

$$\frac{1}{2}m\left(v_{\perp}^2 + v_{\parallel}^2\right) = W_{\perp} + W_{\parallel} = \text{constant}.$$

Then $v_\perp^2 + v_\parallel^2 = \text{constant}$, which implies that if $v_\parallel = \text{constant}$, then $v_\perp = \text{constant}$.

(Note here that this behavior occurs where B is uniform. We will examine later the effects of gradients in B, i.e., ∇B.)

3. In general, we are considering $m\dot{\vec{v}} = q\vec{v} \times \vec{B}$, with $m\dot{v}_\parallel = 0$ and $m\dot{\vec{v}_\perp} = q\vec{v}_\perp \times \vec{B}$.

Also, we can write: $m\ddot{\vec{r}} = q(\dot{\vec{r}} \times \vec{B})$. In this case, we consider: $\vec{B} = B\vec{k}$ where $\vec{k}$ is a unit vector in the z direction. Therefore, v_x and v_y motion is interactive, as:

$$m\ddot{x} = q\dot{y}B_z, \quad \text{and} \quad m\ddot{y} = -q\dot{x}B_z, \quad \text{so:}$$

$$\ddot{x} = \frac{qB_z}{m}\dot{y} = \omega\dot{y}, \quad \text{and} \quad \ddot{y} = -\frac{qB_z}{m}\dot{x} = -\omega\dot{x}.$$

These are coupled equations. We differentiate each with respect to time to get:

$$\dddot{x} = \omega\ddot{y}, \quad \dddot{y} = -\omega\ddot{x},$$

$$\frac{\dddot{x}}{\omega} = -\omega\dot{x}, \quad \frac{\dddot{y}}{\omega} = -\omega\dot{y}$$

These are ordinary differential equations with constant coefficients, second order; the solution is one with simple harmonic motion:

$$\dddot{x} + \omega^2\dot{x} = 0,$$

where, $\dot{x} = Ce^{i\omega t}$, so: $\ddot{x} = i\omega Ce^{i\omega t}$, and $\dddot{x} = -\omega^2 Ce^{i\omega t} = -\omega^2\dot{x}$.

With $\dot{x}(t = 0) = v_\perp \cos\alpha = Ce^{i(0)}$, therefore: $\dot{x} = v_\perp \cos\alpha e^{i\omega t}$.

But with:

$$e^{i\alpha} = \cos\alpha + i\sin\alpha, \; Real\; e^{i\alpha} = \cos\alpha,$$

$$\text{so: } \dot{x} = v_\perp e^{i(\omega t+\alpha)} = v_\perp \cos(\omega t + \alpha).$$

Then we can write: $\dot{y} = \ddot{x}/\omega = -v_\perp \sin(\omega t + \alpha)$, and by integration with $(x = x_0, y = y_0, z = z_0 \text{ at } t = 0)$, we get:

$$x = \frac{v_\perp}{\omega}\sin(\omega t + \alpha) + x_0,$$

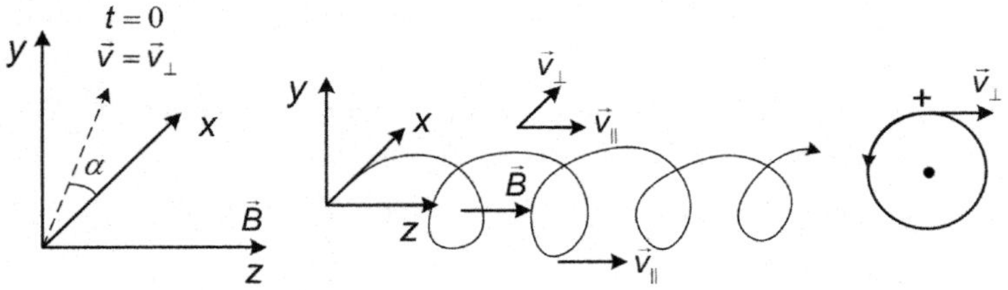

Figure 6.1 Particle motion in constant, uniform B field.

$$y = \frac{v_\perp}{\omega}\cos(\omega t + \alpha) + y_0.$$

This solution is helical motion with velocity v along the z-axis: $z = v_\parallel t + z_0$ (Figure 6.1).

Consider motion in the x-y plane (x_0, y_0 are zero), if we are in a plane moving with velocity, $v_\parallel$, along the z-axis, we get:

$$x^2 + y^2 = \frac{v_\perp^2}{\omega^2} = r^2 = \text{constant},$$

so, we have: $r =$ constant, circular motion in the x, y plane. The center of the circle (the axial position of the particle) is called the *guiding center* of the particle orbit, which in this case is moving with velocity, $v_\parallel$. Note that while the particle may spiral, what is important is the overall "drift" of the particle.

In general, the motion of particles in B and E fields may be represented by cyclic motion with radius r_L and with the center, called a guiding center, in translational motion.

PARTICLE MOTION IN UNIFORM ELECTRIC AND MAGNETIC FIELDS

We consider here a uniform electric field, where $\overrightarrow{E}_\parallel$ will cause uniform accelerated motion of a particle in that direction, and we will examine the particle motion that will result from $\overrightarrow{E}_\perp$. A sketch of the geometry is shown in Figure 6.2.

In a plane $\perp$ to $\overrightarrow{B}$ (the x, y plane): $F_\perp = m\ddot{r}$, but: $q(\overrightarrow{E}_\perp + \overrightarrow{v}_\perp \times \overrightarrow{B}) = m\ddot{\overrightarrow{r}}$. Because of vector properties, something different from the behavior above will happen in this case. Let us take: $\overrightarrow{v}_\perp = \overrightarrow{v} + \overrightarrow{\tilde{v}}$, the sum of circular velocity about the $\overrightarrow{B}$ field and an added component.

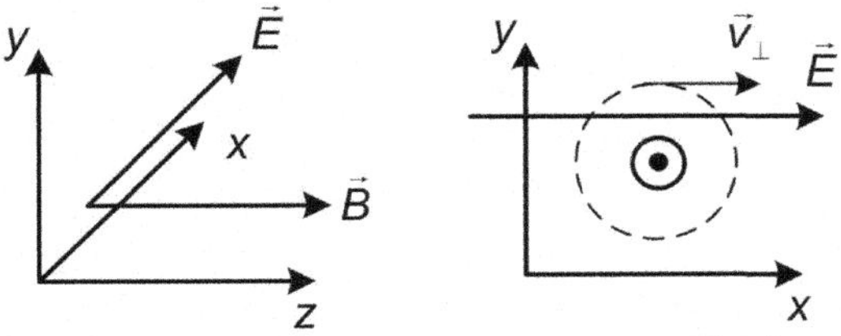

Figure 6.2 Geometry and plane motion in $\vec{E}$, $\vec{B}$ fields.

Then, $\overrightarrow{r} = \overrightarrow{r'} + \overrightarrow{\tilde{r}}$ and substituting into the acceleration term, get:

$$q\left(\overrightarrow{E}_{\perp} + \overrightarrow{v'} \times \overrightarrow{B} + \overrightarrow{\tilde{v}} \times \overrightarrow{B}\right) = m\left(\overrightarrow{\ddot{r}'} + \overrightarrow{\ddot{\tilde{r}}}\right).$$

Now, if we examine the behavior without $\overrightarrow{E}_{\perp}$: $q\left(\overrightarrow{v'} \times \overrightarrow{B}\right) = m\,\overrightarrow{\ddot{r}'}$, where $\overrightarrow{v'} = \overrightarrow{\dot{r}'}$, then $E_{\perp}, B \sim \overrightarrow{\tilde{v}}$, or:

$$q\left(\overrightarrow{E}_{\perp} + \overrightarrow{\tilde{v}} \times \overrightarrow{B}\right) = m\,\overrightarrow{\ddot{\tilde{r}}}.$$

With $\overrightarrow{\tilde{v}} = \text{constant}$, $\overrightarrow{\ddot{\tilde{r}}} = \overrightarrow{\dot{\tilde{v}}} = 0$, then $\overrightarrow{E}_{\perp} = -\overrightarrow{\tilde{v}} \times \overrightarrow{B}$, or $\overrightarrow{E}_{\perp} \times \overrightarrow{B} = -\left(\overrightarrow{\tilde{v}} \times \overrightarrow{B}\right) \times \overrightarrow{B}$,

$$\therefore \overrightarrow{\tilde{v}} = \frac{\overrightarrow{E}_{\perp} \times \overrightarrow{B}}{B^2} = \overrightarrow{v}_{drift} = \overrightarrow{v}_{gc} = \overrightarrow{v}_{E}.$$

So, particles injected into electric and magnetic fields, $\overrightarrow{B}$, $\overrightarrow{E}_{\perp}$, will move as: $\overrightarrow{E}_{\perp} \times \overrightarrow{B}$ (Figure 6.3).

To generalize from this specific case of force application in a magnetic field, what is seen here is a principle: Any force on a particle that acts as $\overrightarrow{E}_{\perp}$ does here will have the same type of vector cross-product interaction with the magnetic field and produce the same kind of motion.

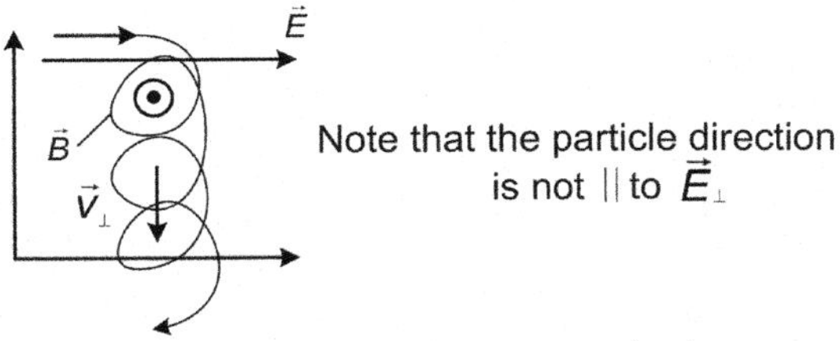

Figure 6.3 Particle drift velocity in $\vec{E}$, $\vec{B}$ fields.

As it is important to understand the type of behavior resulting from linear forces applied simultaneously with magnetic field forces on a particle, the dynamics will also be analyzed using the equations of motion. Again we will consider uniform $\vec{B}$ and $\vec{E}$ fields. The equation of motion for a particle is: $m\frac{d\vec{v}}{dt} = q(\vec{E} + \vec{v} \times \vec{B})$, and along $\vec{B}$: $m\frac{dv_\parallel}{dt} = qE_\parallel$, which has a solution:

$$v_\parallel = \frac{qE_\parallel}{m}t + v_{\parallel_0}(t = 0), \qquad \text{so:}$$

$$x_\parallel = \frac{qE_\parallel}{m}\frac{t^2}{2} + v_{\parallel_0}t + x_{\parallel_0}.$$

This solution represents a behavior of free acceleration along the $\parallel$ direction.

Now, in the direction $\perp$ to $\vec{B}$: $m\frac{d\vec{v}_\perp}{dt} = q(\vec{E}_\perp + \vec{v}_\perp \times \vec{B})$. If $\vec{F}_{E\perp} = -\vec{F}_B$, then $\frac{d\vec{v}_\perp}{dt} = 0$, i.e., the particle moves with constant velocity. But what is the velocity? It must be proportional to $\vec{E}$, $\vec{B}$ and be related in direction. So we take: $\vec{E}_\perp + \vec{v}_\perp \times \vec{B} = 0$, and taking $\vec{B}\times$ gives:

$$\vec{B} \times \vec{E}_\perp + \vec{B} \times \vec{v}_\perp \times \vec{B} = 0,$$

$$\text{and} \quad \vec{B} \times \vec{E}_\perp + \vec{B}\cdot\vec{B}\,(\vec{v}_\perp) - \left(\vec{B}\cdot\vec{v}_\perp\right)\vec{B} = 0,$$

so, $\vec{B} \times \vec{E}_\perp + B^2\vec{v}_\perp = 0$, or: $\vec{v}_\perp = \frac{\vec{E}_\perp \times \vec{B}}{B^2} \equiv \vec{v}_E$. for the case: $\left(\frac{d\vec{v}_\perp}{dt} = 0\right)$. And this is an $\vec{E} \times \vec{B}$ drift motion, $\perp$ to $\vec{B}$ and $\vec{E}$.

Return to the general case; if we break up the motion into velocity components with (*w*) $\vec{E}$ and without (*wo*) $\vec{E}$, we can write: $\vec{v}_\perp = \vec{v}_1 + \vec{v}_{EB}$, where the first term is the motion *wo* $\vec{E}$ and the second is the motion with E and is $\perp$ to $\vec{B}$ and $\vec{E}$. Then we can write:

$$m\frac{d(\vec{v}_1 + \vec{v}_{EB})}{dt} = q\left[\vec{E}_\perp + (\vec{v}_1 + \vec{v}_{EB}) \times \vec{B}\right], \quad \text{and expanding :}$$

$$m\frac{d\vec{v}_1}{dt} + m\frac{d\vec{v}_{EB}}{dt} = q\left(\vec{E}_\perp + \vec{v}_1 \times \vec{B} + \vec{v}_{EB} \times \vec{B}\right),$$

so, $m\frac{d\overrightarrow{v}_{EB}}{dt} = q(\overrightarrow{E}_\perp + \overrightarrow{v}_{EB} \times \overrightarrow{B})$, and $m\frac{d\overrightarrow{v}_1}{dt} = q\overrightarrow{v}_1 \times \overrightarrow{B}$ is the equation of motion without the $\overrightarrow{E}$ field, which case has been solved above. Therefore, the total motion of a particle is a constant drift with $\overrightarrow{v}_{EB}$ plus cyclic motion about $\overrightarrow{B}$.

Note that a particle injected into $\overrightarrow{B}$ and $\overrightarrow{E}_\perp$ will move to $\overrightarrow{E}_\perp \times \overrightarrow{B}$; it will drift $\perp$ to the force $(\overrightarrow{F})$ and $\overrightarrow{B}$. So, in general, we recognize again that any external body force will produce a similar reaction, i.e., an applied $(\overrightarrow{F})$ produces: $\overrightarrow{v}_F = \frac{\overrightarrow{F} \times \overrightarrow{B}}{qB^2}$.

PARTICLE MOTION IN SPATIALLY VARYING (INHOMOGENOUS) MAGNETIC FIELDS

We will next consider motion in magnetic fields that vary very gradually with position. Very gradually means that the variation over one Larmor radius is small: $|(\overrightarrow{r}_L \cdot \nabla)\overrightarrow{B}| \ll \overrightarrow{B}$.

Particle motion is a perturbation from that for a spatially uniform magnetic field, i.e., if we Taylor expand $\overrightarrow{B}$ about some point, $\overrightarrow{r}$, as:

$$\overrightarrow{B}(\overrightarrow{r} + \overrightarrow{\rho}) = \overrightarrow{B}(\overrightarrow{r}) + (\overrightarrow{\rho} \cdot \overrightarrow{\nabla})\overrightarrow{B}(\overrightarrow{r}),$$

where $\overrightarrow{r}$ is the position of the guiding center and ∇ is performed at the guiding center. The spatial variation can be represented in tensor form as:

$$\overrightarrow{\nabla}\overrightarrow{B} = \begin{pmatrix} \frac{\partial B_x}{\partial x} & \frac{\partial B_x}{\partial y} & \frac{\partial B_x}{\partial z} \\ \frac{\partial B_y}{\partial x} & \frac{\partial B_y}{\partial y} & \frac{\partial B_y}{\partial z} \\ \frac{\partial B_z}{\partial x} & \frac{\partial B_z}{\partial y} & \frac{\partial B_z}{\partial z} \end{pmatrix}.$$

Let us consider the physical significance of a simple gradient, $\frac{\partial B_z}{\partial x}$, where B_z varies in the x, y plane. The particle moves in the x, y plane, as (Figure 6.4).

With $\frac{\overrightarrow{F}_B}{q} = \overrightarrow{v}_\perp \times \overrightarrow{B}$, and with B increasing in the $+x$ direction, $B(RHS) > B(LHS)$, so, $\overrightarrow{v}_\perp \times \overrightarrow{B}(RHS) > \overrightarrow{v}_\perp \times \overrightarrow{B}(LHS)$, and as a

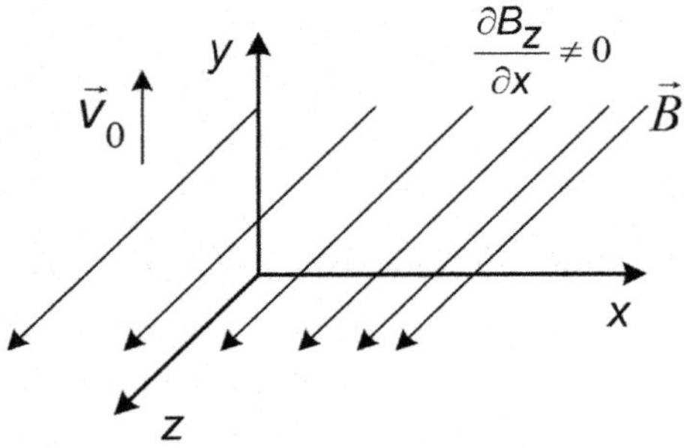

Figure 6.4 Particle motion with $\frac{\partial B_z}{\partial x} \neq 0$.

result, we have a drift of the particle, $v_{drift} = v_y$. Then $\vec{v}_y \times \vec{B}$ (and the force) are in the x-direction. The above has been a discussion based on physical interactions; as before, we now solve the problem as one of particle dynamics.

We will do a simple calculation of this drift effect. For small variations in the magnetic field, we presume equilibrium, and the average force over x, the direction with field gradient, will be taken as zero; so over one cycle, we write:

$$\int_{t_1}^{t_2} F_x dt = 0, \quad \text{where, } F_x = qv_y B_z(x) \approx qv_y \left[B_0 + x\left(\frac{\partial B_z}{\partial x}\right)_0 \right].$$

Then, $\int_{t_1}^{t_2} B_0 v_y dt + \int_{t_1}^{t_2} x\left(\frac{\partial B_z}{\partial x}\right)_0 v_y dt = 0$, and with B_0, $\left(\frac{\partial B_z}{\partial x}\right)_0$ both constant:

$$\int_{t_1}^{t_2} v_y dt = -\frac{1}{B_0}\left(\frac{\partial B_z}{\partial x}\right)_0 \int_{t_1}^{t_2} x v_y dt.$$

But $LHS = y_2 - y_1$, and a weak gradient implies that we approximately have a Larmor orbit, and $v_y \cdot dt \approx dy$, so:

$$\int_{t_1}^{t_2} x v_y dt \approx \int_{t_1}^{t_2} x dy = \int_{t_1}^{t_2} dA_{orbit} = \pi r_L^2$$

for one orbit, and:

$$\therefore \delta y = y_2 - y_1 = -\frac{1}{B_0}\left(\frac{\partial B_z}{\partial x}\right)_0 \pi r_L^2.$$

But:

$$r_L = \frac{mv_\perp}{qB}, \quad \omega = \frac{qB}{m};$$

so:

$$\therefore\ r_L^2 = \frac{mv_\perp}{qB}\cdot\frac{mv_\perp}{qB} = \frac{2\cdot\frac{1}{2}mv_\perp^2}{qB}\cdot\frac{1}{\omega_p} = \frac{2W_\perp}{qB}\cdot\frac{1}{\omega_p}.$$

Since:

$$\delta y = \frac{\text{disp}}{\text{cycle}}, \text{ and } v_{drift}\left(\frac{\text{disp}}{\text{time}}\right) = -\delta y\left(\frac{\text{disp}}{\text{cycle}}\right)\omega_p\left(\frac{\text{rad}}{\text{time}}\right)\left(\frac{\text{cycle}}{2\pi\text{rad}}\right),$$

then:

$$v_{drift}\left(\frac{\partial B_z}{\partial x}\right) = \frac{1}{B_0}\left(\frac{\partial B_z}{\partial x}\right)_0 \pi\cdot\frac{2W_\perp}{qB}\cdot\frac{1}{\omega_p}\cdot\frac{\omega_p}{2\pi} = \frac{W_\perp}{qB_0^2}\left(\frac{\partial B_z}{\partial x}\right)_0,$$

and, we can infer, in general:

$$v_{drift} \rightarrow \frac{W_\perp}{qB_0^3}\left(\vec{B}\times\vec{\nabla}\right)B \rightarrow \frac{W_\perp}{qB_0^3}\left(\nabla B\times\vec{B}\right).$$

So for the field in the z-direction and the gradient in the x-direction, the drift is in the y-direction and the force is in the $-x$-direction.

PARTICLE MOTION WITH CURVATURE OF THE MAGNETIC FIELD LINES

In this case, we consider a velocity, $v_{||}$, along $\vec{B}$, where $\vec{B}$ is curved in the y-z plane. Then, we have the situation presented in Figure 6.5.

As has been developed earlier for the particle drift in a force field: $\vec{v}_F = \frac{\vec{F}\times\vec{B}}{qB^2}$, so the effect of the centrifugal force on the particle will have: $\vec{F}_{CF}(y)\times\vec{B}(z) \Rightarrow v_{drift}(-x)$. Also, due to the gradient in $B_y(z)$:

$$\vec{v}_{curve} = \frac{2W_{||}}{qB_0^3}\left(\vec{B}\times\vec{\nabla}\right)B \rightarrow \frac{2W_{||}}{qB_0^3}\left(\nabla B\times\vec{B}\right).$$

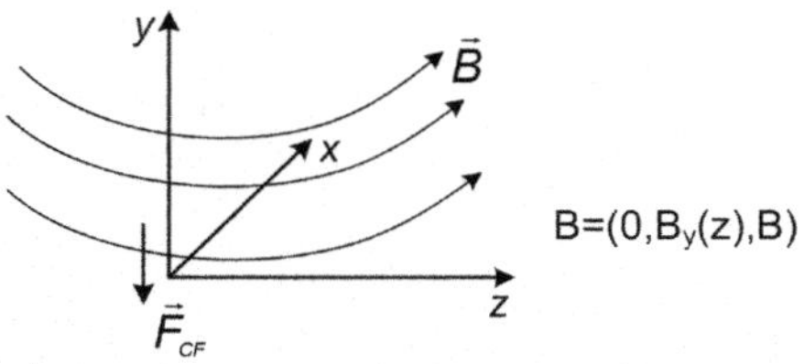

Figure 6.5 Particle motion with field curvature in *y-z* plane.

PARTICLE MOTION IN TIME-VARYING MAGNETIC FIELD

We will consider slowly varying fields, uniform in space, as: $\frac{\dot{B}}{\omega_L} \ll B$. With $\dot{B}$ finite $\Rightarrow \vec{E}$ field, which is $\perp$ to $\vec{B}$ (e.g., $\dot{B}_\theta \Rightarrow j_z$), and $v_\perp$ in the same direction as induced $\vec{E}$ implies that work is being done on the particle and that implies energy transfer.

Now the equation of motion can be written as:

$$m\dot{\vec{v}} = q\left(\vec{E} + \vec{v} \times \vec{B}\right), \text{ and } \vec{E} = \vec{E}_\perp \Rightarrow \vec{v}_\parallel = \text{constant}; \quad \text{so:}$$

$$\vec{v}_\perp \cdot m\dot{\vec{v}}_\perp = q\left[\vec{v}_\perp \cdot \vec{E} + \vec{v}_\perp \cdot \left(\vec{v}_\perp \times \vec{B}\right)\right],$$

$$\text{and} \quad \frac{d}{dt}\left(\frac{1}{2}mv_\perp^2\right) = q\vec{E} \cdot \vec{v}_\perp.$$

Considering one Larmor orbit and taking $d\vec{l} = \vec{v}_\perp dt$, get:

$$\delta W_\perp = q\oint \vec{E} \cdot d\vec{l} = q\int_{A_{r_L}} \left(\nabla \times \vec{E}\right) \cdot d\vec{A}$$

$$= -q\int_{A_{r_L}} \frac{\partial \vec{B}}{\partial t} \cdot d\vec{A}, \quad \text{(from Maxwell's equations)}.$$

(Note that this term will always be positive.) Considering small gradients, $\dot{B} \approx$ constant over one orbit (B changes slowly) then:

$$\delta W_\perp = e\pi r_L^2 \frac{\partial B}{\partial t} = e\pi\left(\frac{mv_\perp}{qB}\right)^2 \frac{\partial B}{\partial t}, \quad \text{and with } W_\perp = \frac{1}{2}mv_\perp^2, \quad \text{we get:}$$

$$\delta W_\perp = W_\perp \frac{2\pi m}{eB} \cdot \frac{1}{B} \cdot \frac{\partial B}{\partial t} = W_\perp \frac{2\pi}{\omega_c B} \cdot \frac{\partial B}{\partial t}.$$

Now since:

$$2\pi\left(\frac{\text{rad}}{\text{rev}}\right)\frac{1}{\omega_c}\left(\frac{\text{time}}{\text{rad}}\right)\frac{\partial B}{\partial t}\left(\frac{\text{Tesla}}{\text{time}}\right)=\frac{\delta B}{B}, \quad \text{then}$$

$$\frac{\delta W_\perp}{W_\perp} = \frac{\delta B}{B}, \text{or: } \delta\left(\frac{W_\perp}{B}\right)=0.$$

So, $\frac{W_\perp}{B}$ is a constant of the motion and is referred to as an invariant of the motion. Placing this in a physical context, slowly varying magnetic fields $\left(\frac{\partial B}{\partial t} \approx 0\right)$ can be taken as adiabatic, i.e., there is no energy addition.

Magnetic Moment

It is important to define the quantity $(W_\perp/B)$, more completely. We begin by examining:

$$\frac{W_\perp}{B}=\frac{1}{2}mv_\perp^2\cdot\frac{1}{B}=\frac{\pi}{2\pi}\cdot\frac{e^2}{e^2}\cdot m\frac{v_\perp^2}{B}\cdot\frac{B}{B}\cdot\frac{m}{m}=\frac{\pi}{2\pi}\cdot\frac{e^2B}{m}\cdot r_L^2=\frac{e\omega_c}{2\pi}\cdot A_{Larmor}.$$

We can also identify that: $e(\text{chg})\cdot\omega\left(\frac{\text{rad}}{\text{time}}\right)\left(\frac{\text{rev}}{2\pi\text{rad}}\right)\cdot A\left(\frac{\text{area}}{\text{rev}}\right) = IA$, i.e., the charge of a particle moving about a loop per unit time is current. So, $W_\perp/B = \text{I}\cdot\text{A}$, and this is the magnetic (dipole) moment, μ, of the particle. We can also recognize that: $\frac{W_\perp}{B}=\frac{e^2}{2\pi m}\cdot\underbrace{\pi r_L^2 B}_{\phi(\text{flux})}$, so, the flux of a particle is invariant, and the flux moves with the particle, i.e., it is "frozen in."

The parameter μ has significance in other particle-field configurations. We will summarize two specific findings.

1. In inhomogeneous fields, μ is invariant

For small gradients in B: $\nabla\cdot\vec{B}=0$, which gives us: $B_r \approx -\frac{r_L}{2}\frac{\partial B}{\partial z}$. From the equation of motion: $m\frac{dv_\parallel}{dt}=q\cdot v_\perp\cdot B_r = q\cdot v_\perp\left(-\frac{r_L}{2}\frac{\partial B}{\partial z}\right)=-\mu\frac{\partial B}{\partial z}$. If we multiply both sides by $v_\parallel$, we can write: $\frac{d}{dt}\left(\frac{1}{2}mv_\parallel^2\right)=v_\parallel\left(-\mu\frac{\partial B}{\partial z}\right)=-\mu\frac{\partial B}{\partial t}$; but $\frac{d}{dt}\left(\frac{1}{2}mv_\perp^2\right)=\frac{d}{dt}(\mu B)$, so, $\frac{d}{dt}(\mu B)-\mu\frac{\partial B}{\partial t}=0$, so we have:

$$\frac{d}{dt}\mu=0, \quad \text{and} \quad \mu=\text{constant}.$$

2. In inhomogeneous fields, μ defines a force relationship.

From above: $m\frac{dv_{\|}}{dt} = -\mu\frac{\partial B}{\partial z}$, so we can write: $\vec{F}_{\|} = -\mu(\nabla B)_{\|}$. We can extend this relationship to the general vector by noting: $\vec{v} = \frac{2W_{\perp}}{qB_0^3}(\nabla B \times \vec{B})$ and also, $\vec{v} = \frac{\vec{F} \times \vec{B}}{qB^2}$, so, $\frac{W_{\perp}}{B_0}\nabla B = \vec{F}$ and we write: $\vec{F}_{\perp,\|} = -\mu\vec{\nabla}_{\perp,\|}B$, which represents a force on the particle, not a drift.

PARTICLE TRAPPING IN MAGNETIC MIRRORS

We consider particle motions with $\vec{E} = 0$. It was noted that in an inhomogeneous field, $\vec{F} = -\mu\nabla B$, where the force is opposite the gradient, i.e., the particle force is away from the increasing $\vec{B}$. In the two directions:

$$\vec{F}_{\|} = -\mu\left(\frac{\partial B_z}{\partial z}\right)\vec{k}, \text{and} \quad \vec{F}_{\perp} = -\mu\left(\frac{\partial B_z}{\partial x}\right)\vec{i}.$$

Here we will consider the motion parallel to the $\vec{B}$ field, i.e., Figure 6.6.

We consider: $m\frac{dv_{\|}}{dt} = -\mu\frac{dB_z}{dz}$, and for particles moving in the $+z$ direction: $dz = v_{\|}dt$, so:

$$mv_{\|}\frac{dv_{\|}}{dz} = -\mu\frac{dB_z}{dz},$$

or integrating:

$$m\left(\frac{v_{\|}^2}{2} - \frac{v_{\|0}^2}{2}\right) = -\mu(B_z - B_{z0}).$$

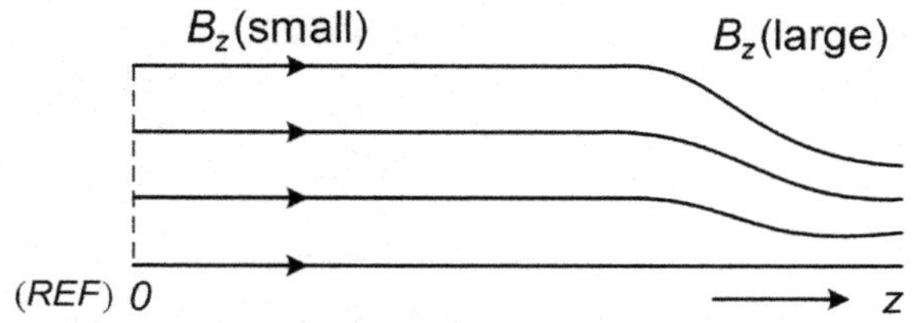

Figure 6.6 Magnetic field geometry for magnetic mirror.

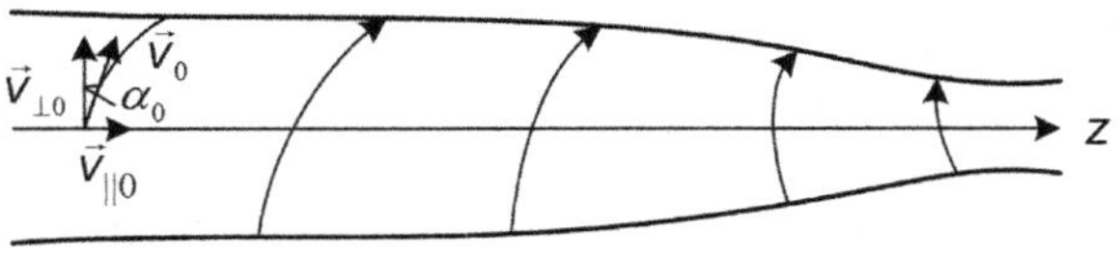

Figure 6.7 Particle motion in magnetic mirror.

So, for B_z increasing, $\Rightarrow$ $v_\parallel$ decreasing; the limit occurs when $v_\parallel = 0$, then:

$$B_{z_{max}}(v_\parallel = 0) = B_{z_0} + \frac{mv_{\parallel_0}^2}{2\mu}.$$

Particles have different energies and velocities; when does a particle stop? Since $\frac{dB_z}{dz} > 0$ where $v_\parallel = 0$, then the particle will *reflect*. What we are considering here is the following motion (Figure 6.7).

We can now consider the different possible motions; we write that: $\frac{v_{\perp 0}}{v_0} = \sin\alpha_0$ and $\frac{v_{\parallel 0}}{v_0} = \tan\alpha_0$. Now, we have the invariants $\mu, \frac{\frac{1}{2}mv_{\perp 0}^2}{B_{z_0}} = \frac{\frac{1}{2}mv_{\perp max}^2}{B_{z max}}$; Energy, $\frac{1}{2}mv_0^2 = \frac{1}{2}mv_{max}^2$; and, at the turn, $\sin\alpha_t = 1$. To stop the particle, $\sin^2\alpha_0 = \frac{v_{\perp 0}^2}{v_0^2} = \frac{v_{\perp 0}^2}{v_{\perp 0}^2 + v_{\parallel 0}^2}$ and $v_0^2 = v_{max}^2 = v_{\perp max}^2 + v_{\parallel max}^2$ $= v_{\perp max}^2$, so, $\sin^2\alpha_0\big|_{stop} = \frac{v_{\perp 0}^2}{v_{\perp max}^2} = \frac{B_{z_0}}{B_{z max}}$, i.e., given B_{z_0} and $B_{z_{max}}$, particles with:

$$\sin^2\alpha_0 = \frac{v_{\perp 0}^2}{v_{\perp 0}^2 + v_{\parallel 0}^2}\begin{cases} > \\ = \frac{B_z}{B_{z max}} \\ < \end{cases}$$

where, $>$ will reflect, $=$ will stop, $<$ will escape.

Accordingly, reflecting particles are trapped in the bottle, while others escape from the bottle. We define: $\frac{B_{max}}{B_0} = R$ as the mirror ratio, so, $\sin^2\alpha_0 > \frac{B_z}{B_{z max}} = \frac{1}{R}$ for reflection. In other words: Large $B_{max} \Rightarrow$ Large $R \Rightarrow$ Small loss.

Note that for particles moving into strong $\vec{B}$: $\frac{v_{\perp 0}^2}{B_0} = \frac{v_\perp^2}{B} \Rightarrow B$ increase, $v_\perp$ increase. However, $v_{\perp_0}^2 + v_{\parallel_0}^2 = v_\perp^2 + v_\parallel^2 \Rightarrow v_\perp$ increase with $v_\parallel$ decrease, so, one way that velocities $v_\parallel, v_\perp$ interchange magnitudes is because of gradients, ∇B.

ADIABATIC INVARIANTS

Particles that do not have their energies changed by E or B fields have a number of properties that are constant; these properties are called adiabatic

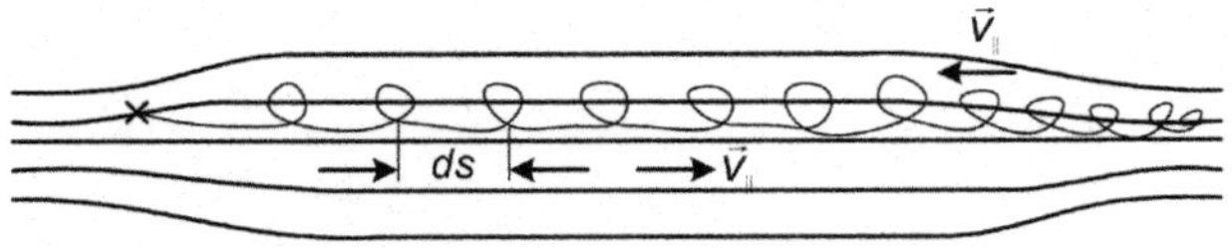

Figure 6.8 Cyclic motion for trapped particles.

invariants. We have already noted such conditions for the following properties to remain constant:

$$W(\textit{Total KE}) = \frac{1}{2}mv_{\perp}^2 + \frac{1}{2}mv_{\parallel}^2,$$

$$\mu(\text{Magnetic moment}) = \frac{W_{\perp}}{B}, \quad \text{or}: \quad \phi_{Larmor}(\text{mag. flux}) = \pi r_L^2 B.$$

There is also an invariant concept that is related to magnetic mirror behavior. Particles in a magnetic mirror captured configuration have the geometry as shown in Figure 6.8.

The particle can be seen to move through a cycle, from one end to the other with velocity $v_{\parallel}$. An invariant of the motion in slowly varying (geometry) fields can be identified. It can be shown that: $\oint v_{\parallel} ds \equiv J_l =$ constant, where J_l is called the longitudinal constant.

An analysis of the effect (Boyd and Sanderson, 1969) uses the relationship: $W = \frac{1}{2}mv_{\parallel}^2 + \mu B$, so that:

$$J_l = \int_{s_1}^{s} \left[\frac{2}{m}(W - \mu B)\right]^{\frac{1}{2}} ds.$$

Taking a function: $\frac{dJ_l}{dt} = \frac{d}{dt}J_l(W, s, t)$ and for small variations, we can show that: $\frac{dJ_l}{dt} = 0$, so J_l is an adiabatic invariant. However, the details of this consideration show the dependence of energy conservation on the length of the bottle region. If the length of the confinement region is changed, the energy of the particle is not conserved; this introduces the notion that there are still-to-be-defined mechanisms for transfer of energy to particles from fields.

REFERENCES

Boyd, T.J.M., Sanderson, J.J., 1969. Plasma Dynamics. Barnes & Noble, New York.

Chen, F.F., 1984. Introduction to Plasma Physics and Controlled Fusion, second ed. Plenum, New York.

Spitzer Jr., L., 1962. Physics of Fully Ionized Gases. Wiley, New York.

Sutton, G.W., Sherman, A., 1965. Engineering Magnetohydrodynamics. McGraw-Hill, New York.

CHAPTER 7

Macroscopic Equations of Plasmas

INTRODUCTION

The previous chapters have dealt with examining different aspects of the behavior of particles in plasmas and differentiating between individual species elements. That treatment is basic to understanding the mechanisms at work in plasmas, as the different species—electrons, ions, and neutral particles—respond differently in electric and magnetic fields. However, in most device applications the collective behavior of the plasma is important, that is, the effects of particle collisions are dominant and the plasma behaves more as a continuum (Shercliffe, 1965). When considering macroscopic behavior, generally the mean free path is short compared to the Debye length and the Larmor radius (Boyd and Sanderson, 1969).

This chapter addresses the mechanisms, laws, and parameters that govern the applications of electric and magnetic fields to ionized gases, and the flow of ionized gases. We take the point of view that the form of the equations can sometimes facilitate not only the analysis, but also the physical understanding of the phenomenology. There is always a choice in the form of the basic equations—one or more dimensions; average or multicomponent species; scalar, vector, or tensor representations—for any particular application, and as this is an introductory work, we present the simplest analytical forms when that is sufficient. However, we present an analysis in the context that further consideration and further research are almost always appropriate.

ELECTROMAGNETIC ENERGY AND MOMENTUM ADDITION TO PLASMAS

We will extend our detailed understanding of particle behavior in electric and magnetic fields to include collective behavior of collisional plasma. (We follow the developments in Jackson, 1963; Sommerfeld, 1964).

Introduction to Plasmas and Plasma Dynamics
ISBN 978-0-12-801661-9

For a single charge, q, the rate of doing work by external electromagnetic fields $\vec{E}$, $\vec{B}$ is given by $q\vec{v}\cdot\vec{E}$ where $\vec{v}$ is the velocity of the charge; the work related to the magnetic field is zero, as $\vec{F_B} = \vec{v} \times \vec{B}$.

For a continuous distribution of charge and current, the total rate of doing work on a finite volume, V, is: $\frac{W}{\text{time}} = \vec{F}\cdot\vec{v} = q\vec{E}\cdot\vec{v}$, and for N particles in volume, V; here, $\vec{J}$ (upper case) represents current density:

$$\frac{W}{\text{time}} = Nq\vec{E}\cdot\vec{v} \rightarrow \int nq\vec{v}\cdot\vec{E}\,dV = \int_V \vec{J}\cdot\vec{E}\,dV.$$

But where is the power transfer coming from/going to? (Do we have electromagnetic energy to mechanical or thermal energy?) To try to clarify component terms we apply Maxwell's equations, as:

$$\int_V \vec{J}\cdot\vec{E}\,dV = \int_V \left(\nabla\times\vec{H} - \frac{\partial\vec{D}}{\partial t}\right)\cdot\vec{E}\,dV,$$

but:

$$\nabla\cdot\left(\vec{E}\times\vec{H}\right) = \vec{H}\cdot\left(\nabla\times\vec{E}\right) - \vec{E}\cdot\left(\nabla\times\vec{H}\right),$$

so:

$$\int_V \vec{J}\cdot\vec{E}\,dV = \int_V \left[\vec{H}\cdot\left(\nabla\times\vec{E}\right) - \nabla\cdot\left(\vec{E}\times\vec{H}\right) - \vec{E}\cdot\frac{\partial\vec{D}}{\partial t}\right]dV,$$

$$= -\int_V \left[\vec{H}\cdot\frac{\partial\vec{B}}{\partial t} + \vec{E}\cdot\frac{\partial\vec{D}}{\partial t} + \nabla\cdot\left(\vec{E}\times\vec{H}\right)\right]dV,$$

and with $\vec{H} = \frac{\vec{B}}{\mu}$, $\vec{D} = \varepsilon\vec{E}$, we get:

$$\int_V \vec{J}\cdot\vec{E}\,dV = -\int_V \left[\frac{\partial}{\partial t}\left(\frac{B^2}{2\mu} + \frac{\varepsilon E^2}{2}\right) + \nabla\cdot\left(\vec{E}\times\vec{H}\right)\right]dV.$$

Then we can define:

$$\mu_M = \frac{B^2}{2\mu} = \frac{\mu H^2}{2}, \quad \mu_E = \frac{\varepsilon E^2}{2}, \quad \mu_{EM} \equiv \mu_E + \mu_M,$$

where these are the magnetic and electric field density (energy/volume), the energies resident in the fields. Further we can write:

$$\int_V \vec{J}\cdot\vec{E}\,dV = -\int_V \left[\frac{\partial \mu_{EM}}{\partial t} + \nabla\cdot\left(\vec{E}\times\vec{H}\right)\right] dV,$$

and for an arbitrary volume:

$$-\vec{J}\cdot\vec{E} = \frac{\partial \mu_{EM}}{\partial t} + \nabla\cdot\left(\vec{E}\times\vec{H}\right),$$

or in words:

Work done by fields on sources in volume = Time rate of change of EM energy in volume + Energy flow out of the volume per unit time

This is a statement of conservation of energy for the volume. We can also define: $\vec{S} = \vec{E}\times\vec{H}$, having units of (energy/time-vol), as the Poynting vector.

If we denote the total energy of the particles in the volume as E_{mech}, and we assume that no particles move out:

$$\frac{dE_{mech}}{dt} = \int_V \vec{J}\cdot\vec{E}\,dV,$$

then:

$$E_{field} = \int_V \left(\frac{B^2}{2\mu} + \frac{\varepsilon E^2}{2}\right) dV,$$

and then:

$$\frac{d\left(E_{mech} + E_{field}\right)}{dt} = -\oint \vec{n}\cdot\vec{S}\,dA.$$

Let us now consider conservation of momentum. The total electromagnetic force on a charged particle is given by: $\vec{F}_{EM} = q\left(\vec{E} + \vec{v}\times\vec{B}\right)$, the Lorentz force, and per unit volume: $\vec{f}_{EM} = \rho_e\vec{E} + \vec{J}\times\vec{B}$. From

Newton's law: $\vec{F}_{EM} = \frac{d\vec{P}_{EM}}{dt}$, where $\vec{P}$ is the momentum and is taken as $\vec{P}_{mech}$, due to mechanical forces on the volume, i.e.:

$$\frac{d\vec{P}_{mech}}{dt} = \int_V \left(\rho_e\vec{E} + \vec{J}\times\vec{B}\right)dV.$$

Now we use:

$$\rho_e = \nabla\cdot\vec{D}, \quad \vec{J} = \nabla\times\vec{H} - \frac{\partial\vec{D}}{\partial t},$$

to get:

$$\frac{d\vec{P}_{mech}}{dt} = \int_V \left[\left(\nabla\cdot\vec{D}\right)\vec{E} + \left(\nabla\times\vec{H} - \frac{\partial\vec{D}}{\partial t}\right)\times\vec{B}\right]dV.$$

But we are considering only field interaction on particles (weak interaction on groups of particles), and so we take $\vec{D}\to\vec{E}$, $\vec{H}\to\vec{B}$ (in Gaussian units : $\mu, \varepsilon\to 1$), then for:

$$\left[\vec{E}\left(\nabla\cdot\vec{E}\right) - \vec{B}\times\left(\nabla\times\vec{B}\right) + \vec{B}\times\frac{\partial\vec{E}}{\partial t}\right],$$

using:

$$\vec{B}\times\frac{\partial\vec{E}}{\partial t} = -\frac{\partial}{\partial t}\left(\vec{E}\times\vec{B}\right) + \vec{E}\times\frac{\partial\vec{B}}{\partial t} \quad \text{and} \quad \vec{B}\left(\nabla\cdot\vec{B}\right) = 0,$$

to get:

$$\left[\vec{E}\left(\nabla\cdot\vec{E}\right) + \vec{B}\left(\nabla\cdot\vec{B}\right) - \vec{E}\times\left(\nabla\times\vec{E}\right) - \vec{B}\times\left(\nabla\times\vec{B}\right) - \frac{\partial}{\partial t}\left(\vec{E}\times\vec{B}\right)\right].$$

But we know that:

$$\frac{1}{2}\nabla\left(\vec{B}\cdot\vec{B}\right) = \left(\vec{B}\cdot\nabla\right)\vec{B} + \vec{B}\times\left(\nabla\times\vec{B}\right),$$

so:

$$\vec{B}(\nabla\cdot\vec{B}) - \vec{B}\times(\nabla\times\vec{B}) = \vec{B}(\nabla\cdot\vec{B}) + (\vec{B}\cdot\nabla)\vec{B} - \frac{1}{2}\nabla B^2.$$

As this is the divergence of a dyadic:

$$\vec{B}(\nabla\cdot\vec{B}) - \vec{B}\times(\nabla\times\vec{B}) = \nabla\cdot\left(\vec{B}\vec{B} - \frac{1}{2}\bar{\bar{I}}B^2\right),$$

and similarly:

where $\bar{\bar{I}}$ is the unit tensor (Jackson, 1963),

$$\vec{E}(\nabla\cdot\vec{E}) - \vec{E}\times(\nabla\times\vec{E}) = \nabla\cdot\left(\vec{E}\vec{E} - \frac{1}{2}\bar{\bar{I}}E^2\right).$$

Then:

$$\frac{d\,\vec{P}_{mech}}{dt} = \int_V \left\{\nabla\cdot\left[\vec{E}\vec{E} + \vec{B}\vec{B} - \frac{1}{2}\bar{\bar{I}}(E^2 + B^2)\right] - \frac{\partial}{\partial t}(\vec{E}\times\vec{B})\right\} dV,$$

so:

$$\frac{d(\vec{P}_{mech} + \vec{P}_{EM})}{dt} = \int_V \nabla\cdot\bar{\bar{T}}\,dV = \oint_A \vec{n}\cdot\bar{\bar{T}}\,dA,$$

where:

$$\bar{\bar{T}} = \vec{E}\vec{E} + \vec{B}\vec{B} - \frac{1}{2}\bar{\bar{I}}(E^2 + B^2),\ \text{is the Maxwell Stress Tensor,}$$

with elements: $T_{ij} = E_iE_j + B_iB_j - \frac{1}{2}\delta_{ij}(E^2 + B^2)$ in Gaussian units. Note that $\vec{n}\cdot\bar{\bar{T}}$ represents momentum flux (per unit area-time) out the volume, V, through the surface, A.

The electromagnetic force acting on an object can be calculated by integrating the Maxwell stress over the closed surface of the body of interest (Stratton, 1941), i.e.:

$$F_i = \int_A T_{ji}dA_j,\ \text{is the total force, and}\ f_{ei} = \frac{\partial T_{ji}}{\partial x_j},\ \text{is the body force,}$$

or:

$$\vec{f} = \rho_e \vec{E} + \vec{J} \times \vec{B} - \frac{\partial}{\partial t}(\vec{D} \times \vec{B}) \approx \rho_e \vec{E} + \vec{J} \times \vec{B},$$

where we have neglected magnetostriction effects in the above discussion.

CONSERVATION EQUATIONS OF MAGNETOFLUID MECHANICS

In the following development we presume: (1) continuum, (2) presence of electric and magnetic fields, and (3) several species that include charged particle species. In order to establish appropriate relationships, we presume that current is conducted: this implies that force is applied and energy is exchanged due to the electric and magnetic fields. This is the multicomponent continuum approach to the equations of conservation (Pai, 1962; Sutton and Sherman, 1965; Bird et al., 1960).

From the macroscopic point of view, for each species (s) there are six variables: temperature (T_s), pressure (p_s), density (ρ_s), and velocity (v_s^i), where $i = 1, 2, 3$ represent orthogonal direction components. The field variables are: E^i, H^i.

So we have:

6n + 6 variables (where n is the number of species),

6 equations for E^i, H^i (the vector Maxwell equations), and

6n fluid equations (3n motion equations, n state, n continuity, n energy).

Note that conservation of charge is included in the species equations.

1. Properties of the plasma

Number density:

$$n = \sum_{s=1}^{n} n_s,$$

Density:

$$\rho = mn = \sum_{s=1}^{n} m_s n_s = \sum_s \rho_s,$$

Pressure:

$$p = \sum_s p_s, \quad \text{Dalton's Law},$$

Temperature:

$$nT = \sum_s n_s T_s,$$

Velocity:

$$\rho u^i = \sum_s \rho_s u_s^i, \quad (i \Rightarrow \text{direction: } 1, 2, 3;\ \text{x, y, z;}\ \text{r, z, }\theta),$$

Charge density:

$$\rho_e = \sum_s e_s n_s = \sum_s \rho_{es},$$

Electrical current density:

$$J^i = \sum_s \rho_{es} u_s^i = \sum_s \left(\rho_{es} w_s^i + \rho_{es} u^i\right), \qquad u_s^i(\text{velocity of s species})$$

$$= J_c^i + \rho_e u^i; \qquad = w_s^i(\text{diffusion vel. wrto. mass avg.}) + u^i(\text{mass avg. vel of mix.}).$$

Note that:

$$\sum_s \rho_s w_s^i = 0.$$

2. Equation of state

$$p_s = n_s k T_s = \rho_s R_s T_s,$$

and:

$$p = \sum_s p_s = nkT = \rho R T.$$

3. Conservation of mass
Conservation of species:

$$\frac{\partial \rho_s}{\partial t} + \frac{\partial}{\partial x^i}\left(\rho_s u^i\right) = \sigma_s,$$

where σ_s is a mass source function; it expresses the creation of species ($m\frac{\partial n}{\partial t}$ due to chemical reactions): $\sum_s \sigma_s = 0$.

For the mixture, sum over species:

$$\frac{\partial \sum \rho_s}{\partial t} + \frac{\partial \sum (\rho_s u^i)}{\partial x^i} = 0,$$

or:

$$\frac{\partial \rho}{\partial t} + \frac{\partial (\rho u^i)}{\partial x^i} = 0.$$

Expressing with species diffusion velocity:

$$\frac{\partial \rho_s}{\partial t} + \frac{\partial}{\partial x^i}\left(\rho_s w_s^i + \rho_s u^i\right) = \sigma_s,$$

or:

$$\frac{\partial \rho_s}{\partial t} + \frac{\partial \left(\rho_s w_s^i\right)}{\partial x^i} + \frac{\partial \left(\rho_s u^i\right)}{\partial x^i} = \sigma_s;$$

and, as above: $\sum_s \rho_s w_s^i = 0$. In standard fashion, we can write for mixture mass conservation:

$$\frac{\partial \rho}{\partial t} + \nabla \cdot (\rho \vec{v}) = 0.$$

4. Conservation of momentum
Species momentum:

$$\frac{\partial \left(\rho_s u_s^i\right)}{\partial t} + \frac{\partial}{\partial x^j}\left(\rho_s u_s^i u_s^j - \tau_s^{ij}\right) = X_s^i + \sigma_s Z_s^i,$$

where τ_s^{ij} represents viscous stress interactions from other species; this will be considered further, below. This term contains collision effects and momentum conservation in collisions.

The term X_s^i is the force per unit volume on the species; $\sigma_s Z_s^i$ is the momentum source per unit volume and is associated with the mass source, $\sum_s \sigma_s Z_s^i = 0$.

Consider summation over all species of the following momentum flux term:

$$\sum_s \left(\rho_s u_s^i u_s^j\right) = \sum_s \rho_s \left(u^i u^j + w_s^i u^j + w_s^j u^i + w_s^i w_s^j\right), \quad \text{or :}$$

$$= \rho u^i u^j + \sum_s \left(\rho_s w_s^i w_s^j\right) = \rho u^i u^j - P_{ij},$$

where P_{ij} is the stress on wall (boundary) at a point. Specifically, we can write for stress in the fluid at a point, i.e.:

$$P_{ij} = -p\delta_{ij} + \tau^{ij},$$

where:

$$\tau^{ij} \approx \mu\left(\frac{\partial u^i}{\partial x^j} + \frac{\partial u^j}{\partial x^i}\right) + \mu_1 \left(\frac{\partial u^k}{\partial x^k}\right)\delta^{ij},$$

and the second coefficient of viscosity, $\mu_1 \approx -\frac{2}{3}\mu$.

In equilibrium, for example,

$$-\frac{\partial p}{\partial x} + \frac{\partial p_{xx}}{\partial x} + \frac{\partial p_{yx}}{\partial y} + \frac{\partial p_{zx}}{\partial z} = F_x,$$

and:

$$-\frac{\partial p}{\partial x} + \left[2\mu\frac{\partial^2 u}{\partial x^2} - \frac{2}{3}\mu\left(\frac{\partial^2 u}{\partial x^2} + \frac{\partial^2 v}{\partial y \partial x} + \frac{\partial^2 w}{\partial z \partial x}\right)\right]$$
$$+ \mu\left(\frac{\partial^2 v}{\partial x \partial y} + \frac{\partial^2 u}{\partial y^2}\right) + \mu\left(\frac{\partial^2 w}{\partial x \partial z} + \frac{\partial^2 u}{\partial z^2}\right) = F_x.$$

With summation of momentum over all species:

$$\frac{\partial(\rho u^i)}{\partial t} + \frac{\partial(\rho u^i u^j)}{\partial x^j} = -\frac{\partial p}{\partial x^j} + \frac{\partial \tau^{ij}}{\partial x^j} + F_{EM}^i + F_g^i,$$

or:

$$\rho\frac{Du^i}{Dt} = -\frac{\partial p}{\partial x^j} + \frac{\partial \tau^{ij}}{\partial x^j} + F_{EM}^i + F_g^i,$$

and written in vector form:

$$\rho\frac{D\vec{v}}{Dt} = -\nabla p + \nabla\cdot\overleftrightarrow{\tau} + \vec{F}_{EM} + \vec{F}_g.$$

5. Conservation of energy
Species energy can be written:

$$\frac{\partial}{\partial t}\left[\rho_s\left(\frac{1}{2}u_s^2 + e_{ms}\right)\right] + \frac{\partial}{\partial x^j}\left[\rho_s\left(\frac{1}{2}u_s^2 + e_{ms}\right)u_s^j\right]$$
$$= -\frac{\partial\left(u_s{}^j p_s\right)}{\partial x^j} + \frac{\partial\left(u_s^i\tau_s^{ij}\right)}{\partial x^j} + \frac{\partial Q_s^i}{\partial x^j} + \boldsymbol{\varepsilon}_s,$$

where Q_s^i is conduction energy flux, $e_{ms} = \int c_{Vs}dT_s$, and we define: $\bar{e}_{ms} = e_{ms} + \frac{1}{2}u_s^2$, and $\bar{e}_s = \rho_s\,\bar{e}_{ms}$; $\boldsymbol{\varepsilon}_s = \vec{E}\cdot\vec{J}_s$. Then:

$$\frac{\partial(\rho_s\bar{e}_{ms})}{\partial t} + \frac{\partial(\rho_s\bar{e}_{ms}u_s^j)}{\partial x^j} = \frac{\partial\bar{e}_s}{\partial t} + \frac{\partial(\bar{e}_s u_s^j)}{\partial x^j}, \quad \text{but,} \quad u_s^j = w_s^j + u^j,$$

so:

$$= \frac{\partial\bar{e}_s}{\partial t} + \frac{\partial}{\partial x^j}\left[\bar{e}_s\left(w_s{}^j + u^j\right)\right].$$

We sum over species to get:

$$\frac{\partial}{\partial t}\sum_s\bar{e}_s + \frac{\partial}{\partial x^j}\sum_s\bar{e}_s w_s^j + \frac{\partial}{\partial x^j}\sum_s\bar{e}_s u^j$$
$$= -\frac{\partial}{\partial x^j}\sum_s u_s^j p_s + \frac{\partial}{\partial x^j}\sum_s u_s^i\tau_s^{ij} + \frac{\partial}{\partial x^j}\sum_s Q_s^i + \sum_s\boldsymbol{\varepsilon}_s,$$

and we take:

$$\sum_s\bar{e}_s = \sum_s\rho_s\bar{e}_{ms} = \rho\bar{e}_m,$$

then:

$$\frac{\partial(\rho\bar{e}_m)}{\partial t} + \frac{\partial(\rho\bar{e}_m u^j)}{\partial x^j} = -\frac{\partial(u^j p)}{\partial x^j} + \frac{\partial(u^i\tau^{ij})}{\partial x^j} + \frac{\partial Q^i}{\partial x^j} + \boldsymbol{\varepsilon},$$

where:

$$Q^i = \sum_s \left(Q_s^i + u^i \rho_s w_s^i w_s^j + \bar{e}_{ms} \rho_s w_s^i + \tau_s^{ij} w_s^j - p_s w_s^j \right),$$

and:

$$\varepsilon = J^i \cdot E^i.$$

So we can write:

$$\frac{\partial(\rho \bar{e}_m)}{\partial t} + \frac{\partial(\rho \bar{e}_m u^j)}{\partial x^j} = \rho \frac{\partial \bar{e}_m}{\partial t} + \left[\bar{e}_m \frac{\partial \rho}{\partial t} + \bar{e}_m \frac{\partial(\rho u^j)}{\partial x^j} \right] + \rho u^j \frac{\partial \bar{e}_m}{\partial x^j},$$

$$\therefore \rho \frac{D\bar{e}_m}{Dt} = -\nabla \cdot (p \vec{v}) + \nabla \cdot (\vec{v} \cdot \overleftrightarrow{\tau}) + \nabla \cdot \vec{Q} + \vec{J} \cdot \vec{E}.$$

To summarize the results of the definition of the equations governing plasma flows, we have written a set of equations that theoretically can be solved for: $T_s, p_s, n_s, u_s^i, E^i, H^i$, but in actual fact this set is quite cumbersome and difficult to use. We have not yet discussed in detail some of the terms involving diffusion velocity, i.e., τ_s^{ij}, Q_s^i, etc. A better approach to a solution is to take a combination of species equations and total equations to get a more tractable set, e.g., assuming fully ionized plasma (2 species), solve for: $T, p, \rho, u^i, E^i, H^i$ and T_e, p_e, n_e, w_e^i, a set of 18 variables. However, the simplest approach is to set up equations for a single fluid and incorporate component species, which will now be outlined.

SINGLE FLUID EQUATIONS OF MAGNETOFLUID MECHANICS

We will consider basic single fluid theory with the following variables: $\vec{v}, p, \rho, T, \vec{H}, \vec{E}, \vec{J}, \rho_e$, a total of 16 unknowns. This set can be solved using the following equations with appropriate boundary conditions (Pai, 1962).

State:

$$p = \rho R T = nkT.$$

Continuity:

$$\frac{\partial \rho}{\partial t} + \nabla \cdot (\rho \vec{v}) = 0.$$

Momentum:

$$\rho\left[\frac{\partial\vec{v}}{\partial t}+(\vec{v}\cdot\nabla)\vec{v}\right]=-\nabla p+\nabla\cdot\overleftrightarrow{\tau}+\rho_e\vec{E}+\vec{J}\times\vec{B},\ \text{where:}$$

$$\tau^{ij}=\mu\left(\frac{\partial u^i}{\partial x^j}+\frac{\partial u^j}{\partial x^i}\right)-\frac{2}{3}\mu(\nabla\cdot\vec{v})\delta_{ij}.$$

Energy:

$$\rho\frac{D\bar{e}_m}{Dt}=-\nabla\cdot(p\vec{v})+\nabla\cdot\left(\vec{v}\cdot\overleftrightarrow{\tau}\right)+\nabla\cdot\vec{Q}+\vec{J}\cdot\vec{E}.$$

But note that:

$$\vec{J}_{cond}=\sigma\left(\vec{E}+\vec{v}\times\vec{B}\right),$$

so:

$$\vec{J}\cdot\vec{E}=\left(\rho_e\vec{v}+\vec{J}_{cond}\right)\cdot\vec{E},$$

$$=\rho_e\vec{v}\cdot\vec{E}+\vec{J}_{cond}\cdot\frac{\vec{J}_{cond}}{\sigma}-\vec{J}_{cond}\cdot\left(\vec{v}\times\vec{B}\right),$$

$$\vec{J}\cdot\vec{E}=\vec{v}\cdot\left(\rho_e\vec{E}+\vec{J}_{cond}\times\vec{B}\right)+\frac{j_{cond}^2}{\sigma}=\vec{v}\cdot\vec{F}_{EM}+\frac{j_c^2}{\sigma}.$$

In words: Energy addition = (Energy) "In" Work + (Joule) Heat (i.e., I^2R).

So:

$$\rho\frac{D}{Dt}\left(e_{int}+\frac{v^2}{2}\right)=-\nabla\cdot(p\vec{v})+\nabla\cdot\left(\vec{v}\cdot\overleftrightarrow{\tau}\right)+\nabla\cdot\vec{Q}+\vec{v}\cdot\vec{F}_{EM}+\frac{j_c^2}{\sigma},$$

$$\text{with } e_{int}=\int c_V\,dT,$$

where $\vec{v}\cdot\vec{F}_{EM}$ can be arrived at by $\vec{v}\cdot$ (*Momentum Equation*), as:

$$\vec{v}\cdot\rho\frac{D\vec{v}}{Dt}=-\vec{v}\cdot\nabla p+\vec{v}\cdot\left(\nabla\cdot\overleftrightarrow{\tau}\right)+\vec{v}\cdot\vec{F}_{EM},$$

which has been referred to as the conservation of mechanical energy. We subtract this to get:

$$\rho\frac{De_m}{Dt} = -p(\nabla\cdot\overrightarrow{v}) + \left(\overleftrightarrow{\tau}\cdot\nabla\right)\cdot\overrightarrow{v} + \frac{j_c^2}{\sigma} + \nabla\cdot\overrightarrow{Q},$$

and:

$$\rho\frac{De_m}{Dt} = (\overleftrightarrow{\tau}_t\cdot\nabla)\cdot\overrightarrow{v} + \frac{j_c^2}{\sigma} + \nabla\cdot\overrightarrow{Q}, \left(\text{where, } \overleftrightarrow{\tau}_t = \tau_{ij}(viscous) - p\delta_{ij}\right),$$

which has been referred to as the conservation of thermal energy.

Along with the above equations, the following relationships govern the plasma behavior.

Ohm's law:

$$\overrightarrow{J} = \rho_e\overrightarrow{v} + \sigma\left(\overrightarrow{E} + \overrightarrow{v}\times\overrightarrow{B}\right).$$

Conservation of charge:

$$\frac{\partial\rho_e}{\partial t} + \nabla\cdot\overrightarrow{J} = 0, \quad \text{or:} \quad \nabla\cdot\overrightarrow{D} = \rho_e.$$

Ampere's law:

$$\nabla\times\overrightarrow{H} = \overrightarrow{J} + \frac{\partial\overrightarrow{D}}{\partial t}, \quad \text{where,} \quad \overrightarrow{D} = \varepsilon\overrightarrow{E}.$$

Faraday's law:

$$\nabla\times\overrightarrow{E} = -\frac{\partial\overrightarrow{B}}{\partial t}, \quad \text{where,} \quad \overrightarrow{B} = \mu\overrightarrow{H}.$$

(Note that $\nabla\cdot\overrightarrow{B} = 0$ is derived from Faraday's law.)

Before considering the solution of this set of equations, it is of interest to consider in detail the electron species momentum equilibrium.

Electron Conservation of Momentum in Current Conduction

We rewrite the species momentum equation as:

$$\frac{\partial\left(\rho_s u_s^i\right)}{\partial t} + \frac{\partial}{\partial x^j}\left(\rho_s u_s^i u_s^j - \tau_s^{ij}\right) = X_s^i,$$

or:

$$\rho_s \frac{\partial u_s^i}{\partial t} + \left[u_s^i \frac{\partial \rho_s}{\partial t} + u_s^i \frac{\partial \left(\rho_s u_s^j\right)}{\partial x^j} \right] + \rho_s u_s^j \frac{\partial u_s^i}{\partial x^j} - \frac{\partial \tau_s^{ij}}{\partial x^j} = X_s^i,$$

and:

$$\rho_s \frac{Du_s^i}{Dt} = \frac{\partial \tau_s^{ij}}{\partial x^j} + X_s^i,$$

but, as we are considering electrons as current carriers:

$$\rho_s = n_e m_e \approx 0,$$

so:

$$\therefore \frac{\partial}{\partial x^j}\left(-p_s \delta_{ij} + \tau_s^{ij}\right) + X_s^i = 0,$$

and:

$$-\frac{\partial\left(p_s \delta_{ij}\right)}{\partial x^j} - n_e e\left(E^i + \varepsilon_{ijk} v^j B^k\right) = -\frac{\partial \tau_s^{ij}}{\partial x^j}.$$

Note that the above equation has dimensions of: $\frac{stress}{dist.} = \frac{force}{area \cdot dist.} = \frac{momentum/time}{area \cdot dist.}$.

Now we will consider a plasma of electron, ion, and neutral (e,i,n) species and turn to vector notation for the momentum equations (following Sutton and Sherman, 1965; Cowling, 1957), as:

$$\nabla p_e + n_e e\left(\vec{E}^* + \vec{w}_e \times \vec{B}\right) = n_e m_e \sum_s \nu_{es} \Delta V_{es}, \text{ and,}$$

$$= n_e m_e\left[\nu_{en}\left(\vec{w}_n - \vec{w}_e\right) + \nu_{ei}\left(\vec{w}_i - \vec{w}_e\right)\right],$$

where subscript n refers to neutral species. Also:

$$J_{cond} = n_e e\left(\vec{w}_i - \vec{w}_e\right) \equiv -n_e e \vec{w}^*, \quad \text{where:} \quad \vec{w}^* = \vec{w}_e - \vec{w}_i,$$

and:

$$m_e n_e \vec{w}_e + m_i n_i \vec{w}_i + m_n n_n \vec{w}_n = 0,$$

so:

$$\vec{w}_n = -\frac{m_e n_e \vec{w}_e + m_i n_i \vec{w}_i}{m_n n_n} = -\frac{m_e n_e(\vec{w}^* + \vec{w}_i) + m_i n_i \vec{w}_i}{m_n n_n}$$
$$= -\frac{m_e n_e \vec{w}^* + (m_e n_e + m_i n_i)\vec{w}_i}{m_n n_n},$$

and with $m_e \ll m_i$, we get: $\vec{w}_n \approx -\frac{m_e n_e}{m_n n_n}\vec{w}^* - \frac{m_i n_i}{m_n n_n}\vec{w}_i$.

Then by substitution (with $\vec{E}^*$, the electric field in coordinates fixed with the particles):

$$\nabla p_e + n_e e(\vec{E}^* + \vec{w}_e \times \vec{B})$$
$$= n_e m_e \left\{ \nu_{en} \left[-\frac{m_e n_e}{m_n n_n}\vec{w}^* - \frac{m_i n_i}{m_n n_n}\vec{w}_i - (\vec{w}^* + \vec{w}_i) \right] - \nu_{ei}\vec{w}^* \right\}$$
$$= -m_e n_e \frac{\vec{w}^*}{\tau_{ei}} - \frac{m_e n_e}{\tau_{en}} \left[\left(1 + \frac{m_e n_e}{m_n n_n}\right)\vec{w}^* + \left(1 + \frac{m_i n_i}{m_n n_n}\right)\vec{w}_i \right]$$

Now with $\vec{J}_i = e n_i \vec{w}_i$ and $\omega_e = \frac{eB}{m_e}$, we can write:

$$\nabla p_e + n_e e\vec{E}^* - (\vec{J} - \vec{J}_i) \times \vec{B}$$
$$= \frac{JB}{\omega_e \tau_{ei}} + \frac{JB}{\omega_e \tau_{en}}\left(1 + \frac{m_e n_e}{m_n n_n}\right) - \frac{J_i B}{\omega_e \tau_{en}}\left(1 + \frac{m_i n_i}{m_n n_n}\right).$$

Now with:

$$\frac{m_e n_e}{m_n n_n} \ll 1, \quad \Omega_{ei} = \omega_e \tau_{ei}, \quad \Omega_{en} = \omega_e \tau_{en},$$

and f is the mass fraction of atoms not ionized, $f \equiv \frac{m_n n_n}{m_i n_i + m_n n_n}$, we get:

$$\nabla p_e + n_e e\vec{E}^* - (\vec{J} - \vec{J}_i) \times \vec{B} = \frac{JB}{\Omega_{ei}} + \frac{JB}{\Omega_{en}} - \frac{J_i B}{\omega_e \tau_{en}}\frac{1}{f}.$$

A similar momentum equation can be derived for the ions, which takes the form:

$$-\nabla\left[\left(\frac{m_i}{m_n} - 2\right)p_i\right] - \vec{J} \times \vec{B} = \frac{JB}{f\Omega_{en}} - \frac{J_i B}{f^2(\Omega_{en} + \Omega_{in})}.$$

Then we can combine these results (i.e., we substitute for $\vec{J_i}$) to get the result:

$$\vec{J} = \frac{n_e e^2}{m_e(\nu_{en} + \nu_{ei})}\left\{\vec{E}^* + \frac{\nabla p_e}{n_e e} - \frac{\vec{J} \times \vec{B}}{n_e e} - \frac{f^2 \tau_{in}}{m_e n_e} \times \left[\left(2 - \frac{m_i}{m_n}\right)\nabla p_i \times \vec{B} + \vec{B} \times (\vec{J} \times \vec{B})\right]\right\}.$$

In words, we are expressing:

J (current density) = (scalar elec. cond.) {(Elec. fields) + (Electron press. Grad.) + (Hall effect) + (Ion slip)}.

This relationship for the current density as a function of fields and species properties is known as the "generalized Ohm's law," and is necessary to describe current flow in collisional plasmas.

We can arrive at some simplifications of the electron equation. If we neglect electron pressure, we get:

$$\vec{J} = \sigma\vec{E}^* - \Omega_{e,avg}\frac{\vec{J} \times \vec{B}}{B} + f^2\Omega_{e,avg}\omega_i\tau_{in}\left[\frac{\vec{B}}{B}\left(\frac{\vec{B}}{B}\cdot\vec{J}\right) - \vec{J}\right]$$

$$= \sigma\vec{E}^* - \beta_2\frac{\vec{J} \times \vec{B}}{B} + \beta_1\left[\frac{\vec{B}}{B}\left(\frac{\vec{B}}{B}\cdot\vec{J}\right) - \vec{J}\right].$$

So we have: $\vec{J} = f_n(\vec{E}, \vec{B}, \vec{J})$, but we want an expression to uniquely determine current as a function of plasma and field properties. So, we begin by taking the dot product, $\vec{J}\cdot\vec{B}$, then:

For $\|$ to B: $\vec{J}_{||} = \sigma\vec{E}_{||}$, where the conductivity is a scalar (no B field effect);

For $\perp$ to B: $\vec{J}_\perp = \sigma\vec{E}^*_\perp - \beta_2\frac{\vec{J}_\perp \times \vec{B}}{B} + \beta_1(-\vec{J}_\perp)$.

To eliminate $J_\perp$ from the right-hand side we take:

$$\vec{J}_\perp \times \vec{B} = \sigma\vec{E}^*_\perp \times \vec{B} - \beta_2\frac{(\vec{J}_\perp \times \vec{B}) \times \vec{B}}{B} - \beta_1\vec{J}_\perp \times \vec{B},$$

$$= \sigma\vec{E}^*_\perp \times \vec{B} + \frac{\beta_2}{B}[(\vec{B}\cdot\vec{J}_\perp)\vec{B} + (\vec{B}\cdot\vec{B})\vec{J}_\perp] - \beta_1\vec{J}_\perp \times \vec{B},$$

and

$$(\vec{J}_\perp \times \vec{B})(1 + \beta_1) = \sigma\vec{E}^*_\perp \times \vec{B} + \beta_2 B\vec{J}_\perp,$$

which can be substituted into the original equation, and:

$$\vec{J}_\perp = \sigma \vec{E}_\perp^* - \frac{\beta_2}{B}\frac{\sigma \vec{E}_\perp^* \times \vec{B} + \beta_2 B \vec{J}_\perp}{1+\beta_1} - \beta_1 \vec{J}_\perp,$$

and rearranging, get:

$$\vec{J}_\perp (1+\beta_1)^2 = \sigma \vec{E}_\perp^* (1+\beta_1) - \beta_2 \sigma \frac{\vec{E}_\perp^* \times \vec{B}}{B} - \beta_2^2 \vec{J}_\perp,$$

and finally:

$$\vec{J}_\perp = \frac{(1+\beta_1)\sigma \vec{E}_\perp^* - \beta_2 \sigma \vec{E}_\perp^* \times \vec{B}/B}{(1+\beta_1)^2 + \beta_2^2}, \qquad \therefore \sigma \text{ is a tensor.}$$

We have: $J_i = \sigma_{ij} E_j$, and for $\vec{B}_z$ (z-direction):

$$\sigma_{ij} = \sigma \begin{bmatrix} \dfrac{1+\beta_1}{(1+\beta_1)^2+\beta_2^2} & -\dfrac{\beta_2}{(1+\beta_1)^2+\beta_2^2} & 0 \\ \dfrac{\beta_2}{(1+\beta_1)^2+\beta_2^2} & \dfrac{1+\beta_1}{(1+\beta_1)^2+\beta_2^2} & 0 \\ 0 & 0 & 1 \end{bmatrix}.$$

But what does this mean? As an example, let $f = 0$ ($n_n = 0$) and $\beta_1 = 0$, then:

$$\vec{J} = \sigma \vec{E}^* - \Omega_e \frac{\vec{J} \times \vec{B}}{B} \Rightarrow \sigma_{ij} = \begin{bmatrix} \dfrac{1}{1+\Omega_e^2} & -\dfrac{\Omega_e}{1+\Omega_e^2} & 0 \\ \dfrac{\Omega_e}{1+\Omega_e^2} & \dfrac{1}{1+\Omega_e^2} & 0 \\ 0 & 0 & 1 \end{bmatrix}.$$

For: $\vec{B} = \vec{B}_z$, $\vec{E}^* = \vec{E}_\perp^* = \vec{E}_x^*$, the applied electric field is in the x-direction, then:

$$J_x = \sigma_{xx} E_x = \frac{\sigma}{1+\Omega_e^2} \vec{E}_x^* < J_x \text{ compared to } \Omega_e \ll 1 \text{ case; and:}$$

$$J_y = \sigma_{yx} E_x = \frac{\sigma \Omega_e}{1+\Omega_e^2} \vec{E}_x^* > J_y \text{ compared to } \Omega_e \ll 1 \text{ case,} \quad \text{where,} \quad \frac{J_y}{J_x} = \Omega_e.$$

Further, to quantify, if $\Omega_e \approx 10$, $\frac{J_x}{J_y} = 0.1$, and $J_y = 10J_x$; in other words, E_x drives J_y.

So the physical interpretation of this result is clear. Depending on the value of Ω, the components of current flow, as well as related values of $\vec{J} \times \vec{B}$ can be very different from the directions of applied electric fields. There are numerous examples of this in experiment. One well-established case is that of magnetohydrodynamics (MHD) generators, where voltages applied across electrodes drive greater current flow axially due to Hall effect, rather than laterally, as intended, and this inhibits the usefulness of the device. A second example is that of a particle collecting biased voltage plane surface on an ionosphere probe, where the particles that enter a collection volume at a distance from the surface will not follow electric field lines, so that the data must be carefully analyzed. Further, in devices such as plasma thrusters, simple models of currents and $\vec{J} \times \vec{B}$ forces that neglect these considerations are not effective in interpreting the complex reality that is evident in experimental data.

THE MHD APPROXIMATIONS

MHD combines the equations of (continuum) fluid mechanics with Maxwell's equations, and so the set of equations is generally difficult to solve. To alleviate this difficulty, we will now consider some approximations to those equations that are based on L, length scale, ω^{-1} time scale, and $L\omega$, velocity and field strengths (Shercliffe, 1965). We make approximations based on the orders of magnitude of terms in the equations.

1. Displacement current:

$$\nabla \times \vec{H} = \vec{J} + \frac{\partial \vec{D}}{\partial t}, \text{ where the ratio,}$$

$$\frac{\partial \vec{D}/\partial t}{\vec{J}} = \frac{\varepsilon \partial \vec{E}/\partial t}{\sigma \vec{E}} \approx \frac{\varepsilon \omega}{\sigma} \ll 1.$$

2. Convection current:

$$\vec{J} = \rho_e \vec{v} + \sigma\left(\vec{E} + \vec{v} \times \vec{B}\right), \text{ where the ratio,}$$

$$\frac{\rho_e \vec{v}}{\sigma \vec{E}} = \frac{\left(\nabla \cdot \vec{D}\right)\vec{v}}{\sigma \vec{E}} = \frac{\varepsilon\left(\nabla \cdot \vec{E}\right)\vec{v}}{\sigma \vec{E}} \approx \frac{\varepsilon\left(\vec{E}/L\right)L\omega}{\sigma \vec{E}} \approx \frac{\varepsilon \omega}{\sigma} \ll 1.$$

3. Electrostatic body force:

$$\vec{F} = \rho_e \vec{E} + \vec{J} \times \vec{B}, \quad \text{where the ratio,}$$

$$\frac{\rho_e \vec{E}}{\vec{J} \times \vec{B}} \approx \frac{\varepsilon E^2/L}{\sigma(E+UB)B} \approx \frac{\varepsilon E^2/L}{\sigma UB^2} \approx \frac{\varepsilon E^2 U}{\sigma U^2 B^2 L} = \frac{\varepsilon\omega}{\sigma} \ll 1.$$

Applying these approximations, we arrive at the following set of (MHD) equations:

Mass:

$$\frac{\partial \rho}{\partial t} + \nabla \cdot (\rho \vec{v}) = 0.$$

Momentum:

$$\rho \frac{D\vec{v}}{Dt} = -\nabla p + \nabla \cdot \overset{\leftrightarrow}{\tau} + \vec{J} \times \vec{B}.$$

Momentum (without viscosity):

$$\rho \frac{D\vec{v}}{Dt} = -\nabla p + \vec{J} \times \vec{B}.$$

Energy (without viscosity and conduction):

$$\rho \frac{De_{int}}{Dt} = -p(\nabla \cdot \vec{v}) + \frac{J^2}{\sigma}.$$

Ohm's:

$$J = \sigma(\vec{E} + \vec{v} \times \vec{B}).$$

Maxwell:

$$\nabla \times \vec{E} = -\frac{\partial \vec{B}}{\partial t}, \quad \nabla \times \vec{H} = J.$$

In this set, we have 14 equations for $\vec{v}$, p, J, $\vec{E}$, $\vec{B}$, e_{int}, T to which we can add:

(1) Either $e_{int} = \frac{3}{2}nkT$ or $\frac{De_{int}}{Dt} = c_V \frac{DT}{Dt}$ and (2) Equation of state: $p = \rho RT$.

The solution of the above set of equations is still difficult. In order to simplify mathematical forms and to facilitate the definition of relationships between variables, it is useful to arrange terms in the equations to combine coefficients in the equations. In this way, we can recognize that the coefficients are ratios of physical terms.

SIMILARITY PARAMETERS

If we take the above set of equations and rearrange them to dimensionless form, we will be able to identify parameters that can help us establish the dominant mechanisms in plasma processes (Sutton and Sherman, 1965). This can allow us to establish orders of magnitudes of terms and so be able to develop approximate, simpler equations in some cases, which can be easier to solve. We neglect viscosity and get:

Momentum:

$$\tilde{\rho}\frac{D\vec{\tilde{v}}}{D\tilde{t}} = -\frac{1}{\gamma_0 M_0^2}\tilde{\nabla}\tilde{p} + S\left(\tilde{\nabla}\times\vec{\tilde{B}}\right)\times\vec{\tilde{B}},$$

where:

$$\tilde{\rho} = \frac{\rho}{\rho_0},\ \tilde{v} = \frac{v}{U_0},\ \gamma_0 = \frac{c_{p0}}{c_{V0}},\ \tilde{B} = \frac{B}{B_0},\ \tilde{t} = \frac{t}{L_0/U_0},\ \tilde{p} = \frac{p}{p_0},\ \frac{\tilde{\nabla}}{L} \equiv \nabla.$$

Energy:

$$\tilde{\rho}\tilde{c}_V\frac{D\tilde{T}}{D\tilde{t}} = -(\gamma_0 - 1)\tilde{p}\left(\tilde{\nabla}\cdot\vec{\tilde{v}}\right) + \frac{\gamma_0(\gamma_0 - 1)}{R_m}SM_0^2\frac{\left(\tilde{\nabla}\times\vec{\tilde{B}}\right)^2}{\mu_0\tilde{\sigma}},$$

where:

$$\tilde{c}_V = \frac{c_V}{c_{V0}},\quad \tilde{T} = \frac{T}{T_0},\quad \tilde{\sigma} = \frac{\sigma}{\sigma_0}.$$

Induction:

$$\frac{\partial\vec{B}}{\partial t} = -\nabla\times\vec{E} = -\nabla\times\left[\frac{\vec{J}}{\sigma} - \vec{v}\times\vec{B} + \frac{\Omega_e}{B\sigma}\left(\vec{J}\times\vec{B}\right)\right]$$

$$= -\nabla\times\left[\frac{\nabla\times\vec{B}}{\mu_0\sigma} - \vec{v}\times\vec{B} + \frac{\Omega_e}{B\sigma}\left(\frac{\nabla\times\vec{B}}{\mu_0}\times\vec{B}\right)\right],$$

or:

$$\frac{\partial \vec{B}}{\partial t} = \nabla \times (\vec{v} \times \vec{B}) - \frac{1}{\mu_0} \frac{\nabla \times \nabla \times \vec{B}}{\sigma} - \nabla \times \left[\frac{\Omega_e}{B\mu_0 \sigma} (\nabla \times \vec{B}) \times \vec{B} \right],$$

to get:

$$\frac{\partial \vec{\tilde{B}}}{\partial \tilde{t}} = \tilde{\nabla} \times (\vec{\tilde{v}} \times \vec{\tilde{B}}) - \frac{1}{R_m} \left\{ \frac{\tilde{\nabla} \times \tilde{\nabla} \times \vec{\tilde{B}}}{\tilde{\sigma}} + \frac{\Omega_e}{\tilde{B}\tilde{\sigma}} \tilde{\nabla} \times \left[(\tilde{\nabla} \times \vec{\tilde{B}}) \times \vec{\tilde{B}} \right] \right\},$$

where:

R_m is magnetic Reynolds number, $R_m = \mu_0 \sigma_0 U_0 L_0$,
M_0 is Mach number, $M_0 = U_0 / \sqrt{\gamma_0 R_0 T_0}$,
Ω_e is Hall parameter, $\Omega_e = \omega \tau$,
S is magnetic force number, $S = \frac{B_0^2}{\mu_0 \rho_0 U_0^2}$.

These are mathematical arrangements, and we can now examine the physical significance of these terms. The names and symbols used here have been defined by earlier works and usage.

Magnetic Reynolds number: This relates induced and allied magnetic fields.

Suppose that $\vec{B} = \vec{B}_0$ (applied) $+ \Delta \vec{B}$ (induced). Then:

$$\vec{J} = \frac{\nabla \times \vec{B}}{\mu_0} \approx \frac{1}{\mu_0} \frac{\Delta B}{\Delta x},$$

so:

$$J\mu_0 L \approx \Delta B,$$

with $J = \sigma UB$, $\Delta B \approx \sigma \mu_0 UBL$, then:

$$\frac{\Delta B}{B} = \frac{\text{Induced B field}}{\text{Applied B field}} = \sigma \mu_0 UL = R_m.$$

Magnetic force number: This relates applied field energy density to flow energy density.

$$S = \frac{B_0^2 / 2\mu_0}{\frac{1}{2}\rho_0 U_0^2} = \frac{\text{Magnetic energy density (pressure)}}{\text{Inertia energy density (stress)}};$$

also:

$$S = \frac{B_0^2}{\mu_0 \rho_0 U_0^2} = \frac{B_0^2/\rho_0 \mu_0}{U_0^2} \equiv \frac{V_A^2}{U_0^2},$$

$$\text{where Alfven speed,} \quad V_A = \sqrt{B_0^2/\rho_0 \mu_0},$$

and Alfven Mach number for plasma flow is $M_A = \frac{U_0}{V_A}$.

Hall parameter: This relates the effect of cyclotron time due to B to collision time.

$$\omega_e \left(\frac{rad}{time}\right) \tau \left(\frac{time}{coll.}\right) = \Omega_e = \frac{\frac{\sigma \Omega_e}{1+\Omega_e^2} E_x}{\frac{\sigma}{1+\Omega_e^2} E_x}.$$

Magnetic force interaction parameter:

$$I = \frac{\text{Magnetic Body Force/vol}}{\text{Inertia Force/vol}} = \frac{JB}{\rho_0 U_0^2/L_0} \approx \frac{(\sigma U_0 B_0) B_0}{\rho_0 U_0^2/L_0} = \frac{\sigma B_0^2 L_0}{\rho_0 U_0}.$$

Beta (parameter): This relates random thermal energy density to magnetic energy density.

$$\beta = \frac{nkT}{B_0^2/2\mu_0} = \frac{\text{Thermal energy density}}{\text{Magnetic energy density}}.$$

REFERENCES

Bird, R.B., Stewart, W.E., Lightfoot, E.N., 1960. Transport Phenomena. Wiley, New York.

Boyd, T.J.M., Sanderson, J.J., 1969. Plasma Dynamics. Barnes & Noble, New York.

Cowling, T.G., 1957. Magnetohydrodynamics. Interscience, New York.

Jackson, J.D., 1963. Classical Electrodynamics. Wiley, New York.

Pai, S.I., 1962. Magneto gas dynamics and plasma dynamics. Prentice-Hall, New York.

Shercliffe, J.A., 1965. A Textbook of Magnetohydrodynamics. Pergamon, London.

Sommerfeld, A., 1964. Electrodynamics; Lectures on Theoretical Physics, vol. III. Academic, New York.

Stratton, J.A., 1941. Electromagnetic Theory. McGraw-Hill, New York.

Sutton, G.W., Sherman, A., 1965. Engineering Magnetohydrodynamics. McGraw-Hill.

CHAPTER 8

Hydromagnetics—Fluid Behavior of Plasmas

INTRODUCTION

Many applications in which there are plasma interactions involve densities where the medium behaves largely as a fluid. The basic behavior in terms of energy input, force applications, and the influence of magnetic and electric fields is that of a continuum fluid (Shercliff, 1965). So this description and interpretation can serve as a starting point to understand the complex behavior of plasmas; we then can improve the modeling and experiments to clarify uniqueness of regimes and interactions so as to finally achieve understanding of the plasma and so enable efficient device operation. In order to develop this important description, we will use the magneto-hydrodynamics (MHD) equations to allow us to understand and quantify plasma behavior.

BASIC EQUATIONS OF CONTINUUM PLASMA DYNAMICS

Mass conservation:

$$\frac{\partial \rho}{\partial t} + \nabla \cdot (\rho \vec{v}) = 0;$$

Momentum conservation:

$$\rho \frac{D\vec{v}}{Dt} = -\nabla p + \vec{J} \times \vec{B};$$

Energy conservation:

$$\rho \frac{De_{int}}{Dt} = -p(\nabla \cdot \vec{v}) + \frac{J^2}{\sigma};$$

Ohm's law:

$$\vec{J} = \sigma(\vec{E} + \vec{v} \times \vec{B});$$

Introduction to Plasmas and Plasma Dynamics
ISBN 978-0-12-801661-9

Ampere's law:

$$\nabla \times \vec{H} = \vec{J}, \quad \vec{H} = \frac{\vec{B}}{\mu};$$

Faraday's law:

$$\nabla \times \vec{E} = -\frac{\partial \vec{B}}{\partial t}.$$

The variables of the problem are ρ, $\vec{v}$, p, $\vec{J}$, $\vec{E}$, $\vec{B}$, e, σ where E and B are generally specified by the properties of the device or problem of interest.

TRANSPORT EFFECTS IN PLASMAS AND PLASMA DEVICES

Diffusion of Particles Across Magnetic Field Lines

General Treatment of Transient Diffusion of Gas

To establish a framework for the discussion of transport, we look at a general analytical treatment for a static fluid in a (−) semi-infinite volume of a species of number density, n, which at time $t = 0$ begins diffusing into a (+) semi-infinite region. In this case, the general equation for the diffusion of density is $\frac{\partial n}{\partial t} = D\nabla^2 n$, which can be solved analytically depending on the boundary conditions (Wylie, 1966).

If we consider one dimension and take the boundary and initial conditions, (1) $n(x,t) = 0$ for $t \leq 0$ and (2) $n(0,t) = n_0$ for $t > 0$, we can get the solution: $n(x,t) = n_0\, erfc\left(\frac{x}{\sqrt{4Dt}}\right)$, in terms of the mathematical complementary error function: $\left[erfc(\eta) = \frac{2}{\sqrt{\pi}} \int_\eta^\infty e^{-\zeta^2} d\zeta\right]$. This is referred to as a similarity solution and displays the functional combination $\frac{x}{\sqrt{4Dt}}$, or $\frac{x^2/D}{t}$, which identifies a time relevant to diffusion (x^2/D).

If we also look at other diffusion geometries with boundary conditions that enable solution by application of separation of variables, we then take $n(x,t) = T(t)X(x)$ and we get:

$\frac{1}{T}\frac{dT}{dt} = -\frac{1}{\tau}$, and also $\frac{D}{X}\nabla^2 X(x) = -\frac{1}{\tau}$, with solutions:

$T = T_0 e^{-t/\tau}$, where T_0 is the value of n at $t = 0$, and with:

$X = \cos\frac{\pi x}{2L}$, for boundary conditions on n at $x = 0$, L.

So we have $n \sim n_0\, e^{-t/\tau}$, where $\tau \sim \frac{L^2}{D}$ is a characteristic time for diffusion.

Steady State Diffusion

Now, let us consider the macroscopic equations for the steady state diffusion of a fully ionized plasma (Spitzer, 1962; Chen, 1984); for behavior "following the fluid," i.e., $\frac{D}{Dt} = 0$, then:

$$0 = -\nabla p + \vec{J} \times \vec{B}.$$

For charged particles moving with $\vec{v} = \vec{v}_\perp + \vec{v}_\parallel$, with respect to the direction of the magnetic field, then:

$$\vec{J} = \sigma(\vec{E} + \vec{v}_\parallel \times \vec{B} + \vec{v}_\perp \times \vec{B}) = \sigma(\vec{E} + \vec{v}_\perp \times \vec{B}).$$

We take:

$$\vec{J} \times \vec{B} = \sigma_\perp[\vec{E} \times \vec{B} + (\vec{v}_\perp \times \vec{B}) \times \vec{B}],$$

but from momentum:

$$\vec{J} \times \vec{B} = \nabla p,$$

so:

$$\nabla p = \sigma_\perp[\vec{E} \times \vec{B} + (\vec{v}_\perp \times \vec{B}) \times \vec{B}] = \sigma_\perp(\vec{E} \times \vec{B} - \vec{v}_\perp B^2),$$

or:

$$\vec{v}_\perp = -\frac{\nabla p}{\sigma_\perp B^2} + \frac{\vec{E} \times \vec{B}}{B^2}.$$

In words, the particle velocity across the magnetic field lines is the sum of (1) the diffusion velocity in the $-\nabla p$ direction and (2) the drift velocity in the $\vec{E} \times \vec{B}$ direction (zero if $\vec{E} = 0$).

The particle flux across the field lines can be associated with a velocity, as:

$$\Gamma(\text{Particles/time} - \text{area}) = nv_\perp = -\frac{n\nabla p}{\sigma_\perp B^2} = -\frac{nkT\nabla n}{\sigma_\perp B^2}.$$

Recall that in continuum, diffusion is phenomenologically described by Fick's law, $\Gamma = -D\nabla n$; accordingly, $D_\perp = \frac{nkT}{\sigma_\perp B^2}$ is called the classical diffusion coefficient.

It can be seen that this diffusion is $\sim\frac{1}{B^2}$, n, T, and ∇n. Again, we are describing the motion of charged particles across uniform magnetic field lines.

Ambipolar Diffusion

As was noted above, electrons will diffuse because of density gradients. As plasmas have equal number densities of ions and electrons ($n_+ = n_e$), the ions will diffuse, but at a lower rate because of a lower diffusion coefficient, D_i. However, faster drift of electrons would lead to charge separation and the formation of an electric field $(\overrightarrow{E})$ to pull the ions along. So the particles are tied together and drift as pairs; this is called ambipolar diffusion. As the flux of ions and electrons must be the same:

$\Gamma_{amb}\left(\frac{p}{\text{cm}^2-\text{s}}\right) = -D_i \frac{dn}{dx} + \mu_i\ n\ E$ with $\mu_{i,e}\, E = v_{drift}^{+,-}$, and $\mu_{i,e}$ are $(+, e)$ species mobility. We solve for $E = -D_e \frac{dn}{dx} - \mu_e \cdot n \cdot E$, and substitute into:

$$\Gamma_{amb}\left(\frac{p}{\text{cm}^2-\text{s}}\right) = -D_i \frac{dn}{dx} + \mu_+\ n\ E \equiv -D_{amb}\frac{dn}{dx}.$$

So:

$$D_{amb} = \frac{D_i\, v_{drift}^- + D_e\, v_{drift}^+}{v_{drift}^- + v_{drift}^+}$$

but

$$v_{drift}^- \gg v_{drift}^+$$

and then:

$$D_{amb} \approx D_i + D_e \frac{v_{drift}^+}{v_{drift}^-} = D_i + D_e \frac{\mu_+}{\mu_e},$$

and using the Einstein relationships for $\mu \sim D$: $D_{amb} = D_i\,(1 + T_e/T_i)$.

Equilibration of Species Energy

The concept of temperature for a species assumes that an equilibrium distribution of random velocities or energies exists in a species or a species energy component. As energy can be preferentially delivered to one species (mostly to electrons) in a plasma, and then transferred from that species to another (mostly to ions) by collision processes, it is important to establish an order of magnitude for such transfer times. For groups of charged particles that are in collision, the encounters are primarily those related to fields of the particles or Coulomb collisions, and these occur "at a distance" with an impact parameter, p. The impact parameter minimum can be expressed as

$p_{imp-min} = \frac{e^2}{4\pi\epsilon_0 T_e}$, and in terms of collision time, $\tau_c \sim \frac{1}{np^2 v_c}$, where v_c is the collision frequency and $\tau_c \sim \frac{\epsilon_0^2 \sqrt{m} T^{3/2}}{e^4 n}$. When the minimum impact parameter is set equal to the maximum encounter parameter (the Debye length), we have:

$$\frac{3}{2ZZ_f e^3}\left(\frac{k^3 T^3}{\pi n_e}\right)^{1/2} \equiv \Lambda,$$

where Z_f is the charge of "field" particles or particles that test particles collide with. The time for two species, test and field particles with temperatures and Maxwellian (distributions) T and T_f, to reach equipartition of energy has been derived and expressed to be (Spitzer, 1962):

$$\frac{dT}{dt} = \frac{T_f - T}{t_{eq}}, \quad \text{with:} \quad t_{eq} = \frac{3mm_f k^{3/2}}{8(2\pi)^{1/2} n_f Z^2 Z_f^2 e^4 \ln \Lambda}\left(\frac{T}{m} + \frac{T_f}{m_f}\right)^{3/2}.$$

The term $\ln \Lambda$ is referred to as the Coulomb logarithm; it does not vary widely when $n > 10^{15}\ \text{cm}^{-3}$, and so it can be approximated by $\ln \Lambda \cong 6.6 - 0.5 \ln n + 1.5 \ln T_e$, where n is in units of $10^{20}\ \text{m}^{-3}$ and T_e (eV).

Independent from transfer between species, particularly in intense heating schemes, there may also be the need to identify time for equilibration of energy within a species; this will be considered later with respect to discussion of applications.

Transport Properties—Magnetic Field Effects

The mechanisms to transfer momentum, energy, and mass in plasmas can become quite complex due to interactions of charged particles with magnetic fields. Very basic is the cyclotron motion of ions and electrons; when the particle interactions include collision effects, the transfer processes are quite different from elastic, billiard-ball models.

To develop expressions for plasma transport, the equations of mass, momentum, and energy of charged species are fundamental. Transport coefficients differ for directions relative to the magnetic field; components are specified as $\perp$ or $\|$ to magnetic field directions. Expressions for transport are given in terms of cyclotron frequencies, ω_e and ω_i, and collision times for electrons and ions (τ_e and τ_i), as:

$$\tau_{ce} = 3.44 \times 10^5 \cdot \frac{T_e^{3/2}}{n \ln \Lambda}, \quad \text{and} \quad \tau_{ci} = 2 \times 10^7 \cdot \frac{T_i^{3/2}}{n \ln \Lambda}\ (\text{s}).$$

Following are expressions that have been derived (Braginskii, 1965) for transport coefficients in the different regimes of behavior that can be differentiated by the parameter, $\omega\tau$.

For $\omega_i\tau_i$, $\omega_e\tau_e \ll 1$:

The electrical conductivity is:

$$\sigma_{\parallel} = 1.96\frac{ne^2\tau_e}{m_e}.$$

The electron and ion thermal conductivities are:

$$K_{\parallel}^{e} = 3.2\frac{n\tau_e T_e}{m_e},$$

and:

$$K_{\parallel}^{i} = 3.9\frac{n\tau_i T_i}{m_i}.$$

The electron and ion viscosities are:

$$\eta_0^e = 0.73n\tau_e T_e,$$

and:

$$\eta_0^i = 0.96n\tau_i T_i.$$

For $\omega_i\tau_i$, $\omega_e\tau_e \gg 1$ (strong magnetic fields), the corresponding expressions are:

$\sigma_{\parallel}$ (same as above)

$$\sigma_{\perp} = 0.51\sigma_{\parallel} = \frac{ne^2\tau_e}{m_e}.$$

$K_{\parallel}^{e}$, $K_{\parallel}^{i}$ (same as above)

$$K_{\perp}^{e} = 4.7\frac{nT_e}{m_e\omega_e^2\tau_e}, \quad K_{\perp}^{i} = 2\frac{nT_i}{m_i\omega_i^2\tau_i}.$$

$\eta_{\parallel}^{e,i}$ (same as above)

$$\eta_{\perp}^{e} = 0.51\frac{nT_e}{\omega_e^2\tau_e}, \quad \eta_{\perp}^{i} = 0.3\frac{nT_i}{\omega_i^2\tau_i}.$$

The conditions under which the above equations can be considered exact are quite specific and detailed. So the use of macroscopic coefficients to represent inherently atomic-scale interactions is problematic, in general, in plasmas. A general discussion of one important aspect of such behavior is now noted.

Anomalous Transport

As the understanding of plasma heating and compression techniques were developing to guide experimental design, evidence of enhanced transport losses in various modes of energy delivery has become apparent (Liewer, 1985; Papadopoulos, 1985). Largely, this enhanced transport is due to atomic-scale microinstabilities whose occurrence is related to quite specific field, species, density, and temperature conditions.

This behavior has been given the name anomalous transport in a general description of a wide range of physical conditions (Connor and Wilson, 1994). So, for any given experiment, concern must be given to the possibility of the occurrence of such microinstabilities and their effects. The understanding of plasma wave-type behavior as related to microinstabilities has advanced considerably (Klages et al., 2008). In particular, the advances in a number of experimental achievements, as in controlled magnetic fusion, are largely due to control of local instabilities that can destroy confinement and heating mechanisms. Specific examples of anomalous transport behavior will be noted in the sections below that describe specific experiments and experimental results.

KINEMATICS (AND DYNAMICS) OF MAGNETIC FIELDS IN PLASMAS

Depending on the order of magnitude of magnetic field, plasma conductivity, and flow properties, the magnetic field interaction with the plasma can demonstrate aspects of diffusion and convection (Cowling, 1957; Boyd and Sanderson, 1969). To examine this behavior, we consider the governing equations and some limiting cases.

We had seen above that Faraday's law can be expressed as:

$$\frac{\partial \vec{B}}{\partial t} = -\nabla \times \vec{E} = -\nabla \times \left[\frac{\vec{J}}{\sigma} - \vec{v} \times \vec{B} + \frac{\Omega_e}{B\sigma}(\vec{J} \times \vec{B})\right], \quad \text{and}$$

$$\frac{\partial \vec{B}}{\partial t} = -\nabla \times \left[\frac{\nabla \times \vec{B}}{\mu\sigma} - \vec{v} \times \vec{B} + \frac{\Omega_e}{B\sigma}(\vec{J} \times \vec{B})\right].$$

Now let us take: $\vec{J} \times \vec{B} \approx 0$. Then we have: $\frac{\partial \vec{B}}{\partial t} = -\nabla \times \left(\frac{\nabla \times \vec{B}}{\mu\sigma}\right) + \nabla \times (\vec{v} \times \vec{B})$, so we have: $\vec{B} = \vec{B}(\mu, \sigma, \vec{v})$, and we can examine the effects of flow velocity ($\vec{v}$) on B.

We write the above equation in more detail, as:

$$\nabla \times (\nabla \times \vec{B}) = \nabla(\nabla \cdot \vec{B}) - (\nabla \cdot \nabla)\vec{B} = -\nabla^2 \vec{B}, \text{ then:}$$

$$\frac{\partial \vec{B}}{\partial t} = \frac{1}{\mu\sigma} \nabla^2 \vec{B} + \nabla \times (\vec{v} \times \vec{B}) = \lambda \nabla^2 \vec{B} + \nabla \times (\vec{v} \times \vec{B}),$$

where we take σ as uniform and adopt the usage that $\lambda = \frac{1}{\mu\sigma}$, the magnetic diffusivity.

In words, the above equation shows that: Rate of change of B ~ (Diffusion + Convection).

Diffusion of B in Plasmas (We Take $\vec{v} = 0$ and Examine the Effects of Diffusion)

Diffusion of Magnetic Fields into Conducting Medium (Order of Magnitude)

We begin by establishing the phenomenology of diffusion (progression of a property, due to a gradient, with time through space) of magnetic fields through uniform composition conductor or plasma. In devices such as the theta- or Z-pinch, pulsed discharges create magnetic fields and plasmas that are in near proximity to conductor walls. The subsequent behavior is categorized by the characteristic time for the process of diffusion of fields to occur.

We take the simple case of a plane conductor, which at time $t = 0$ is suddenly brought in contact with a uniform magnetic field, $\vec{B}_0$ (Jackson, 1963). The governing diffusion equation for magnetic fields in a conductor can be written (as above): $\frac{\partial \vec{B}}{\partial t} = \lambda \nabla^2 \vec{B}$, where $\lambda = 1/\mu\sigma$. Here, to simply get the timescale, we approximate: $\nabla^2 \Rightarrow \frac{1}{L^2}$, so, $\frac{\partial B}{\partial t} = -\frac{1}{\mu\sigma}\frac{B}{L^2}$, which has a solution: $B = B_0 e^{-t/\tau}$, where, $B_0 = B$ at $t = 0$ and $\tau = \mu\sigma L^2$ is the characteristic time for diffusion. (The time for 1-kG fields to diffuse through 2.5 cm of copper ($\sigma = 0.5 \times 10^6$/ohm cm) is 0.6 ms).

Diffusion of Magnetic Fields (Diffusion Velocity)

We would like to identify some concepts so that we can envision the physical behavior. Again: $\frac{\partial B}{\partial t} = \lambda \nabla^2 \vec{B}$, the diffusion equation, where $\lambda = 1/\mu\sigma$ is magnetic diffusivity. (Note that for the limiting condition: $\sigma = \infty$, there is no diffusion.) If we begin by specifying a steady state $\left(\frac{\partial B}{\partial t} = 0\right)$, so, $\lambda \nabla^2 \overrightarrow{B} = 0$; then, in one dimension, $0 = \lambda \frac{d^2 B}{dx^2}$, which can be integrated to give $\frac{dB}{dx}$ = constant. We can write for this process:

$$\text{Flux} = -\text{diffusion coefficient} \times \text{property gradient, i.e.} : \Gamma = -D\frac{dB}{dx}.$$

This physical behavior is analogous to steady-state heat flow $\left(q = -k\frac{dT}{dx}\right)$, where there is energy flux, and we can write:

Energy per unit time per unit area = (Energy per unit volume) × (velocity or rate of flow), and the latter term defines a flux velocity. Similarly, we can recast the field diffusion as:

$$\left(\frac{\overrightarrow{B}\left[\frac{\mathrm{W}}{\mathrm{m}^2}\right]}{dist.}\right) \cdot \overrightarrow{v}_{diff}\left(\frac{dist.}{time}\right) = \frac{dB}{dx} \cdot v_{diff} = \lambda \nabla^2 \overrightarrow{B},$$

where $\overrightarrow{v}_{diff}$ is a velocity of diffusion for the magnetic field. So a constant gradient in magnetic field exhibits a dynamic flux behavior in plasma.

Convection (Dynamics) of B in Plasmas (We Neglect Diffusion; Effect occurs with $\sigma = \infty$—Infinite Conductivity Plasma)

Taking the above equation for $\overrightarrow{B}$ without diffusion, we can write:

$$\frac{\partial \overrightarrow{B}}{\partial t} = \nabla \times (\overrightarrow{v} \times \overrightarrow{B}) = \overrightarrow{v}(\nabla \cdot \overrightarrow{B}) + (\overrightarrow{B} \cdot \nabla)\overrightarrow{v} - \overrightarrow{B}(\nabla \cdot \overrightarrow{v}) - (\overrightarrow{v} \cdot \nabla)\overrightarrow{B},$$

$$\text{and} \quad \frac{\partial \overrightarrow{B}}{\partial t} + (\overrightarrow{v} \cdot \nabla)\overrightarrow{B} = (\overrightarrow{B} \cdot \nabla)\overrightarrow{v},$$

$$\text{or} \quad \frac{D\overrightarrow{B}}{Dt} = (\overrightarrow{B} \cdot \nabla)\overrightarrow{v}.$$

Note that this functional form is similar to that for the fluid equation for the convection of vorticity: $\frac{D\overrightarrow{\Omega}}{Dt} = (\overrightarrow{\Omega} \cdot \nabla)\overrightarrow{v}$ (Karamcheti, 1966). That equation expresses the change of vorticity of an element of ideal fluid under irrotational force, $\nabla \times \overrightarrow{v} = 0$. So the concepts for understanding the transport of vorticity in ordinary fluid mechanics (OFM) will provide insight into magnetic field transport in plasmas.

We now consider further some aspects of the motion of an infinite conductivity plasma. While being an ideal, limiting value for conductivity, this assumption is an effective artifice for theoretical study of plasma in order to simplify equations. Accordingly: $\frac{\vec{J}}{\sigma} = \vec{E} + \vec{v} \times \vec{B} \rightarrow 0 \Rightarrow \vec{E} = -\vec{v} \times \vec{B}$, there is no effective current flow, and with $v \sim E/B$, there is no diffusion. We now consider two important principles for this plasma.

1. Magnetic flux through any closed contour moving with a perfectly conducting fluid is constant.

The magnetic flux is: $\varphi = \int_S \vec{B} \cdot d\vec{A}$. As: $\frac{D\varphi}{Dt} = 0$, this state is defined as: "frozen-in flux" that travels with the loop in the plasma flow. Further, with $\frac{\partial \vec{B}}{\partial t} = \nabla \times (\vec{v} \times \vec{B})$, get $\frac{D}{Dt}\left(\frac{\vec{B}}{\rho}\right) = \left(\frac{\vec{B}}{\rho} \cdot \nabla\right)\vec{v}$, or if we integrate with respect to time following the element, $\frac{\vec{B}}{\rho} = \left(\frac{\vec{B}_0}{\rho_0} \cdot \nabla_0\right)\vec{r}$, where the subscript 0 refers to initial (reference) conditions, so:

$$\frac{B}{\rho} = \frac{B_0}{\rho_0}\frac{dr}{dr_0}$$

and,

$$\frac{B}{\rho dr} = \frac{B_0}{\rho_0 dr_0} = \text{constant}.$$

For the second principle, it is important to establish direction with respect to the magnetic field lines, as the above relationship is scalar. We had integrated with respect to time following an element of fluid, with $v_{ref.} = 0$, so using the above vector relationship: $\vec{B}_0 \parallel d\vec{r}_0 \Rightarrow \frac{\vec{B}_0}{\rho_0} = C d\vec{r}_0$, and the left-hand side (LHS) defines the $\vec{r}$ direction as parallel to $\vec{B}$. We can then sketch an element as shown in (Figure 8.1).

We can interpret $d\vec{r}$ as an element of length along $\vec{B}$, so we can state:

$$d\vec{r} = (d\vec{r}_0 \cdot \nabla)\vec{r} = \left(\frac{\vec{B}_0}{C\rho_0} \cdot \nabla\right)\vec{r} = \frac{1}{C}\frac{\vec{B}}{\rho}, \quad \therefore d\vec{r} \parallel \vec{B}.$$

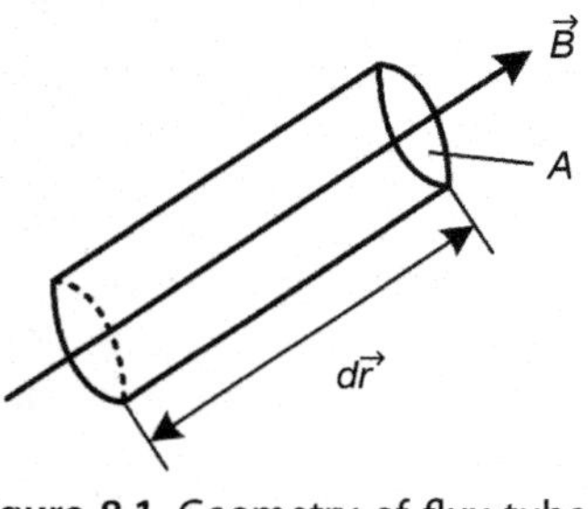

Figure 8.1 Geometry of flux tubes.

So we state the second principle:

2. Fluid elements on magnetic lines of force continue to lie on magnetic lines of force.

Further, $\frac{B}{\rho dr} = \text{constant} \Rightarrow$ with a dr increase (stretching element), then B increases. However, since flux ($\varphi = B \cdot A$) must be conserved, the cross-section area of the cylindrical element must decrease. As a corollary (Cowling, 1957) of the above statement, fluid motion transverse to $\overrightarrow{B}$ field lines transports them (B field lines) along with the fluid; motion along the field lines does not affect them. This concept is important when we discuss wave motion in plasmas with magnetic fields.

Definition of Parameters for Convection and Diffusion

To develop parameters to simplify equations and establish relevance of mechanisms, we again consider the basic equation for fields and fluids:

$$\frac{\partial \overrightarrow{B}}{\partial t} = \lambda \nabla^2 \overrightarrow{B} + \nabla \times (\overrightarrow{v} \times \overrightarrow{B}).$$

We define the following variables to place the equation in dimensionless form:

$$\overrightarrow{\tilde{B}} = \frac{\overrightarrow{B}}{B_0}, \quad \tilde{\nabla} = \frac{d}{dr/L} = L\nabla, \quad \overrightarrow{\tilde{v}} = \frac{\overrightarrow{v}}{U}, \quad \tau = \frac{t}{L/U}, \text{ then:}$$

$$\frac{\partial \overrightarrow{\tilde{B}}}{\partial \tau} = \left(\frac{\lambda}{UL}\right) \tilde{\nabla}^2 \overrightarrow{\tilde{B}} + \tilde{\nabla} \times (\overrightarrow{\tilde{v}} \times \overrightarrow{\tilde{B}}), \text{ and}$$

we represent the coefficient as: $\frac{\partial \overrightarrow{\tilde{B}}}{\partial \tau} = \frac{1}{R_m} \tilde{\nabla}^2 \overrightarrow{\tilde{B}} + \tilde{\nabla} \times \left(\overrightarrow{\tilde{v}} \times \overrightarrow{\tilde{B}}\right).$

We define the parameter $UL/\lambda \equiv R_m$ as the magnetic Reynolds number in analogy to the definition in fluid mechanics of where $R \equiv UL/\nu$ and ν is the viscous diffusivity. For:

Large R_m, neglect B diffusion (convection is dominant);

Small R_m, neglect B convection (B diffusion dominant).

In this section, we have ignored the influence of forces, considering only the effects of flow on $\overrightarrow{B}$. In the next section, we will neglect flow and consider only force effects.

MAGNETOHYDROSTATICS

Analogous to hydrostatics in OFM, relating pressure and force, we are here interested in determining the interrelationship of force, pressure, and magnetic field (Boyd and Sanderson, 1969). Again we begin by writing the momentum equation:

$$\rho\frac{D\vec{v}}{Dt} = -\nabla p + \vec{J}\times\vec{B}, \text{or:}$$

$$\rho\left[\frac{\partial\vec{v}}{\partial t} + (\vec{v}\cdot\nabla)\vec{v}\right] = -\nabla p + \left(\frac{\nabla\times\vec{B}}{\mu}\right)\times\vec{B}, \text{ and for steady state;}$$

$$\rho(\vec{v}\cdot\nabla)\vec{v} = -\nabla p + (\vec{B}\cdot\nabla)\frac{\vec{B}}{\mu} - \frac{\nabla B^2}{2\mu} = -\nabla\left(p + \frac{B^2}{2\mu}\right) + (\vec{B}\cdot\nabla)\frac{\vec{B}}{\mu},$$

$$\text{and: } (\vec{v}\cdot\nabla)\vec{v} = -\frac{1}{\rho}\nabla\left(p + \frac{B^2}{2\mu}\right) + (\vec{B}\cdot\nabla)\frac{\vec{B}}{\rho\mu}.$$

If $\vec{v} = 0$, $\rho =$ constant, then $\nabla\left(p + \frac{B^2}{2\mu}\right) - \rho(\vec{V}_A\cdot\nabla)\vec{V}_A = 0$, where: $(B^2/\mu\rho)^{1/2} \equiv \vec{V}_A$.

So, the fluid pressure term has an equivalence to $(B^2/2\mu)$, which is termed magnetic pressure.

The term, $\vec{V}_A$, which has units of velocity, is referred to as Alfven velocity, and it is a quantity of considerable significance in plasma dynamics; it will be considered again later.

The third term in the above equation is in analogy to OFM where a comparable momentum equation can be written:

$$\nabla p + \rho(\vec{v}\cdot\nabla)\vec{v} = 0.$$

The second term can be referred to as an inertia gradient. So we can see that magnetic pressure plays an important role in the fluid dynamics of plasmas, and that demonstrates the interrelationships of fluid and magnetic properties.

Magnetic Pressure in Plasma Fluids

First, we look at the mathematical relationships that follow from the above equation involving magnetic pressure. Taking $\vec{v} = 0$ in the momentum equation,

$$\nabla\left(p + \frac{B^2}{2\mu}\right) = (\vec{B}\cdot\nabla)\frac{\vec{B}}{\mu} = \frac{1}{\mu}\nabla\cdot(\vec{B}\,\vec{B}),$$

where the right-hand side (RHS) is a dyad, so, $\nabla P - \nabla\cdot\left(\frac{\vec{B}\,\vec{B}}{\mu}\right) = 0.$

This is a vector equation, but: $\overrightarrow{B}\overrightarrow{B} = B_i B_k$, and $\nabla \cdot \left(\overrightarrow{B}\overrightarrow{B}\right) = \frac{\partial}{\partial r_k}(B_i B_k)$. Then, an equivalent form of this equation is (Jackson, 1963):

$$\frac{\partial}{\partial r_k}(P\delta_{ik}) - \frac{\partial}{\partial r_k}T_{ik} = 0,$$

where:

$$T_{ik} = \frac{1}{\mu}\left(B_i B_k - \frac{\delta_{ik}}{2}B^2\right),$$

so:

$$\frac{\partial}{\partial r_k}\left[\left(p + \frac{B^2}{2\mu}\right)\delta_{ik} - \frac{B_i B_k}{\mu}\right] = 0,$$

where the δ_{ik} term is the diagonal component of a total stress tensor, $\overleftrightarrow{\tau}_m$.

For the case where $\overrightarrow{B} = B_z = B\overrightarrow{k}$ and $i = k$ only for z, the tensor becomes:

$$\overleftrightarrow{\tau}_m = \begin{bmatrix} p + \frac{B^2}{2\mu} & 0 & 0 \\ 0 & p + \frac{B^2}{2\mu} & 0 \\ 0 & 0 & p - \frac{B^2}{2\mu} \end{bmatrix}.$$

This tensor represents the physical occurrence of a compression pressure $(B^2/2\mu)$ transverse to $\overrightarrow{B}$ and tension pressure $(-B^2/2\mu)$ along $\overrightarrow{B}$ lines.

Fluid-Plasma Equilibrium Configurations

The physical significance of "magnetic pressure" becomes evident when we examine plasma configurations that are in force equilibrium. To express the relative magnitudes of each of these terms, we define: Beta Parameter, $\beta \equiv \frac{p}{B^2/2\mu} = \frac{\text{Fluid pressure}}{\text{Magnetic pressure}}$.

As a first physical example (compatible with $\sigma = \infty$), we consider a cylindrical geometry, which has properties that are uniform in the axial direction and a radial configuration that has uniform plasma pressure, p, and $B = 0$ from $r = 0$ to $r = r_0$; and from $r > r_0$, $p = 0$ and B compatible with Maxwell's equations (Figure 8.2).

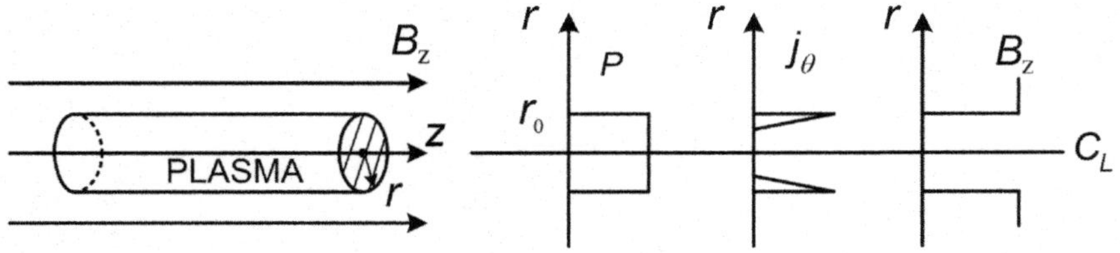

Figure 8.2 Cylindrical plasma/field equilibrium configuration.

From momentum:

$$\nabla\left(p+\frac{B^2}{2\mu}\right)=(\vec{B}\cdot\nabla)\frac{\vec{B}}{\mu}=\left(B_z\cdot\frac{\partial}{\partial z}\right)\frac{B_z}{\mu}=0,$$

then:

$$\frac{d}{dr}\left(p+\frac{B^2}{2\mu}\right)=0,$$

so:

$$p(r)+\frac{B^2(r)}{2\mu}=\text{constant}.$$

This means: $p(0-r_0)=p$, $p(>r_0)=0$; $B(0-r_0)=0$, $B(>r_0)=B_0$, or:

$$p\text{ (Fluid pressure in plasma)}=\frac{B_{out}^2}{2\mu}\text{(Magnetic pressure outside } r_0).$$

Another way to understand the force mechanism is as follows. If we take $\sigma=\infty$,

$$\oint\vec{B}\cdot d\vec{l}=\mu\int\vec{J}\cdot dA=\mu I.$$

So in the layer between plasma and magnetic field, using Maxwell's equation:

$$\frac{\partial B_z}{\partial r}=\textit{finite},\quad\text{and}\quad\nabla\times\frac{\vec{B}}{\mu}\Rightarrow\vec{J}_\theta,\quad\text{an azimuthal current, as:}$$

$$\nabla\times\vec{B}=\left(\frac{1}{r}\frac{\partial B_z}{\partial\theta}-\frac{\partial B_\theta}{\partial z}\right)\vec{r}+\left(\frac{\partial B_r}{\partial z}-\frac{\partial B_z}{\partial r}\right)\vec{\theta}+\left[\frac{1}{r}\frac{\partial(rB_\theta)}{\partial r}-\frac{1}{r}\frac{\partial B_r}{\partial\theta}\right]\vec{z},$$

which has the one component: $\overrightarrow{J}_\theta > 0$ on a surface at r_0, and creates: $\overrightarrow{J}_\theta \times \overrightarrow{B}_z = \overrightarrow{f}_r$. This physical behavior is inherent in magnetic confinement devices. The present example (r,θ,z) is the geometry of a theta pinch device.

As a second physical example, let us consider a cylindrical geometry where there is a uniform current density, J_z, constant from $r = 0$ to $r = r_0$ and the current is flowing in a plasma of finite conductivity, which is in pressure equilibrium with magnetic fields outside $r = r_0$ (Figure 8.3).

We want to define current density, fluid pressure, and magnetic field as a function of r consistent with fluid and Maxwell's equations up to $r = r_0$. First, taking Maxwell's equation $\nabla \times \overrightarrow{B} = \mu \overrightarrow{J}$, integrating both sides over surface area and applying Stoke's theorem we get: $\oint \overrightarrow{B} \cdot d\overrightarrow{l} = \mu \int \overrightarrow{J} \cdot dA$, which when applied to this geometry gives us $B_\theta \cdot 2\pi r = \mu\pi r^2 J_z$. Then $B_\theta = \mu \frac{r}{2} J_z$, and so the azimuthal magnetic field increases linearly from zero on the axis and is related to the magnitude of enclosed current within the cross-sectional area.

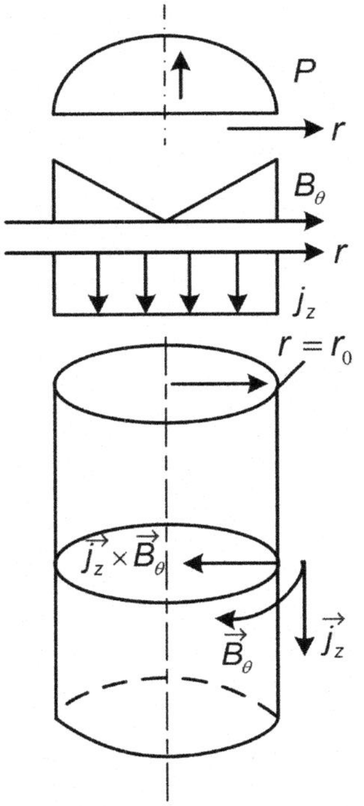

Figure 8.3 Linear (uniform current) pinch.

Next, we consider the momentum equation: $\nabla p = \vec{J} \times \vec{B}$, which for this case gives us:

$$\frac{\partial p}{\partial r} = -J_z B_\theta = -\mu \frac{r}{2} J_z^2,$$

or:

$$dp = -\mu \frac{J_z^2}{2} r dr,$$

and:

$$p_2 - p_1 = -\frac{\mu}{4} J_z^2 \left(r_2^2 - r_1^2\right).$$

So we get:

$$p + \frac{\mu}{4} J_z^2 r^2 = p + \frac{B^2}{\mu} = \text{constant},$$

and the pressure is zero at r_0 and is parabolic in shape. Calculating the total current flowing in the column:

$$I = \int_0^{r_0} J \cdot 2\pi r dr = \frac{\oint \vec{B} \cdot d\vec{l}}{\mu},$$

so:

$$I = \frac{2\pi r_0 B_\theta(r_0)}{\mu}.$$

It can be shown that: $I^2 \sim \underbrace{NKT}_{p \cdot \pi r_0^2}$, where N is the total number of particles per unit length, and the current required to contain the plasma is proportional to the product of pressure and cross-sectional area of the plasma column. This example illustrates the force equilibrium inherent in the linear (Z-) pinch, which we consider in application, later.

HYDROMAGNETIC STABILITY

We have introduced the notion of confinement of plasma by the force density: $\vec{J} \times \vec{B} \Leftrightarrow \frac{B^2}{2\mu} = p$. But the real process of force application in experiment involves force generation and initiation, which is unsteady.

There is then the related question of stability of an equilibrium in response to time-dependent perturbations. A system is stable if the forces acting tend to restore equilibrium. A system is unstable if the forces acting tend to take the system away from equilibrium. There are two types of instabilities: (1) macroscopic, which are hydromagnetic and involve physical displacement of the boundary; and, as we discussed above with respect to anomalous transport, (2) microscopic, which involve local density–field interactions and these involve wave perturbations and wave–particle interactions (Klages et al., 2008). We are concerned here with macroscopic MHD stability of confinement.

Physical Considerations of MHD Stability

Let us examine the linear (Z-) pinch plasma described above for high conductivity ($\sigma = \infty$); physically we have a current carrying cylindrical plasma surrounded by a magnetic field configuration in B_θ, as shown in Figure 8.3.

From the momentum equation, $\nabla p = \vec{J} \times \vec{B}$, and we must consider boundary conditions for the interface between plasma column and external field. At that interface, we have:

$$\frac{B_\theta^2}{2\mu} = p, \quad \text{where} \quad B_\theta \cdot 2\pi r = \mu \pi r^2 j_z, \quad \text{then:} \quad B_\theta = \frac{\mu I}{2\pi r},$$

where I is total current enclosed within radius, r, here taken to be the boundary radius. We will consider perturbations with the configurations shown in Figures 8.4 and 8.5.

First, consider a symmetrical perturbation of the boundary (to r'); let it be azimuthally (in θ) uniform (in or out) in radius. Then $B'_\theta = \frac{\mu I}{2\pi r'}$, where $r' < r \therefore B'_\theta > B_\theta$, and $\frac{B'^2_\theta}{2\mu} \gg \frac{B^2_\theta}{2\mu}$, but $p' = p$. The configuration is unstable to the perturbation, and it is called the sausage instability.

This instability can be inhibited by an applied B_z field. In the plasma, the flux: $\varphi = B_z \cdot A = B_z \pi r^2 = \text{constant}$. With an applied field, $B_z = \frac{C}{\pi r^2}$, at the boundary: for $r' < r$, $B'_z \gg B_z \Rightarrow p + \frac{B'^2_z}{2\mu} > \frac{B^2_\theta}{2\mu}$, and this is now a stable configuration.

Second, consider an asymmetric perturbation of the boundary, r'; let it be a left or right displacement from the original axis, as in Figures 8.5 and 8.6. Then $B_\theta = \frac{\mu I}{2\pi r'}$, but $\varphi = B \cdot A$ and $B = \frac{\varphi}{A}$. At the perturbation, $A_{LHS} < A_{RHS}$, $\varphi_{LHS} = \varphi_{RHS}$; therefore: $B_{LHS} > B_{RHS}$ and $\frac{B^2}{2\mu}(LHS) + p > \frac{B^2}{2\mu}(RHS)$. This is unstable, and it is called the kink instability.

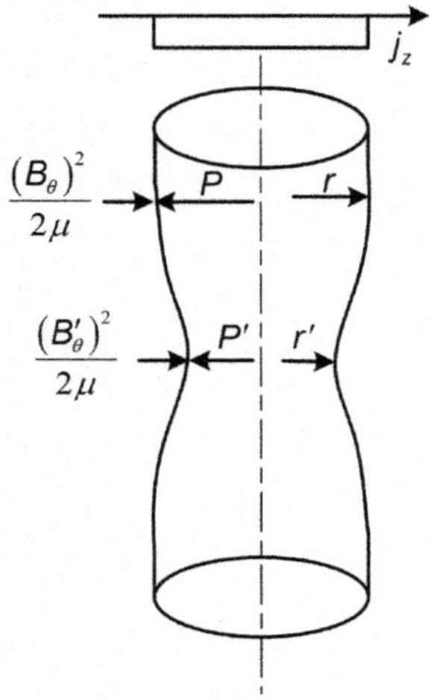

Figure 8.4 Boundary (symmetrical) perturbation linear pinch.

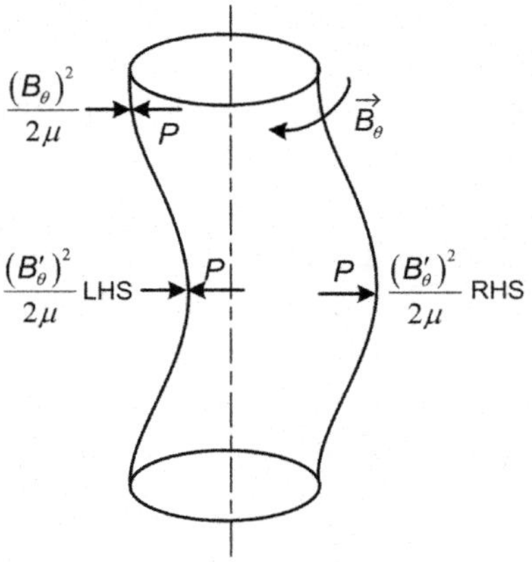

Figure 8.5 Boundary (asymmetrical) perturbation linear pinch.

Analysis of MHD Perturbations

There are several techniques for analyzing the stability of static plasma configurations:

Energy minimum principle (E_{Mag}, E_{plasma})—thermodynamics based (Spitzer, 1962);

Orbit theory—particle drift velocities (Chandrasekhar, 1960);

Perturbed hydromagnetic equations—small perturbation theory (Cowling, 1957).

We will apply the method of perturbed hydromagnetic equations (following Boyd and Sanderson, 1969). An outline of the method is presented first. We assume perfect conductivity ($\sigma = \infty$).

1. Write equilibrium equations (take flow velocity, $U_0 = 0$).
2. Introduce perturbations ($p = p_0 + p'$, etc.) and drop second-order terms,

$$\text{e.g.,} \quad \rho \frac{D\vec{v}}{Dt} = -\nabla p + \vec{J} \times \vec{B};$$

$$\text{get:} \quad \rho_0 \frac{D\vec{v'}}{Dt} = -\nabla p' + \vec{J'} \times \vec{B}_0 + \vec{J}_0 \times \vec{B'}.$$

3. Reduce the number of equations and unknowns, get v', p', B' expressed in three equations: induction, momentum, and continuity.

4. Introduce the displacement vector:
$\vec{\xi}(\vec{r}, t) = \vec{r} \cdot \vec{r}_0$, with $\vec{v} = \frac{D\vec{r}}{Dt} = \frac{\partial \vec{\xi}}{\partial t}$, the velocity of boundary.
Then $\frac{\partial^2 \vec{\xi}}{\partial t^2} = f(\rho_0, p_0, B_0)$; this is linear and needs two boundary conditions.
5. We assume a periodic solution and perform a Fourier analysis, i.e., we take the form for solution: $\vec{\xi}(\vec{r}, t) = \sum_n \vec{\xi}(\vec{r}, \omega)e^{i\omega_n t}$, where ω_n is the normal mode frequency, and substitute.
6. We specify and incorporate the boundary conditions (BC) on fields; force equilibrium at boundary: $p_0 + \frac{B_{0i}^2}{2\mu} = \frac{B_{0e}^2}{2\mu}$ (i is internal, e is external to the plasma).

Stability of the Linear Pinch

In this analysis (we summarize that of Boyd and Sanderson, 1969), we will assume: $\sigma = \infty$, so we have a thin layer of (skin) current and frozen-in fields. We take an applied axial field and examine the effect on stability, as: $\vec{B}_{0i} = (0,\ 0,\ B_{0z})$, and also $\vec{B}_{0e} = (0,\ B_\theta(r),\ 0)$. We have: $\vec{\xi}(\vec{r}, t) = \vec{\xi}(\vec{r})e^{i\omega t}$, and we take: $\vec{\xi}(\vec{r}) = \vec{\xi}(r)e^{im\theta + ikz}$ for cylindrical symmetry. We consider the $m = 0$ mode, with no angular dependence. We get:

$$\frac{d^2\xi_z}{dr^2} + \frac{1}{r}\frac{d\xi_z}{dr} - \overbrace{f(k,\ c_s,\ \omega,\ V_A)}^{K^2}\xi_z = 0.$$

This is the Bessel equation with solution $\xi_z = I_0(Kr)$; we have a similar solution for ξ_r.

We apply the boundary condition to get the dispersion relationship, $\omega = \omega(k)$,

$$\omega^2 = k^2 V_A^2 - \frac{B_\theta^2}{B_{0z}^2}\frac{K}{r}\frac{I_0'(Kr)}{I_0(Kr)} = \frac{k^2 B_{0z}^2}{4\pi\rho_0} - \frac{B_\theta^2}{4\pi\rho_0 r^2}\frac{KrI_0'(Kr)}{I_0(Kr)}.$$

Now recall: $\vec{\xi}(\vec{r}, t) = \vec{\xi}(\vec{r})e^{i\omega t}$; for $\omega^2 < 0, -\omega^2 > 0$ and $i\omega > 0$ and $e^{i\omega t}$ grows with time and the equilibrium is unstable. For conservative systems, ω^2 must be real to evidence no damping or instability. In the above case, $\omega^2 < 0$, means:

$$k^2 B_{0z}^2 < \frac{B_\theta^2}{r^2}\frac{KrI_0'(Kr)}{I_0(Kr)}, \quad \text{or:} \quad B_{0z}^2 < \frac{B_\theta^2}{(kr)^2}\frac{KrI_0'(Kr)}{I_0(Kr)},$$

which is indicative of an unstable configuration. However, if $B_{0z}^2 > RHS$, the configuration can be made stable.

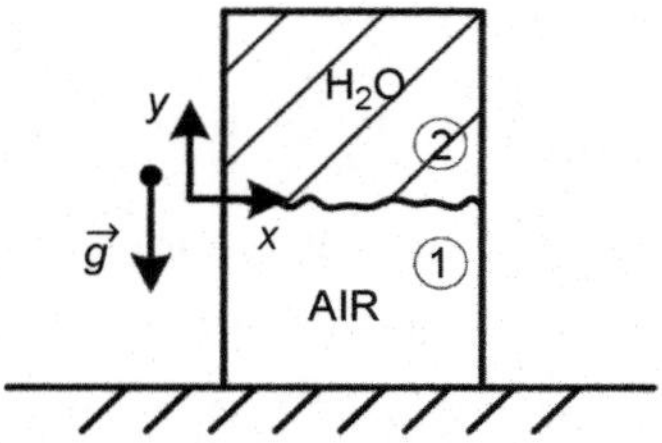

Figure 8.6 Hydrodynamic Rayleigh–Taylor instability.

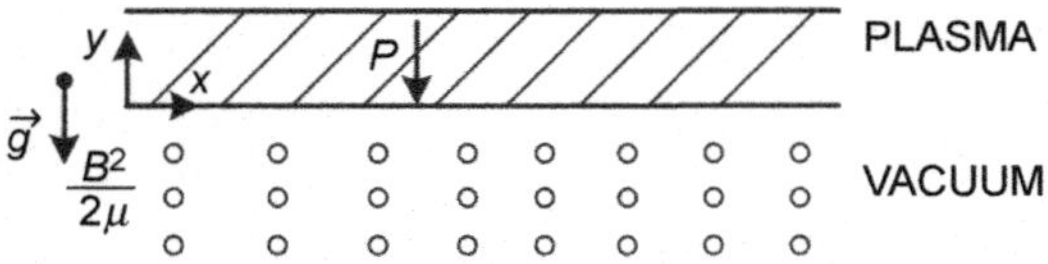

Figure 8.7 Plasma Rayleigh–Taylor instability.

Rayleigh–Taylor Instability

This instability occurs in a region where a fluid has an interface with a force field; this can occur with a liquid and a gravity force field (Figure 8.6) or with plasmas and magnetic fields (Figure 8.7) where the fluid interface occurs with a region of lower density or with lower (fluid or magnetic) pressure.

Ordinary Fluid Mechanics

Consider a fluid of high density supported by fluid of low density under the action of a gravitational field. (This would occur if we instantly inverted a glass half-filled with water. We would have a plane interface between water (top) and air (bottom)).

$$\text{Basic equations:} \quad \frac{\partial \rho}{\partial t} + \nabla \cdot (\rho \vec{v}) = 0,$$

$$\text{and} \quad \rho \frac{D\vec{v}}{Dt} = -\nabla p + \rho \vec{g}.$$

We set $v = v_0 + \tilde{v}$, where we have a perturbation: $\tilde{v} = v(y)e^{i(kx-\omega t)}$. We apply the boundary conditions in a normal mode analysis to get a dispersion relation:

$\omega^2 = -gk\frac{(\rho_2-\rho_1)}{\rho_2+\rho_1} < 0$, which is unstable (perturbation growth) for $\rho_2 > \rho_1$.

Plasma

In this case, we examine a (dense) plasma in a gravity field supported by vacuum and magnetic fields. The basic momentum equation (with force equilibrium) is:

$$\rho \frac{D\vec{v}}{Dt} = -\nabla p + \vec{J} \times \vec{B} + \rho \vec{g}.$$

For straight, uniform fields with the geometry shown in Figure 8.7, we have:

$$-\nabla p + \frac{\nabla \times \overrightarrow{B}}{\mu} \times \overrightarrow{B} + \rho \overrightarrow{g} = 0,$$

or:

$$-\nabla \left(p + \frac{B^2}{2\mu} \right) - \rho g = 0.$$

If $p \neq f(y)$:

$$-\frac{1}{2\mu} \frac{\partial B^2}{\partial y} - \rho g = 0,$$

so, in the fluid:

$$\frac{B^2(y)}{2\mu} = \frac{B^2(0)}{2\mu} + \rho g y,$$

and B must decrease with y for equilibrium. The boundary condition at $y = 0$ (again, $p \neq f(y)$) is:

$$\frac{B^2_{y<0}}{2\mu} = \frac{B^2(0)}{2\mu} + p.$$

To examine the stability, consider the basic equations:

$$\frac{\partial \rho}{\partial t} + \nabla \cdot (\rho \overrightarrow{v}) = 0,$$

$$\rho \frac{D\overrightarrow{v}}{Dt} = -\nabla \left(p + \frac{B^2}{2\mu} \right) + \rho \overrightarrow{g}.$$

Again, we take $v = v_0 + \tilde{v}$, where $\tilde{v}$ is a (small) perturbation, $\tilde{v} = v(y)e^{i(kx - \omega t)}$. We assume that the fluid is incompressible, $\nabla \cdot \tilde{v} = 0$, and the equation of plasma equilibrium becomes:

$$\frac{d^2\tilde{v}_y}{dy^2} + \left(\frac{1}{\rho} \frac{d\rho}{dy} \right) \frac{d\tilde{v}_y}{dy} - k^2 \left(1 + \frac{g}{\omega^2} \frac{1}{\rho} \frac{d\rho}{dy} \right) \tilde{v}_y = 0.$$

This has a solution of the form: $\tilde{v} = \tilde{v}_y = K\tilde{v}(y)e^{-i\omega t}$, and $\tilde{v}(y) = \widehat{v}e^{-ky}$.

At $y = 0$, the differential equation for the perturbation is: $\frac{d\tilde{v}_y}{dy} - k^2 \frac{g}{\omega^2} \tilde{v}_y = 0$, and as before $\omega^2 = -kg_0$, and the perturbation and system equilibrium are unstable.

WAVES IN PLASMA—PROPAGATION OF PERTURBATIONS

Introduction

Plasmas are unique in that there is a coupling of fluid and electromagnetic field properties. Accordingly, the medium can support wave motions with a combination of fluid and field perturbations. While a comprehensive discussion of plasma waves is not the intent in this introductory work, the importance of this unique physical behavior requires that basic descriptions and analysis of the type of unique wave motions possible be presented here.

Many of the important aspects of wave behavior have been introduced in fundamental treatments of electromagnetic waves in gases; these are familiar to all readers. Here, the presentation will review these basic ideas for emphasis and for a foundation to the discussion of plasma waves. Light waves or radio waves travel through a nonconducting medium with the functional behavior (for plane waves), as (Jackson, 1963):

$$\vec{E}(\vec{x}, t) = \vec{\epsilon}_1 E_0 e^{i(\vec{k}\cdot\vec{x}-\omega t)}; \quad \vec{B}(\vec{x}, t) = \vec{\epsilon}_2 B_0 e^{i(\vec{k}\cdot\vec{x}-\omega t)},$$

where E_0, B_0 are complex amplitudes; $\vec{\epsilon}_1, \vec{\epsilon}_2$ are directional vectors perpendicular to the direction of wave propagation, $\vec{k}$; $\vec{\epsilon}_2 = \frac{\vec{k} \times \vec{\epsilon}_1}{\vec{k}}$; $B_0 = \sqrt{\mu\epsilon} E_0$. The wave number, $k = \frac{\omega}{v} = \sqrt{\mu\epsilon}\ \frac{\omega}{c} = n\frac{\omega}{c}$ where ω is the frequency, v is the wave velocity, c is the velocity of light in a vacuum and n is the index of refraction. Electromagnetic waves are transverse waves, i.e., the perturbed values are perpendicular to the direction of propagation. The velocity of a point on the wave is the phase velocity, $v_{\text{ph}} = \frac{\omega}{k}$; this can exceed the speed of light. The wave pulse travels undistorted in shape with a group velocity, $v_{\text{g}} = \frac{d\omega}{dk}$. The energy density associated with the wave travels with the group velocity; this cannot exceed the speed of light. In a vacuum (where $\vec{J} = 0$), $v_{\text{ph}} \equiv c = (\epsilon_0\mu_0)^{-1/2}$.

Dispersive Media and Cutoff of Electromagnetic Waves

In a plasma medium, electrons can respond to the electric field in the electromagnetic wave and be driven into motion, as $\vec{J} = \vec{J}_e = -n_e e \vec{v}_e$.

(In a vacuum, $\overrightarrow{J} = 0$.) If we have an effectively cold plasma, ($kT_e = 0$) then $\overrightarrow{v}_e = \frac{e\overrightarrow{E}_0}{im_e\,\omega}$, and the dispersion relationship,

$$\omega(k), \quad \text{is:} \quad \omega^2 = \omega_p^2 + c^2k^2, \quad \text{or:} \quad k^2 = \frac{\omega^2}{c^2}\left(1 - \frac{\omega_p^2}{\omega^2}\right).$$

$$\text{So,} \quad v_{ph}^2 = \frac{\omega^2}{k^2} = c^2 + \frac{\omega_p^2}{k^2} > c^2,$$

$$\text{and} \quad v_g = \frac{d\omega}{dk} = \frac{c^2}{v_{ph}} \quad \text{with } v_g < c, \text{ when } v_{ph} > c.$$

If the index of refraction is a function of the frequency, the electromagnetic wave in a medium is dispersive. The dispersion relationship predicts an important property for passage and propagation of electromagnetic waves: when $\omega < \omega_p$, k cannot be real, so the wave cannot propagate. Only waves with $\omega > \omega_p = \left(\frac{n_e e^2}{m\epsilon_0}\right)^{1/2}$ can pass, so we have "cutoff or blackout" of certain wavelengths for a given electron density. As a practical example, during reentry of an Earth orbiting spacecraft, the plasma that forms around the vehicle shape interferes with transmission of microwave wavelength bands that are normally used for communication.

Ion (Sound) Waves

In un-ionized gases, particles transmit perturbations by collisions. The result is longitudinal waves (the perturbation is in the direction of the wave motion), and they move with (sound speed): $c_s = \frac{\omega}{k} = \left(\frac{\gamma k_B T}{m_i}\right)^{1/2}$, where m_i is the ion mass, and k_B is the Boltzmann constant.

In plasmas ($n_i = n_e$) without significant (close encounter) interparticle collisions, perturbations in ion density can propagate due to the formation of an electric field. It is presumed that particle species have a thermal energy (Maxwellian) distribution. For small perturbations we find that sound (longitudinal) waves will travel at (Chen, 1984):

$$c_s = \frac{\omega}{k} = \left(\frac{k_B T_e + \gamma k_B T_i}{m_i}\right)^{1/2}.$$

This is the dispersion relationship for ion acoustic waves. Note that in plasmas with $T_i \ll T_e$, wave (sound) speed is proportional to electron temperature ($c_s \sim T_e$), not ion temperature.

Longitudinal Electron Oscillations Perpendicular to $\overrightarrow{B}$

We presume an applied magnetic field, $\overrightarrow{B_0}$, and specify a plasma where ions do not move to form a background for electrons that will respond to forces; charge separation will create electric fields (Chen, 1984). We neglect thermal motion of electrons; we have perturbation values of electron density (n'_e), velocity (v'_e), and electric field ($\overrightarrow{E'}$); we have $\overrightarrow{k} \parallel \overrightarrow{E'}$ (the electron wave moves in the x-direction and the $\overrightarrow{E'}$field is in x-direction) and the governing equations are:

$$m_e \frac{\partial \overrightarrow{v'_e}}{\partial t} = -e\left(\overrightarrow{E'} + \overrightarrow{v'_e} \times \overrightarrow{B}_0\right),$$

$$\frac{\partial \overrightarrow{n'_e}}{\partial t} + n_0 \nabla \cdot \overrightarrow{v'_e} = 0,$$

$$\epsilon_0 \nabla \cdot \overrightarrow{E'} = -e\overrightarrow{n'_e}.$$

We take:

$$\overrightarrow{v'} = v' \overrightarrow{x} e^{i(kx-\omega t)}, \quad n'_e = n'_e \overrightarrow{x} e^{i(kx-\omega t)}, \quad \overrightarrow{E'} = E' \overrightarrow{x} e^{i(kx-\omega t)}.$$

Considering a geometry with $\overrightarrow{k} = \overrightarrow{k}_x$, $\overrightarrow{B}_0 = \overrightarrow{z} B_0$, $\overrightarrow{E'} = \overrightarrow{x} E'$, from the x,y,z equations of motion, we have: $v'_z = 0$, and $v'_x = \frac{eE'/i\omega m_e}{1-\omega_c^2/\omega^2}$, where: $\omega_c = \frac{eB_0}{m_e}$. Now, substituting into Mass, as: $n'_e = \frac{k}{\omega} n_0 v'_x$, with v'_x from above, and n'_e from Field $\epsilon_0 \frac{\partial E'}{\partial x} = en'_e$, we get:

$$\epsilon_0 E' ik = -e\frac{k}{\omega} n_0 \cdot \frac{eE/i\omega m_e}{1-\omega_c^2/\omega^2} \Rightarrow \omega^2 = \omega_p^2 + \omega_c^2 \equiv \omega_{uh}^2,$$

so this wave behavior indicates a functional form as a combination of plasma frequency and cyclotron frequency. In fact, the analysis due to Spitzer (1962) identified this frequency of a longitudinal electron wave as one of two combined wave resonances in the plasma and termed it the upper hybrid frequency.

Longitudinal Ion Oscillations Perpendicular to $\overrightarrow{B}$

In attempting to identify basic wave behavior, in this case with ions, we allow ion longitudinal waves generated with electric fields by charge separation (from electrons) due to the magnetic field and make the assumption: $T_i \approx 0$; this is called the cold plasma limit (Boyd and Sanderson, 1969). Waves are presumed to move in the x-direction, and we take: n_0, $\vec{B}_0 = \vec{x} B_0$ as constant and $\vec{v}_0$, $\vec{E}_0$ are both zero. With electrostatic waves and $\vec{k} \times \vec{E} = 0$, we have $\vec{E} = -\nabla\varphi$. We use the primed notation (′) for (small) perturbation values, as above. The equation of motion for the ions (Chen, 1984) is then written: $m_i \frac{\partial \vec{v_i'}}{\partial t} = -e\nabla\varphi' + e\vec{v_i'} \times B_0$.

Writing the equations of motion for the x,y directions and solving for perturbation velocity in the direction of propagation, we get:

$$v'_{ix} = \frac{ek}{m_i\omega}\varphi'\left(1 - \frac{\Omega_i^2}{\omega^2}\right),$$

where, $\Omega_i = \frac{eB_0}{m_i}$, the ion cyclotron frequency.

We take: $n_i = n_e$, and assuming that electrons are also perturbed with the field creation (as above) we get: $n' = n_0 \frac{e\varphi'}{kT_e}$. We substitute in mass conservation: $n'_i = n_0 \frac{k}{\omega} v'_{ix}$ to get: $\omega^2 = \Omega_i^2 + k^2 \frac{kT_e}{m_i}$, as the dispersion frequency of this longitudinal ion wave; the wave is termed the ion cyclotron oscillation, for motion perpendicular to $\vec{B}$.

Alfven Waves—Propagation of Magnetic Perturbations Along $\vec{B}_0$

Introduction

We have been examining the fluid characteristics of plasma behavior. In OFM, the fluid wave that is most evident is the sound wave. It is a longitudinal wave that emanates from a pressure difference and that propagates outward at a speed that is related to the random speed of the molecules. A sound wave also involves perturbations in density; if a fluid is incompressible ($\rho =$ constant), the analytically indicated speed of this wave would be infinite. In practical experiment, higher density fluids do evidence higher sound speed.

Now we will examine a wave that is based on a perturbation of the magnetic field; it is of considerable significance for plasma dynamics. As we have seen above, there are physical laws that connect perturbations in magnetic field with perturbations in density. This wave is distinctly different from a sound wave; however, the two behaviors can be coupled in a plasma. Eventually we will introduce the complexity that can develop because of the simultaneous existence of both of these types of waves. First, we examine the physical situation that is predicted when we ignore the existence of sound waves.

Propagation of Magnetic Perturbations in Plasma

We had noted above that for plasma in the presence of magnetic field:

$$\nabla\left(p+\frac{B^2}{2\mu}\right) - \rho(\overrightarrow{V_A}\cdot\nabla)\overrightarrow{V_A} = 0,$$

which was derived from the momentum equation with $v_{flow} = 0$ and where $V_A = (B^2/\mu\rho)^{1/2}$. But in OFM, the momentum equation written with flow velocity, $\overrightarrow{v}$, gives:

$$\nabla p + \rho(\overrightarrow{v}\cdot\nabla)\overrightarrow{v} = 0.$$

So in plasma, with the magnetic pressure effectively interrelating with the kinetic pressure, V_A, Alfven speed, takes on a role (with negative sign) in momentum relationships. So, clearly, it is important to understand the Alfven speed and its relationship to physical behaviors that influence the broader spectrum of fluid and field interactions.

The derivation of the Alfven speed and discussion of its physical significance was discussed early by Cowling (1957) and Spitzer (1962). Here, for simplicity, we take a limiting case that assumes an incompressible (ρ = constant), perfectly conducting ($\sigma = \infty$) fluid and apply a perturbation analysis (following Boyd and Sanderson, 1969). The governing equations are as follows:

Mass:

$$\frac{\partial \rho}{\partial t} + \nabla\cdot(\rho\overrightarrow{v}) = 0,$$

And for steady flow:

$$\nabla\cdot\overrightarrow{v} = 0.$$

Momentum:

$$\rho\frac{D\overrightarrow{v}}{Dt} = -\nabla p + \overrightarrow{J}\times\overrightarrow{B} = -\nabla p - \overrightarrow{B}\times\frac{\nabla\times\overrightarrow{B}}{\mu}.$$

Inductance:

$$\frac{\partial\overrightarrow{B}}{\partial t} = \nabla\times(\overrightarrow{v}\times\overrightarrow{B}).$$

We take perturbed field and fluid conditions: $\vec{B} = \vec{B}_0 + \vec{B}'$; $\vec{v} = \vec{v}'$ $(\vec{v}_0 = 0)$; and $p = p_0 + p'$. Substituting these in momentum, and retaining first-order terms, we get:

$$\rho\left[\frac{\partial \vec{v}'}{\partial t} + (\vec{v}' \cdot \nabla)\vec{v}'\right] = -\nabla p' - (\vec{B}_0 + \vec{B}') \times \frac{\nabla \times (\vec{B}_0 + \vec{B}')}{\mu},$$

so:

$$\rho\left[\frac{\partial \vec{v}'}{\partial t}\right] = -\nabla p' - \frac{\vec{B}_0}{\mu} \times (\nabla \times \vec{B}'),$$

or:

$$\rho\left[\frac{\partial \vec{v}'}{\partial t}\right] = -\nabla p' - \frac{\nabla(\vec{B}_0 \cdot \vec{B}')}{\mu} + \frac{(\vec{B}_0 \cdot \nabla)\vec{B}'}{\mu}.$$

First, we take $\nabla \cdot$ equation to get: $\rho \frac{\partial}{\partial t}(\nabla \cdot \vec{v}') = -\nabla^2\left(p' + \frac{\vec{B}_0 \cdot \vec{B}'}{\mu}\right) + \frac{(\vec{B}_0 \cdot \nabla)(\nabla \cdot \vec{B}')}{\mu}$, and as $\nabla \cdot \vec{v}' = 0$, $\nabla \cdot \vec{B}' = 0$: $\therefore \nabla^2\left(p' + \frac{\vec{B}_0 \cdot \vec{B}'}{\mu}\right) = 0$.

Physically, this means that perturbations diminish to zero at a distance away from the source, and mathematically, the term must be zero everywhere, so, for momentum: $\rho \frac{\partial \vec{v}'}{\partial t} = \frac{(\vec{B}_0 \cdot \nabla)\vec{B}'}{\mu}$.

We want to solve for either B' *or* v', so we begin with the induction relationship:

$\frac{\partial \vec{B}}{\partial t} = \nabla \times (\vec{v} \times \vec{B}) = (\vec{B} \cdot \nabla)\vec{v} - (\vec{v} \cdot \nabla)\vec{B}$, and with perturbations:

$$\frac{\partial(\vec{B}_0 + \vec{B}')}{\partial t} = \left[(\vec{B}_0 + \vec{B}') \cdot \nabla\right]\vec{v}' - (\vec{v}' \cdot \nabla)(\vec{B}_0 + \vec{B}'),$$

and retaining first order terms: $\frac{\partial \vec{B}'}{\partial t} = (\vec{B}_0 \cdot \nabla)\vec{v}'$.
In order to combine, we take the derivative of momentum, as:

$$\rho \frac{\partial^2 \vec{v}'}{\partial t^2} = \frac{(\vec{B}_0 \cdot \nabla)}{\mu}\frac{\partial \vec{B}'}{\partial t} = \frac{(\vec{B}_0 \cdot \nabla)}{\mu}(\vec{B}_0 \cdot \nabla)\vec{v}' = \frac{(\vec{B}_0 \cdot \nabla)^2}{\mu}\vec{v}'.$$

If we take $\vec{B}_0$ along the z-axis, we get: $\frac{\partial^2 \vec{v}'}{\partial t^2} = \frac{B_0^2}{\rho\mu}\frac{\partial^2 \vec{v}'}{\partial z^2} = V_A^2 \frac{\partial^2 \vec{v}'}{\partial z^2}$.

Figure 8.8 Propagation of hydromagnetic wave perturbations with Alfven speed.

This is the wave equation for disturbances $\left(\overrightarrow{v'}\right)$ propagating in the z-direction with velocity, V_A.

To clarify the nature of the disturbances in $B'$and v', from the induction equation we take:

$$\frac{\partial \overrightarrow{B'}}{\partial t} = \nabla \times \left(\overrightarrow{v'} \times \overrightarrow{B_0}\right) \Rightarrow \overrightarrow{v'} \perp \overrightarrow{B_0} \text{ where } \overrightarrow{B'} \perp \overrightarrow{B_0},$$

and from the combination of induction and fluid equations:

$$\frac{\partial \overrightarrow{B'}}{\partial t} = (\overrightarrow{B}_0 \cdot \nabla)\overrightarrow{v'} \Rightarrow \overrightarrow{B'} \parallel \overrightarrow{v'}, \overrightarrow{v'} \perp \overrightarrow{B}_0.$$

This describes a transverse wave. So if we envisage magnetic field lines as strings in tension, the physical behavior is comparable to striking a string and creating propagating sinusoidal oscillations, as shown in Figure 8.8.

Recall that we have assumed an incompressible fluid (where we have noted that sound waves would have very high speed), $\rho' = 0$, and so these magnetic perturbation waves moving with Alfven speed would not have any interaction with compression wave effects:

$$v_{\mathrm{A}} \ll v_{\mathrm{s}}, \quad \text{and} \quad \frac{B^2}{\mu e} \ll \gamma \frac{p}{\rho}.$$

This physical situation would occur with weak fields or dense fluids.

FLUID WAVES AND SHOCK WAVES IN PLASMA

Waves in a Compressible Plasma Medium

Because of the possibilities of fluid and electromagnetic force interactions, there is a broader range of possible fluid waves in plasma with applied magnetic fields (Cowling, 1957; Spitzer, 1962; Jeffrey, 1966). By introducing compressibility into the equations, we take a step to allow this coupling to become evident in the analysis. As above, we take the basic set of equations and generally follow the procedures of Boyd and Sanderson (1969):

Continuity:

$$\frac{\partial \rho}{\partial t} + \nabla \cdot (\rho \vec{v}) = 0;$$

Momentum:

$$\rho \frac{D\vec{v}}{Dt} = -\nabla p + \vec{J} \times \vec{B};$$

Maxwell:

$$\nabla \times \vec{B} = \mu \vec{J},$$

and:

$$\nabla \times \vec{E} = -\frac{\partial \vec{B}}{\partial t};$$

Ohm's law:

$$\frac{\vec{J}}{\sigma} = \vec{E} + \vec{v} \times \vec{B}.$$

The set of equations for the properties is completed by an energy relationship; as a simplifying condition we assume isentropic processes, as: $p\rho^{-\gamma} = \text{constant}$.

We begin with Maxwell's equations, as:

$$\nabla \times \vec{B} = \mu \vec{J}, \quad \text{and we take } \frac{\partial}{\partial t} \text{ to get:} \quad \nabla \times \frac{\partial \vec{B}}{\partial t} = \mu \frac{\partial \vec{J}}{\partial t},$$

and then substitute to get: $\nabla \times \left(\nabla \times \vec{E}\right) = -\mu \frac{\partial \vec{J}}{\partial t}$.

Then for: $\sigma \approx \infty$, we have: $\frac{\vec{J}}{\sigma} = \vec{E} + \vec{v} \times \vec{B} \approx 0$.

We again will carry out a solution with perturbation variables ($'$) for properties, as:

$$\vec{v} = \vec{v'},\ \vec{J} = \vec{J'},\ \vec{E} = \vec{E'},\ \vec{B} = \vec{B}_0 + \vec{B'},\ p = p_0 + p',\ \rho = \rho_0 + \rho',$$

to get:

$$\frac{\partial \rho'}{\partial t} + \rho_0 \nabla \cdot \vec{v'} = 0;$$

$$\nabla \times \left(\nabla \times \overrightarrow{E'}\right) = -\mu \frac{\partial \overrightarrow{J'}}{\partial t};$$

$$\rho_0 \frac{\partial \overrightarrow{v'}}{\partial t} = -\nabla p' + \overrightarrow{J'} \times \overrightarrow{B_0};$$

$$\overrightarrow{E'} + \overrightarrow{v'} \times \overrightarrow{B_0} = 0.$$

Now, we take plane wave variations in the $\overrightarrow{x}$-direction, where $\overrightarrow{x}$ can be viewed as arbitrary, and $\overrightarrow{k}$ is the wave number for waves in the $\overrightarrow{k}$ direction. So we have:

$$\overrightarrow{v'}(\overrightarrow{x}, t) = \overrightarrow{v'} e^{i \overrightarrow{k} \cdot \overrightarrow{x} - i\omega t} \equiv \overrightarrow{v'} \varphi, \text{ and similarly,}$$

$$\overrightarrow{J'}(\overrightarrow{x}, t) = \overrightarrow{J'} \varphi; \quad \overrightarrow{E'}(\overrightarrow{x}, t) = \overrightarrow{E'} \varphi; \quad p'(\overrightarrow{x}, t) = p' \varphi; \quad \rho'(\overrightarrow{x}, t) = \rho' \varphi.$$

Then on substituting into Maxwell's equations:

$$\nabla \times \left[\nabla \times \left(\overrightarrow{E'} \varphi\right)\right] = \nabla\left[\nabla \cdot \left(\overrightarrow{E'} \varphi\right)\right] - \nabla^2\left(\overrightarrow{E'} \varphi\right) = -\mu \frac{\partial\left(\overrightarrow{J'} \varphi\right)}{\partial t},$$

$$\nabla\left[\varphi \nabla \cdot \overrightarrow{E'} + \overrightarrow{E'} \nabla \varphi\right] - \nabla^2\left(\overrightarrow{E'} \varphi\right) = -\mu \frac{\partial\left(\overrightarrow{J'} \varphi\right)}{\partial t}, \text{ and with:}$$

$$\nabla \varphi = \nabla e^{i \overrightarrow{k} \cdot \overrightarrow{x} - i\omega t} = \varphi \nabla\left(i \overrightarrow{k} \cdot \overrightarrow{x} - i\omega t\right),$$

and:

$$\nabla(\overrightarrow{k} \cdot \overrightarrow{x}) = (\overrightarrow{k} \cdot \nabla)\overrightarrow{x} + (\overrightarrow{x} \cdot \nabla)\overrightarrow{k} + \overrightarrow{k} \times (\nabla \times \overrightarrow{x}) + \overrightarrow{x} \times (\nabla \times \overrightarrow{k})$$

$$= k \cdot \frac{d}{dx} x \overrightarrow{i} = \overrightarrow{i} k = \overrightarrow{k}.$$

So:

$$\nabla(\overrightarrow{E'} \varphi \cdot i \overrightarrow{k}) - \overrightarrow{E'} \varphi (ik)^2 = i\omega\mu \overrightarrow{J'} \varphi,$$

and:

$$i \overrightarrow{k} (\overrightarrow{E'} \varphi \cdot i \overrightarrow{k}) + k^2 \overrightarrow{E'} \varphi = i\omega\mu \overrightarrow{J'} \varphi,$$

then:

$$-\overrightarrow{k}(\overrightarrow{E'} \cdot \overrightarrow{k}) + k^2 \overrightarrow{E'} = i\omega\mu \overrightarrow{J'}.$$

Now substituting into momentum, we get:

$$\rho_0 \frac{\partial}{\partial t}(\overrightarrow{v'}\varphi) = -\nabla(p'\varphi) + (\overrightarrow{J'}\varphi) \times \overrightarrow{B_0},$$

$$\rho_0(-i\omega)\overrightarrow{v'}\varphi = -i\overrightarrow{k}p'\varphi + \varphi\overrightarrow{J'} \times \overrightarrow{B_0},$$

$$-i\omega\rho_0\overrightarrow{v'} = -i\overrightarrow{k}p' + \overrightarrow{J'} \times \overrightarrow{B_0}.$$

So we have: $\overrightarrow{v'} = fn\,(p', \overrightarrow{J'})$, but we would like to eliminate p' and get $\overrightarrow{v'} = fn\,(\overrightarrow{J'})$. First, substituting into Ohm's law: $\overrightarrow{E'}\varphi + (\overrightarrow{v'}\varphi) \times \overrightarrow{B_0} = 0$, or:

$$\overrightarrow{E'} + \overrightarrow{v'} \times \overrightarrow{B_0} = 0.$$

We also have the isentropic energy relationship: $p\rho^{-\gamma} = \text{constant}$. We take ∇ (equation) to get:

$$\rho^{-\gamma}\nabla p - p\gamma\rho^{-\gamma-1}\nabla\rho = 0,$$

or:

$$\rho\nabla p = \gamma p\nabla\rho;$$

then:

$$(\rho_0 + \rho')\nabla(p_0 + p') = \gamma(p_0 + p')\nabla(\rho_0 + \rho'),$$

to get:

$$\frac{\nabla p'}{p_0} = \gamma\frac{\nabla\rho'}{\rho_0}.$$

Finally, with continuity $\frac{\partial(\rho'\varphi)}{\partial t} = -\rho_0\nabla\cdot(\overrightarrow{v'}\varphi)$, get:

$$-i\omega\rho'\varphi = -\rho_0\overrightarrow{v'}\cdot(i\overrightarrow{k})\varphi.$$

We then have the relationship: $\rho' = \rho_0\frac{\overrightarrow{v'}\cdot\overrightarrow{k}}{\omega}$, which is a scalar with no time or space variation.

So we evaluate: $\nabla\rho' = \rho' i\overrightarrow{k}\varphi = i\overrightarrow{k}\rho_0\frac{\overrightarrow{v'}\cdot\overrightarrow{k}}{\omega}\varphi$, and, as:

$$\nabla p' = \gamma\frac{p_0}{\rho_0}\nabla\rho',$$

we get:

$$i\vec{k}p'\varphi = \gamma\frac{p_0}{\rho_0}i\vec{k}\rho_0\frac{\vec{v'}\cdot\vec{k}}{\omega}\varphi,$$

and so:

$$i\vec{k}p' = i\gamma p_0\frac{\vec{k}(\vec{v'}\cdot\vec{k})}{\omega}.$$

We then substitute into the momentum equation to get:

$$-i\omega\rho_0\vec{v'} = -i\vec{k}p' + \vec{J'}\times\vec{B_0},$$

and:

$$i\omega\rho_0\vec{v'} = i\gamma p_0\frac{\vec{k}(\vec{v'}\cdot\vec{k})}{\omega} - \vec{J'}\times\vec{B_0},$$

or:

$$\vec{v'} = \frac{\gamma p_0}{\rho_0}\frac{\vec{k}(\vec{v'}\cdot\vec{k})}{\omega^2} + i\frac{\vec{J'}\times\vec{B_0}}{\omega\rho_0} = c_s^2\frac{\vec{k}(\vec{v'}\cdot\vec{k})}{\omega^2} + i\frac{\vec{J'}\times\vec{B_0}}{\omega\rho_0},$$

where:

$$c_s^2 = \frac{\gamma p_0}{\rho_0}.$$

As a first case, we choose $\vec{k}$ (direction of wave propagation) in the x-direction (Figure 8.9), as:

Then, from above: $\vec{v'} = \frac{c_s^2\vec{k_x}(kv_x)}{\omega^2} + i\frac{\vec{J'}\times\vec{B_0}}{\omega\rho_0}$, and for this geometry:

$$v_x = \frac{c_s^2k^2v_x}{\omega^2} + \frac{i}{\omega\rho_0}J_yB_0\sin\theta,$$

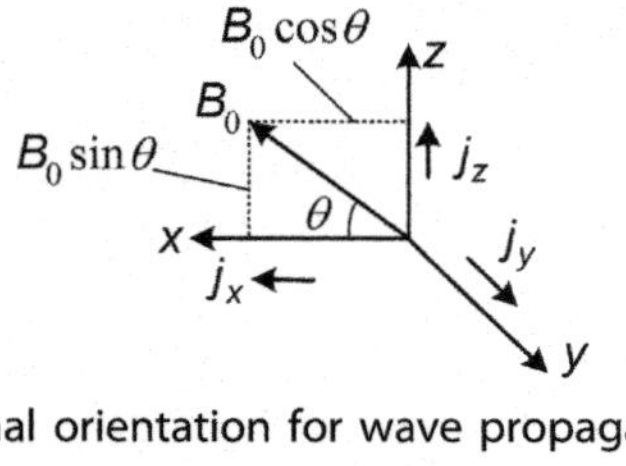

Figure 8.9 Directional orientation for wave propagation in x-direction.

or:

$$v_x\left(1-\frac{c_s^2k^2}{\omega^2}\right)=\frac{i}{\omega\rho_0}J_yB_0\sin\theta,$$

so:

$$v_x=\frac{iB_0}{\omega\rho_0}\frac{J_y\sin\theta}{1-\frac{c_s^2k^2}{\omega^2}}.$$

Similarly:

$$v_y=\frac{iB_0}{\omega\rho_0}(J_z\cos\theta-J_x\sin\theta);$$

$$v_z=-\frac{iB_0}{\omega\rho_0}J_y\cos\theta.$$

Then from Maxwell's equations, $k^2\overrightarrow{E'}-\overrightarrow{k}(\overrightarrow{k}\cdot\overrightarrow{E'})=i\omega\mu\overrightarrow{J'}$, and for this geometry:

$$k^2\overrightarrow{E'}-\overrightarrow{k}(kE_x)=i\omega\mu J',\quad \text{so for each direction:}$$

$$k^2E_x-k^2E_x=i\omega\mu J_x=0,\quad \text{so:}\quad J_x=0;$$

$$k^2E_y=i\omega\mu J_y,\quad \text{so,}\quad E_y=\frac{i\omega\mu J_y}{k^2};$$

$$k^2E_z=i\omega\mu J_z,\quad \text{so,}\quad E_z=\frac{i\omega\mu J_z}{k^2}.$$

Substitute into Ohm's law for: $\overrightarrow{E'}+\overrightarrow{v'}\times\overrightarrow{B_0}=0$, which for each direction gives:

X. $E_x+v_yB_z-v_zB_y=0$, or:

$$E_x+v_yB_0\sin\theta=0.$$

Y. $E_y-v_xB_z+v_zB_x=0$, and substituting:

$$\frac{i\omega\mu J_y}{k^2}-v_xB_0\sin\theta+v_zB_0\cos\theta=0,$$

so:

$$\frac{i\omega\mu J_y}{k^2}-\left(\frac{iB_0}{\omega\rho_0}\frac{J_y\sin\theta}{1-\frac{c_s^2k^2}{\omega^2}}\right)B_0\sin\theta+\left(-\frac{iB_0}{\omega\rho_0}J_y\cos\theta\right)B_0\cos\theta=0,$$

and:

$$\left[\frac{\omega\mu}{k^2} - \frac{B_0^2}{\omega\rho_0}\left(\frac{\sin^2\theta}{1-\frac{c_s^2k^2}{\omega^2}} + \cos^2\theta\right)\right]J_y = 0.$$

Z. $E_z + v_xB_y - v_yB_x = 0$, and substituting:

$$\frac{i\omega\mu J_z}{k^2} - \frac{iB_0}{\omega\rho_0}(J_z\cos\theta - J_x\sin\theta)B_0\cos\theta = 0,$$

so:

$$\left(\frac{\omega\mu}{k^2} - \frac{B_0^2}{\omega\rho_0}\cos^2\theta\right)J_z = 0.$$

First, we examine the relationship for the z-direction; we have:

$$\frac{\omega}{k^2} - \frac{B_0^2}{\omega\mu\rho_0}\cos^2\theta = 0, \quad \text{but as:} \quad \frac{B_0^2}{\mu\rho_0} = V_A^2, \quad \text{the Alfven velocity,}$$

then:

$$\frac{\omega}{k} = \pm V_A\cos\theta.$$

This is an oblique Alfven wave (x direction) with θ as the angle between $\vec{k}$ and $\vec{B}_0$. Further, notice that as $\cos\theta \leq 1.0$, the speed is less than the Alfven speed. Recall that we have the geometry (Figure 8.10):

With $V_A\cos\theta = \frac{(B_0\cos\theta)}{(\mu\rho)^{1/2}}$ and as $(B_0\cos\theta)$ is the component of B_0 in the direction of propagation, for:

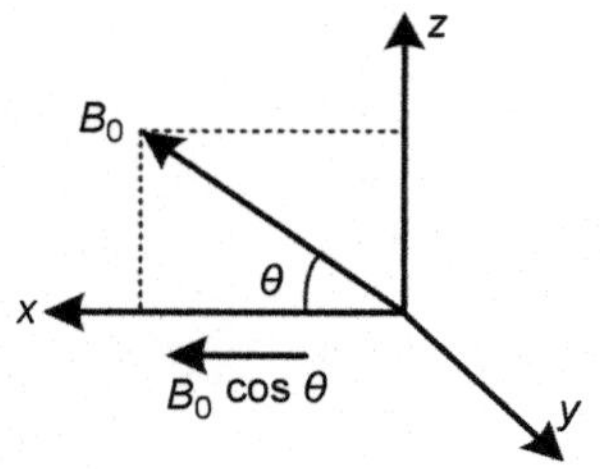

Figure 8.10 Geometry for wave propagation (angle θ) referred to magnetic field direction, $\vec{B}_0$.

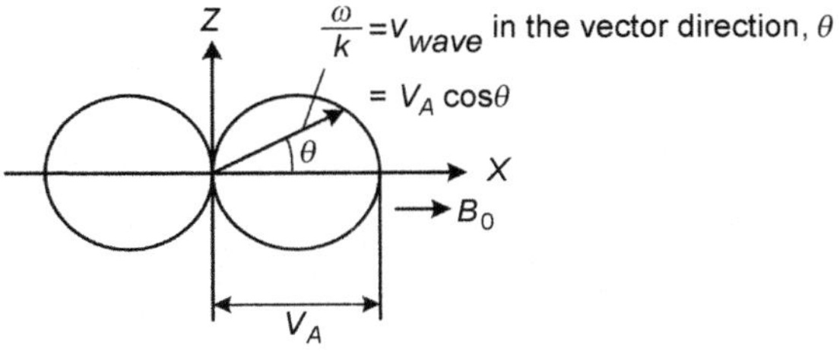

Figure 8.11 Geometry of oblique Alfven wave propagation in B_0 plane.

$$\theta = 90^\circ, \quad \cos\theta = 0, \quad \text{so,} \quad \frac{\omega}{k} = 0, \quad \text{no wave propagation;}$$

$$\theta = 0^\circ, \quad \cos\theta = 1, \quad \text{so,} \quad \frac{\omega}{k} = V_A, \quad \text{Alfven speed propagation.}$$

Alternatively, if we take the $\overrightarrow{B}$ field in the x-direction, but we consider propagation with arbitrary $\overrightarrow{k}$ (Figure 8.11), we have $\frac{\omega}{k} = v_{wave}$ in the vector direction, θ; so the wave travels at a speed that is a component of Alfven speed. Note that from: $E_z - v_y B_x = 0$ where B_x is applied, the wave involves E_z, v_y, transverse to B_x.

Second, let us now examine the relationship for the y-direction; for that we write, as above:

$$\frac{\omega}{k^2} - \frac{B_0^2}{\omega\mu\rho_0}\left(\frac{\sin^2\theta}{1-\frac{c_s^2k^2}{\omega^2}} + \cos^2\theta\right) = \frac{\omega^2}{k^2} - V_A^2\left(\frac{\sin^2\theta}{1-\frac{c_s^2k^2}{\omega^2}} + \cos^2\theta\right) = 0,$$

and:

$$\frac{\omega^2}{k^2}\left(1-\frac{c_s^2k^2}{\omega^2}\right) - V_A^2\sin^2\theta - V_A^2\left(1-\frac{c_s^2k^2}{\omega^2}\right)\cos^2\theta = 0,$$

so:

$$\frac{\omega^2}{k^2} - c_s^2 - V_A^2\sin^2\theta - V_A^2\cos^2\theta + V_A^2\frac{c_s^2k^2}{\omega^2}\cos^2\theta = 0,$$

to get:

$$\frac{\omega^4}{k^4} - \frac{\omega^2}{k^2}\left(c_s^2 + V_A^2\right) + c_s^2V_A^2\cos^2\theta = 0.$$

We can apply the quadratic formula to get:

$$\frac{\omega^2}{k^2} = \frac{1}{2}\left(c_s^2 + V_A^2\right) \pm \left[\frac{\left(c_s^2 + V_A^2\right)^2}{4} - c_s^2 V_A^2 \cos^2\theta\right]^{1/2},$$

or:

$$= \frac{1}{2}\left(c_s^2 + V_A^2\right)\left\{1 \pm \left[1 - \frac{4c_s^2 V_A^2 \cos^2\theta}{\left(c_s^2 + V_A^2\right)^2}\right]^{1/2}\right\}.$$

Now for: $\left(\frac{c_s V_A \cos\theta}{c_s^2 + V_A^2} \ll 1,\ c_s \ll V_A\right)$ or $(c_s \gg V_A,\ \cos\theta \ll 1)$, we get:

$\frac{\omega^2}{k^2} = \frac{1}{2}\left(c_s^2 + V_A^2\right)\left\{1 \pm \left[1 - \frac{2c_s^2 V_A^2 \cos^2\theta}{\left(c_s^2 + V_A^2\right)^2}\right]\right\}$, with two ($\pm$) solutions. We have two waves:

$\left.\frac{\omega^2}{k^2}\right|_+ = \frac{1}{2}\left(c_s^2 + V_A^2\right)\left\{1 + \left[1 - \frac{2c_s^2 V_A^2 \cos^2\theta}{\left(c_s^2 + V_A^2\right)^2}\right]\right\} = V_A^2 + c_s^2,\ (> V_A^2)$, a "fast" wave;

$\left.\frac{\omega^2}{k^2}\right|_- = \frac{1}{2}\left(c_s^2 + V_A^2\right)\frac{2c_s^2 V_A^2 \cos^2\theta}{\left(c_s^2 + V_A^2\right)^2} = \frac{c_s^2 V_A^2 \cos^2\theta}{c_s^2 + V_A^2} = \left(V_A^2\right)\frac{\cos^2\theta}{1 + \frac{V_A^2}{c_s^2}},\ (< V_A^2)$, a "slow" wave;

$\left.\frac{\omega^2}{k^2}\right|_{Alfven} = V_A^2$, is termed an "intermediate" wave.

The "fast" and "slow" waves involve: $E_y - v_x B_z + v_z B_x = 0$, where the second term is longitudinal, and the third term is transverse, so these waves have mixed components.

For $B_z = 0$, there is no v_x term so the wave is transverse only.

For $B_x = 0$, $\cos\theta = 0$, there is no v_z term so it is a longitudinal wave only.

Let us now consider the general case of wave direction relative to the $\vec{B}$ field. We have:

$\left.\frac{\omega}{k}\right|_+ = \pm\sqrt{c_s^2 + V_A^2} \neq f(\theta)$, constant in all directions, even $\perp$ to $\vec{B}$; and:

$$\left.\frac{\omega}{k}\right|_- = \pm\frac{c_s V_A}{\sqrt{c_s^2 + V_A^2}}\cos\theta = f(\theta).$$

The geometry of these waves is illustrated on the following diagram (Figure 8.12), called a Friedrich's diagram (Jeffrey, 1966).

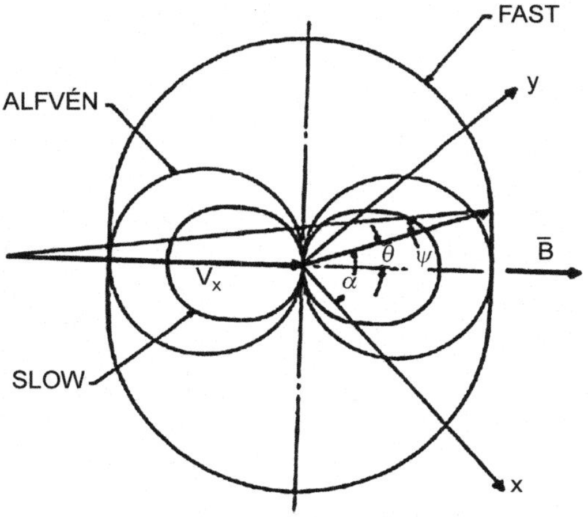

Figure 8.12 Phase diagram for fast, slow and (intermediate) Alfven plasma waves. *Anderson (1963), Figure 3.11, page 72, with permission.*

We can make the following observations with respect to these plasma waves:

1. Fast and slow waves have: $\frac{\omega}{k} = f(c_s, V_A)$, so they combine sound and Alfven speeds; these waves are termed magnetoacoustic or magneto-sonic waves.
2. For a fast wave: $v_{Fast} = \sqrt{c_s^2 + V_A^2}$, and for $V_A \gg c_s$, then the wave travels in all directions, perpendicular or parallel to $\overrightarrow{B}$, with this high speed.
3. For $\overrightarrow{k} \parallel \overrightarrow{B}_0$ there are three separate wave modes, two uncoupled and one coupled.

But what are these waves? (What is propagating?) The Alfven wave travels at intermediate speeds and is transverse. There is a density perturbation induced here, so is it related to shock wave formation? Magnetosonic waves involve Δp, and therefore these waves will steepen and can form MHD shocks, which have longitudinal and transverse components.

These questions regarding properties of plasma waves have important ramifications in all high-speed plasma flows. In OFM, there is a relatively simple physics: flows will behave differently below the speed of sound than above the speed of sound, and sonic flow introduces a singular, limiting condition. So, for plasmas, if there are different characteristic speeds for the propagation of disturbances, the determination of definable, limiting ("choking") conditions due to wave motion is an important consideration. In fact, in an attempt to interpret experimental results from a number of plasma devices, consideration has been given to the possibility of just such limiting speeds. However, in order to make a definitive determination of limiting speed in any configuration, very careful attention must be paid to the specific modes of wave propagation with respect to magnetic field orientation that are possible; only then is it reasonable to propose a correct analytical relationship for such a speed and predict experimental behavior. Unfortunately, there are few cases with direct experimental verification of limiting plasma speeds with clear reference to field and flow properties. This topic will be discussed further below and in the following chapter where data are discussed for specific experiments.

Shock Wave Formation and Plasma Flow Effects

As plasmas are formed in high-temperature gases, those gases possess high energy and enthalpy. The property of high enthalpy is characteristic of source conditions in high-speed flows encountered as an aspect of space flight. Earth reentry was one of the first physical experiences that involved hypersonic flow and the inherent problem of strong shock waves; the flow about the vehicle generated ionized plasma, and that flow had to be understood to control reentry.

As a separate problem, the existence of planetary shock waves as part of a magnetosphere structure created by the solar wind is another example of the occurrence and complexity of fluid and electromagnetic phenomena that is a natural component of the solar system environment. Some applied aspects of this topic will be discussed later; at this point, the physical problem of plasma shock wave formation will be considered, and basic analysis to recover shock wave relationships will be outlined.

Shock Waves in Ordinary Fluid Flow

Pressure disturbances propagate at sound speed. Strong local compressions emit waves that will steepen until the effects of viscosity and heat conduction establish an equilibrium of stresses that occurs across a shock wave (Courant and Friedrichs, 1948). As fluid passes through a shock wave, pressure, temperature, and density will increase; velocity will decrease. As the normal shock wave presents a one-dimensional flow configuration, it is an ideal phenomenon through which to study transport processes and flow

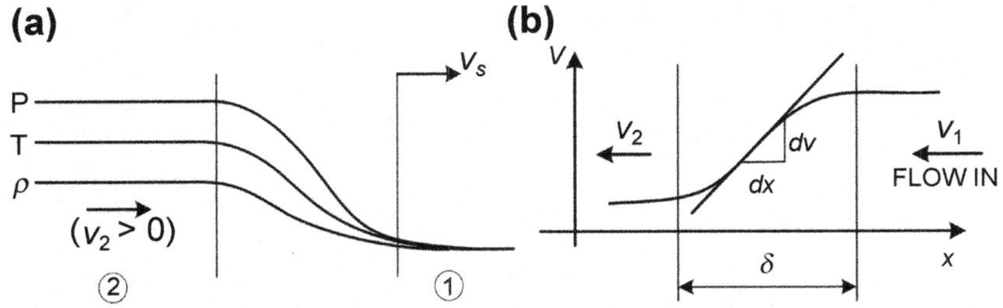

Figure 8.13 Fluid mechanical shock wave property transitions. (a) Propagating shock wave. (b) Flow through shock wave.

behavior. As the shock discontinuity is thin, velocity and temperature gradients are high and approach limiting values.

In steady, one-dimensional flow, steepening of waves due to pressure difference and inertia will be balanced by dispersion due to viscosity and heat conduction. In Figure 8.13(a), a shock wave propagating with speed V_s into a gas in state (1) induces changes in properties to state (2). In Figure 8.13(b), supersonic ($v_1 > a$, sound speed) flow from right to left encounters a normal shock wave and experiences a reduction in velocity (to v_2) across a distance, δ.

In the phenomenology, we can see:

$$\frac{dV}{dx} \rightarrow \frac{\Delta V}{0} \rightarrow \infty, \text{ and } \tau = \mu \frac{dV}{dx} \rightarrow \infty;$$

$$\frac{dT}{dx} \rightarrow \frac{\Delta T}{0} \rightarrow \infty, \text{ and } q = k \frac{dT}{dx} \rightarrow \infty.$$

The shock wave formation is driven by the pressure difference: $\Delta p = \frac{\dot{m}}{A} \Delta V$; the shock thickness is defined as $\delta \equiv \frac{|v_2 - v_1|}{(dv/dx)_{max}}$. We can express the steepening pressure gradient as:

$\frac{dp}{dx} = \rho V^* \left(\frac{dV}{dx}\right)$ (where V^* is wave velocity).

The dispersion due to stress is: $\frac{d\tau}{dx} = \frac{d}{dx}\left(\mu \frac{dV}{dx}\right)$, with $\tau_{xx} = \mu \frac{dV_x}{dx}$, so: $\rho V^* \left(\frac{dV}{dx}\right) = \mu \frac{d}{dx}\left(\frac{dV}{dx}\right)$, and integrating across the shock gradient region, we get shock thickness: $\delta = \frac{\mu}{\rho V^*}$, or: $\frac{\rho V^* \delta}{\mu} = 1$.

Now, kinetic theory showed: $\mu = \frac{1}{2} mn\bar{c}\lambda$, where, λ is mean free path and $\bar{c} \approx c_s$ (sound speed), so we have: $\delta = \frac{c_s \lambda}{2V^*} = \frac{1}{2} \frac{\lambda}{M^*}$, and so the thickness of a shock wave in ordinary fluids is on the scale of one mean free path (λ).

Shock Waves in Plasmas; Magnetic Field Effects

We have noted the existence of three types of waves in plasma with magnetic fields. These waves propagate as a fluid wave and may possibly steepen to form shock waves as in OFM. In the plasma medium with magnetic fields present we need to determine the following: (1) What are the speeds of (propagation or flow speeds where there is formation of) shock waves and (2) What are the changes in properties that occur across the shock waves in plasmas. In the following basic considerations, for simplicity we will assume one-dimensional flow: $f(x) \neq 0$, but $f(y) = f(z) = 0$.

1. Hydromagnetic shock equations

We consider the following orthogonal coordinate situation (Figure 8.14) with incoming flow in the x-direction $\perp$ to the shock and the flow downstream of the shock having possible components $\perp$ and $\|$ to the shock; the magnetic field upstream is generally taken to be $\perp$ or $\|$ to the shock and can have downstream components in all directions.

The basic equations of plasma dynamics and Maxwell's equations for this simplest situation are, as above, taken to be:

$$\frac{\partial \rho}{\partial t} + \nabla \cdot (\rho \vec{v}) = 0,$$

$$\rho \frac{D\vec{v}}{Dt} = -\nabla p + \vec{J} \times \vec{B},$$

$$\nabla \times \vec{B} = \mu \vec{J},$$

$$\nabla \times \vec{E} = -\frac{\partial \vec{B}}{\partial t},$$

$$\frac{\vec{J}}{\sigma} = \vec{E} + \vec{v} \times \vec{B}.$$

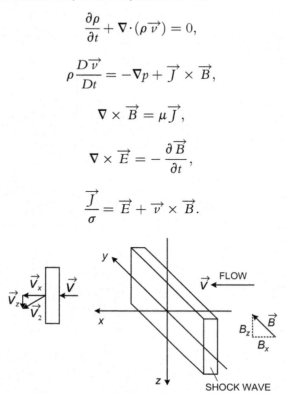

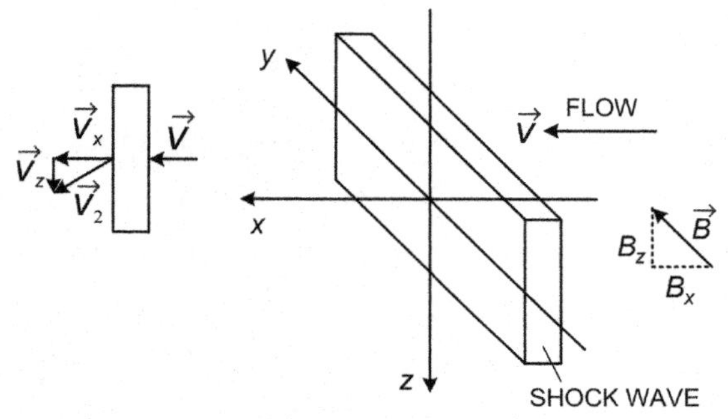

Figure 8.14 Coordinates and flow components for shock discontinuities.

The momentum equation above is not valid through the shock because of viscosity effects, but it is valid across the shock.

Here, we will follow the standard calculation formalism (Boyd and Sanderson, 1969; Sutton and Sherman, 1965). First, for Maxwell's equations, for steady flow we have ($\overrightarrow{\epsilon}_{y,z}$ are unit direction vectors):

$$\nabla \times \overrightarrow{E} = -\frac{\partial \overrightarrow{B}}{\partial t} = 0, \text{ or: } \overrightarrow{e}_y\left(-\frac{\partial E_z}{\partial x}\right) + \overrightarrow{e}_z\left(\frac{\partial E_y}{\partial x}\right) = 0$$

$$\Rightarrow \frac{\partial E_z}{\partial x} = 0, \ \frac{\partial E_y}{\partial x} = 0.$$

From Faraday (1–D): $-\frac{\partial B_z}{\partial x} = \mu j_y$, $\frac{\partial B_y}{\partial x} = \mu j_z$, and $\nabla \cdot \overrightarrow{B} = 0 \Rightarrow \frac{\partial B_x}{\partial x} = 0$.

Conservation of mass is: $\frac{\partial \rho}{\partial t} + \nabla \cdot (\rho \overrightarrow{v}) = 0$, and for steady flow is: $\frac{d(\rho v_x)}{dx} = 0$, which can be integrated to provide a jump relationship.

Conservation of momentum can be written: $\frac{\partial \overrightarrow{G}}{\partial t} = -\nabla \cdot \overleftrightarrow{\Pi}$, where: $\overrightarrow{G} = \rho \overrightarrow{v} + \overrightarrow{E} \times \overrightarrow{B}$;

$\overleftrightarrow{\Pi}$ is a tensor, as $\overleftrightarrow{\Pi} = \Pi_{ij} = \rho v_i v_j + p\delta_{ij} - \frac{1}{\mu_0} T_{ij}$, with $T_{ij} = B_i B_j - \frac{1}{2}\delta_{ij} B^2$.

For steady flow: $\frac{d\Pi_{xy}}{dx} = 0$.

Conservation of energy can be written: $\frac{\partial W}{\partial t} = -\nabla \cdot \overrightarrow{S}$, where:

$$W = \frac{3}{2}nkT + \frac{1}{2}\rho v^2 + \frac{1}{2}(E^2 + B^2),$$

and:

$$\overrightarrow{S} = \frac{\overrightarrow{E} \times \overrightarrow{B}}{\mu} + \overleftrightarrow{p} \cdot \overrightarrow{v} + \overrightarrow{v}\left(\frac{3}{2}nkT + \frac{1}{2}\rho v^2\right),$$

with:

$$\overleftrightarrow{p} = p\delta_{ij}, \quad \overleftrightarrow{Q} = 0.$$

For steady flow, energy becomes: $\frac{dS_x}{dx} = 0$.

In the solution, generally we assume that all conditions upstream of the shock (1) are known. Based on the above, the equations for the change of properties across the shock are:

$$\frac{dB_x}{dx} = 0, \quad B_{x1} = B_{x2};$$

$$\frac{dE_y}{dx} = \frac{d}{dx}(v_z B_x - v_x B_z) = 0, \qquad (v_z B_x - v_x B_z)_1 = (v_z B_x - v_x B_z)_2;$$

$$\frac{dE_z}{dx} = \frac{d}{dx}\left(v_x B_y - v_y B_x\right) = 0, \qquad \left(v_x B_y - v_y B_x\right)_1 = \left(v_x B_y - v_y B_x\right)_2;$$

$$\frac{d(\rho v_x)}{dx} = 0, \quad (\rho v_x)_1 = (\rho v_x)_2;$$

$$\frac{d\Pi_{xx}}{dx} = 0, \quad \left[\rho v_x^2 + p - \frac{1}{\mu_0}\left(B_x^2 - \frac{1}{2}B^2\right)\right]_1 = [\ ]_2, \text{ and}$$

$$\left[\rho v_x^2 + p + \frac{1}{\mu_0}\left(\frac{1}{2}B_y^2 - \frac{1}{2}B_z^2\right)\right]_1 = [\ ]_2;$$

$$\frac{d\Pi_{xy}}{dx} = 0, \quad \left(\rho v_x v_y - \frac{B_x B_y}{\mu_0}\right)_1 = \left(\rho v_x v_y - \frac{B_x B_y}{\mu_0}\right)_2;$$

$$\frac{d\Pi_{xz}}{dx} = 0, \quad \left(\rho v_x v_z - \frac{B_x B_z}{\mu_0}\right)_1 = \left(\rho v_x v_z - \frac{B_x B_z}{\mu_0}\right)_2;$$

and $[S_x]_1 = [S_x]_2$, where:

$$S_x = \frac{1}{\mu}\left(E_y B_z - E_z B_y\right) + p v_x + v_x\left(\frac{3}{2}nkT + \frac{1}{2}\rho v^2\right).$$

As: $\frac{\text{Energy}}{\text{particle}} = \frac{1}{2}m\bar{v}^2 = \frac{3}{2}kT$; $\frac{\text{Energy}}{\text{vol.}} == \frac{3}{2}nkT$; $I = \frac{\text{Energy}}{\text{Mass}} = \frac{1}{2}\bar{v}^2 = \frac{3}{2}\frac{nkT}{\rho}$, so:

$$I\rho = \frac{3}{2}nkT = \frac{3}{2}p.$$

If we have random translation energy only: $C_V = \frac{3}{2}R$, and, $I\rho = \frac{pC_V}{R} = \frac{pC_V}{C_p - C_V} = \frac{p}{\gamma - 1}$.

From Ohm's law: $\frac{\vec{J}}{\sigma} = \vec{E} + \vec{v} \times \vec{B}$; the first term is taken zero by assuming no current flow except in the shock. Therefore: $\vec{E} = -\vec{v} \times \vec{B}$, and, to recover terms in the energy equation:

$$E_y = -(-v_x B_z + v_z B_x) = v_x B_z - v_z B_x;$$

$$E_z = -\left(v_x B_y - v_y B_x\right) = -v_x B_y + v_y B_x.$$

Then, multiplying each:

$$E_y B_z = v_x B_z^2 - v_z B_x B_z,$$

and:

$$E_z B_y = -v_x B_y^2 + v_y B_x B_y,$$

so, subtracting:

$$E_y B_z - E_z B_y = v_x B_z^2 - v_z B_x B_z + v_x B_y^2 - v_y B_x B_y,$$

so:

$$= v_x\left(B_y^2 + B_z^2\right) - B_x\left(v_y B_y + v_z B_z\right).$$

So the energy equation provides the final equation needed for solution, as:

$$\left[\frac{v_x}{\mu}\left(B_y^2 + B_z^2\right) - \frac{B_x}{\mu}\left(v_y B_y + v_z B_z\right) + p v_x + v_x \rho I + \frac{1}{2}\rho v^2 v_x\right]_1 = [\ ]_2.$$

With all the initial conditions at (1) given, we have eight equations for the eight variables (with one to two across the shock), as:

$$\vec{B}\left(B_x, B_y, B_z\right); \quad \vec{v}\left(v_x, v_y, v_z\right); \quad p; \quad \rho.$$

First, we recall from OFM that a Hugoniot condition can be written:

$$(I_2 - I_1) + \frac{1}{2}(p_1 + p_2)\left(\frac{1}{\rho_2} - \frac{1}{\rho_1}\right) = 0,$$

which is a statement of: (change in internal energy ~ average work done). So we can write:

$$\frac{1}{\gamma - 1}\left(\frac{p_2}{\rho_2} - \frac{p_1}{\rho_1}\right) + \frac{1}{2}(p_1 + p_2)\left(\frac{1}{\rho_2} - \frac{1}{\rho_1}\right) = 0,$$

which expresses: $(p_2,\rho_2) = f(p_1,\rho_1)$.

It can be shown (Boyd and Sanderson, 1969) that there is an equivalent relationship for MHD shocks, as:

$$(I_2 - I_1) + \frac{1}{2}(p_1 + p_2)\left(\frac{1}{\rho_2} - \frac{1}{\rho_1}\right) + \frac{1}{4\mu}\left(B_2^2 - B_1^2\right)\left(\frac{1}{\rho_2} - \frac{1}{\rho_1}\right) = 0.$$

This clearly demonstrates the interrelationship of the fluid (plasma) and magnetic field properties. This result allows the recovery of the form for ordinary fluid mechanics when: $B = 0$. Also, if $B_1 = B_{1x}$, $B_2 = B_{2x}$, i.e., if we say that there is no tangent field after the shock, from Maxwell's equation, $B_{2x} = B_{1x}$; therefore, if $B_2 = B_{2x}$ the equation reduces to that for OFM. So the possibility of the transformation of magnetic field components in the plasma shock wave processes will prove to be a unique characteristic of plasma interactions.

The above equation means that for shocks with $\vec{k} \parallel \vec{B}$, the result will be different in plasma. As in *OFM*, shocks will steepen until a balance with dissipation is reached, i.e., $S_2 \geq S_1$. Also, it can be shown for MHD shocks that: $p_2 \geq p_1$, i.e., shocks are compressive.

2. Shock propagation parallel to magnetic field.

In this case, the plasma flow and magnetic fields are aligned in the same direction, normal to the shock wave. The geometry in this case is shown in Figure 8.15.

We choose the y, z axes to define the plane of the shock, as:

$$\vec{B}_1 = B_1\vec{i} + 0\vec{j} + 0\vec{k}; \quad \vec{B}_2 = B_x\vec{i} + B_y\vec{j} + 0\vec{k};$$

$$\vec{v}_1 = v_1\vec{i} + 0\vec{j} + 0\vec{k} = -v\vec{i}; \quad \vec{v}_2 = v_x\vec{i} + v_y\vec{j} + v_z\vec{k}.$$

Now the equations for the x, y directions are:

$$B_{x2} = B_{x1} \quad \text{or:} \quad B_x = B_1;$$

$$-v_{x1}B_{y1} - v_{y1}B_{x1} = v_{x2}B_{y2} - v_{y2}B_{x2},$$

or:

$$0 = v_xB_y - v_yB_x,$$

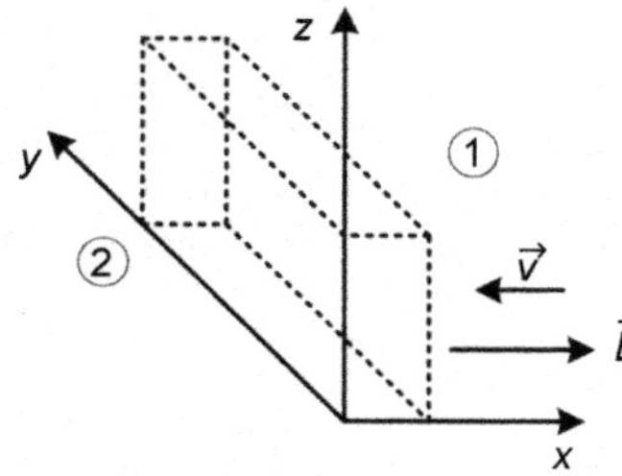

Figure 8.15 Geometry and field orientation for shock perpendicular to $\vec{B}_0$.

and:

$$v_x B_y = v_y B_x.$$

With momentum: $\rho_1 v_{x1} v_{y1} - \frac{B_{x1} B_{y1}}{\mu} = \rho_2 v_{x2} v_{y2} - \frac{B_{x2} B_{y2}}{\mu}$, or: $0 = \rho_2 v_x v_y - \frac{B_1 B_y}{\mu}$.

Then substituting: $\rho_2 v_x \left(\frac{v_x B_y}{B_1}\right) - \frac{B_1 B_y}{\mu} = 0$, we get: $\left(v_x^2 - \frac{B_1^2}{\mu \rho_2}\right) B_y = 0$.

This has two possible solutions: (a) $B_y = 0$, or: (b) $v_x^2 = \frac{B_1^2}{\mu \rho_2}$.

1. We consider the first possible solution: $B_{y2} = B_y = 0$.
Then $\overrightarrow{B}_2 = B_{x2} \overrightarrow{i} + 0 \overrightarrow{j} + 0 \overrightarrow{k} = B_{x1} \overrightarrow{i} + 0 \overrightarrow{j} + 0 \overrightarrow{k} = B_1$.
From x,y momentum: $0 = \rho_2 v_x v_y - \frac{B_1 B_y}{\mu} = \rho_2 v_x v_y$,
and as $\rho_2 v_x \neq 0$, $\therefore v_y = 0$.
From x,z momentum: $0 = \rho_2 v_{x2} v_{z2} - \frac{B_{x2} B_{z2}}{\mu}$, but: $B_{z2} = 0$, so:

$$\rho_2 v_{x2} v_{z2} = 0, \quad \text{but: } \rho_2 v_{x2} \neq 0$$

$$\therefore v_z = 0, \quad \overrightarrow{v}_2 = v_x \overrightarrow{i}$$

So, in this case, we get the following equations to govern the changes across the shock:

Continuity: $(\rho v_x)_1 = (\rho v_x)_2$;
x,x momentum: $(\rho v_x^2 + p)_1 = (\rho v_x^2 + p)_2$;
Energy: $\left(p v_x + \rho v_x I + \frac{1}{2} \rho v^2 v_x\right)_1 = \left(p v_x + \rho v_x I + \frac{1}{2} \rho v^2 v_x\right)_2$.
These are the equations of a shock wave in *OFM* with:

$$v_1 > a_1, \quad v_2 < v_1, \quad p_2 > p_1.$$

2. We now consider the second possible solution: $B_y \neq 0$, and $v_x^2 = \frac{B_1^2}{\mu \rho_2}$.
Since: $B_{x1} = B_{x2} = B_1$, then $\overrightarrow{B}_2 = B_1 \overrightarrow{i} + B_y \overrightarrow{j} + 0 \overrightarrow{k}$.
With the x,y equations again, we have:

Momentum: $\rho_1 v_{x1} v_{y1} - \frac{B_{x1} B_{y1}}{\mu} = \rho_2 v_{x2} v_{y2} - \frac{B_{x2} B_{y2}}{\mu}$, and as: $v_{y1} = 0$, $B_{y1} = 0$,

$$0 = \rho_2 v_x v_y - \frac{B_1 B_y}{\mu},$$

but:

$$\rho_2 v_x = \rho_1 v_1,$$

and we substitute to get: $v_y = \frac{B_1 B_y}{\mu \rho_2 v_x} = \frac{B_1 B_y}{\mu \rho_1 v_1} = f(B_y) \neq 0.$

To solve for B_y we take the following:
x,x momentum: $\rho_1 v_{x1}^2 + p_1 = \rho_2 v_{x2}^2 + p_2 + \frac{1}{\mu}\frac{B_{y2}^2}{2}$;
Energy:
$p_1 v_{x1} + \rho_1 v_{x1} I_1 + \frac{1}{2}\rho_1 v_{x1}^3 = \frac{v_{x2} B_y^2}{\mu} - \frac{v_{y2} B_x B_y}{\mu} + p_2 v_{x2} + \rho_2 v_{x2} I_2 + \frac{1}{2}\rho_2 v_2^2 v_{x2}$, or:

$$v_1\left(p_1 + \frac{p_1}{\gamma - 1}\right) + \frac{1}{2}\rho_1 v_1^3 = v_x\left(p_2 + \frac{p_2}{\gamma - 1}\right) + \frac{v_x B_y^2}{\mu} - \frac{v_y B_x B_y}{\mu} + \frac{1}{2}\rho_2 v_2^2 v_x,$$

and:

$$\frac{\gamma p_1}{(\gamma - 1)\rho_1} + \frac{1}{2}v_1^2 = \frac{\gamma p_2}{(\gamma - 1)\rho_2} + \frac{B_y^2}{\mu \rho_2} - \frac{v_y B_x B_y}{\mu \rho_2 v_x} + \frac{1}{2}\left(v_x^2 + v_y^2\right).$$

We solve for p_2 from the momentum equation, as:
$p_2 = \rho_1 v_1^2 + p_1 - \rho_2 v_x^2 - \frac{1}{\mu}\frac{B_{y2}^2}{2}$, to substitute into energy,
and using: $\rho_1 v_1 = \rho_2 v_x$; $v_x^2 = \frac{B_1^2}{\mu \rho_2}$; $v_y = \frac{B_1 B_y}{\mu \rho_1 v_1}$, we combine, to get:

$$B_y^2 = 2\left(\frac{\rho_2}{\rho_1} - 1\right)B_1^2\left[\frac{(\gamma + 1) - \frac{\rho_2}{\rho_1}(\gamma - 1)}{2} - \frac{\mu \gamma p_1}{B_1^2}\right],$$

and:

$$B_y = \pm\left[2\left(\frac{\rho_2}{\rho_1} - 1\right)B_1^2\right]^{1/2}\left[\frac{(\gamma + 1) - \frac{\rho_2}{\rho_1}(\gamma - 1)}{2} - \frac{\mu \gamma p_1}{B_1^2}\right]^{1/2}.$$

Taking the real part of the radical, where, $\frac{(\gamma + 1) - \frac{\rho_2}{\rho_1}(\gamma - 1)}{2} - \frac{\mu \gamma p_1}{B_1^2} > 0$,

we get a solution only when: $\frac{(\gamma + 1) - \frac{\rho_2}{\rho_1}(\gamma - 1)}{2} > \frac{\mu \gamma p_1}{B_1^2} = \frac{c_s^2 (sound\ speed)^2}{V_A^2\ (Alfven\ speed)^2}$.

So, physically, there is a solution only when: V_A (*Alfven speed*) $> c_s$ (*sound speed*).

Further: $v_x^2 = \frac{B_1^2}{\mu_0 \rho_2} \cdot \frac{\rho_2}{\rho_2} \Rightarrow v^2 = \frac{B_1^2 \rho_2}{\mu_0 \rho_1 \rho_1} = V_A^2 \frac{\rho_2}{\rho_1}$, and since for a "real" shock wave: $\frac{\rho_2}{\rho_1} > 1$,

$$\therefore v^2 > V_A^2 > c_s^2.$$

The result that we have derived mathematically can be described physically as follows:

$$\overrightarrow{B_1} = B_{x1}\overrightarrow{i} + 0\overrightarrow{j} + 0\overrightarrow{k}; \quad \overrightarrow{B}_2 = B_{x2}\overrightarrow{i} + B_{y2}\overrightarrow{j} + 0\overrightarrow{k};$$

and:

$$\overrightarrow{v_1} = v_{x1}\overrightarrow{i} + 0\overrightarrow{j} + 0\overrightarrow{k}; \quad \overrightarrow{v_2} = v_{x2}\overrightarrow{i} + v_{y2}\overrightarrow{j} + 0\overrightarrow{k}.$$

Graphically, this unique behavior is shown in Figure 8.16.

In this case, velocity and magnetic field components perpendicular to the upstream conditions are "switched on" by the normal shock wave, and the MHD waves are referred to as "switch-on" shocks. Recognize that for such waves to occur physically there must be current flow in the shock wave, as: $J_z = \frac{\partial B_y}{\partial x}$. So currents must be induced to flow by the external physical structure or Maxwell stresses must be actively involved. Under appropriate conditions and geometry, there are comparable solutions for "switch-off" shocks.

3. Shock propagation perpendicular to magnetic field.

In this case, we have the following geometry (Figure 8.17) with a shock wave positioned at the coordinates origin and magnetic field direction parallel to the shock:

Take: $\overrightarrow{B_1} = \overrightarrow{B_y} = 0\overrightarrow{i} + B\overrightarrow{j} + 0\overrightarrow{k}$, and with $B_{x1} = B_{x2}$, we have: $\overrightarrow{B_2} = 0\overrightarrow{i} + B_y\overrightarrow{j} + B_z\overrightarrow{k}$.

Now with $\rho_1 v_{x1} = \rho_2 v_{x2}$, then $\rho_1 v_1 = \rho_2 v_x$, so, $v_x = \frac{\rho_1}{\rho_2} v_1 \equiv \left(\frac{1}{\alpha}\right) v_1$, and with $\overrightarrow{v_1} = v_1\overrightarrow{i} + 0\overrightarrow{j} + 0\overrightarrow{k}$ $(v_1 = -v)$, we write: $\overrightarrow{v_2} = v_x\overrightarrow{i} + v_y\overrightarrow{j} + v_z\overrightarrow{k}$.

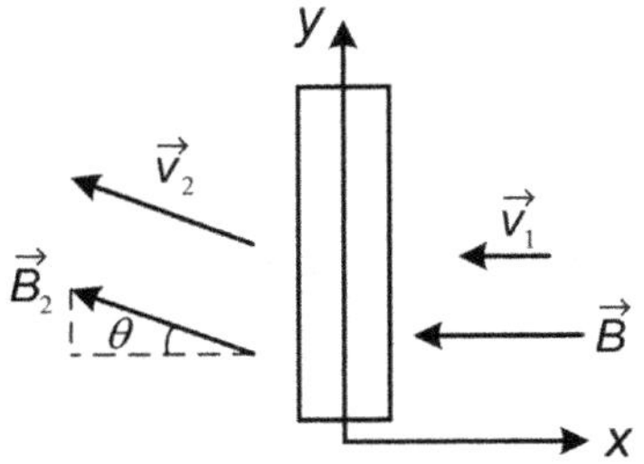

Figure 8.16 Flow and field orientations with magnetic field "switch on" across the shock wave.

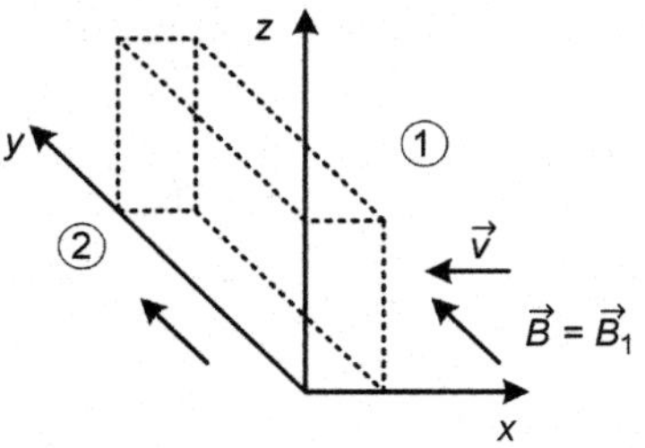

Figure 8.17 Shock geometry with magnetic field perpendicular to flow.

Further:

$$(v_x B_z - v_z B_x)_1 = 0 = v_x B_z - v_z B_x; \quad \therefore B_z = 0.$$

$$\left(\rho v_x v_y - \frac{B_x B_y}{\mu}\right)_1 = 0 = \rho v_x v_y - \frac{B_x B_y}{\mu}; \quad \therefore v_y = 0.$$

$$\left(\rho v_x v_z - \frac{B_x B_z}{\mu}\right)_1 = 0 = \rho v_x v_z - \frac{B_x B_z}{\mu}; \quad \therefore v_z = 0.$$

Also: $(v_x B_y - v_y B_x)_1 = v_x B_y - v_y B_x$; or:

$$v_1 B_1 = v_x B_y; \quad \text{so,} \quad B_y = \left(\frac{v_1}{v_x}\right) B_1 = (\alpha) B_1;$$

$$\therefore \vec{v}_2 = \frac{v_1}{\alpha}\vec{i} + 0\vec{j} + 0\vec{k}; \quad \vec{B}_2 = 0\vec{i} + \alpha B_1 \vec{j} + 0\vec{k}.$$

We combine momentum and energy, as:
Momentum:

$$\rho_1 v_1^2 + p_1 + \frac{B_1^2}{2\mu} = \rho_2 v_2^2 + p_2 + \frac{B_2^2}{2\mu}, \quad \text{so:}$$

$$\rho_1 v_1^2 + p_1 + \frac{B_1^2}{2\mu} = \frac{\rho_1 v_1^2}{\alpha^2} + p_2 + \frac{\alpha^2 B_1^2}{2\mu},$$

then divide by p_1:

$$\frac{\rho_1 v_1^2}{p_1} + 1 + \frac{B_1^2}{2\mu p_1} = \frac{\rho_1 v_1^2}{\alpha^2 p_1} + \frac{p_2}{p_1} + \frac{\alpha^2 B_1^2}{2\mu p_1},$$

and with:

$$\frac{\rho_1}{p_1} = \frac{\gamma}{c_s^2}, \quad \beta = \frac{p_1}{B_0^2/2\mu}, \quad R \equiv \frac{p_2}{p_1};$$

we get:

$$\gamma M_1^2\left(1 - \frac{1}{\alpha^2}\right) = (R-1) + \frac{1}{\beta}(\alpha^2 - 1).$$

For energy, as above:

$$\frac{\gamma p_1}{(\gamma-1)\rho_1} + \frac{v_1^2}{2} + \frac{B_1^2}{\mu\rho_1} = \frac{\gamma p_2}{(\gamma-1)\rho_2} + \frac{v_2^2}{2} + \frac{B_2^2}{\mu\rho_2}$$

$$= \frac{\gamma p_2}{(\gamma-1)\rho_2} + \frac{v_1^2}{2\alpha^2} + \frac{\alpha^2 B_1^2}{\mu\rho_2}, \text{ we multiply by } 2\rho_1/p_1,$$

then:

$$\frac{2\gamma}{(\gamma-1)} + \frac{\rho_1}{p_1}v_1^2 + \frac{2B_1^2}{\mu p_1} = \frac{2\gamma p_2\rho_1}{(\gamma-1)\rho_2 p_1} + \frac{\rho_1}{p_1}\frac{v_1^2}{\alpha^2} + \frac{2\rho_1}{p_1}\frac{\alpha^2 B_1^2}{\mu\rho_2},$$

and:

$$\frac{2\gamma}{(\gamma-1)} + \gamma M_1^2 + 4\frac{1}{\beta} = \frac{2\gamma}{\gamma-1}\frac{R}{\alpha} + \frac{\gamma M_1^2}{\alpha^2} + 4\frac{\alpha}{\beta},$$

rearrange to get:

$$\gamma M_1^2\left(1 - \frac{1}{\alpha^2}\right) = \frac{2\gamma}{\gamma-1}\left(\frac{R}{\alpha} - 1\right) + \frac{4}{\beta}(\alpha - 1).$$

We eliminate R in momentum and energy equations to get:

$$\frac{1}{\beta}(2-\gamma)\alpha^2 + \left[\gamma\left(\frac{1}{\beta} + 1\right) + \frac{1}{2}\gamma(\gamma-1)M^2\right]\alpha - \frac{1}{2}\gamma(\gamma+1)M^2 = 0.$$

This is a quadratic functional form: $a\alpha^2 + b\alpha + c = 0$, and as $\alpha > 0$, has a solution:

$$\therefore \alpha = \frac{-b + \sqrt{b^2 - 4ac}}{2a},$$

with:

$$\frac{ac}{b^2} \sim \frac{M^2}{M^4} = M^{-2},$$

so:

$$\alpha = \frac{-b + b\sqrt{1 - \frac{4ac}{b^2}}}{2a} \approx \frac{-b + b\left(1 - \frac{2ac}{b^2}\right)}{2a} = \frac{-\frac{2ac}{b}}{2a} = -\frac{c}{b},$$

so:

$$\alpha = \frac{\frac{1}{2}\gamma(\gamma+1)M^2}{\gamma\left(\frac{1}{\beta}+1\right) + \frac{1}{2}\gamma(\gamma-1)M^2}, \quad \text{and} \quad v_x = \left(\frac{1}{\alpha}\right)v_1.$$

Therefore, as $\alpha > 1$:

$$\frac{1}{2}\gamma(\gamma+1)M^2 > \gamma\left(\frac{1}{\beta}+1\right) + \frac{1}{2}\gamma(\gamma-1)M^2, \text{ and:}$$

$$\gamma M^2 > \gamma\left(\frac{1}{\beta}+1\right) \approx 2\left(\frac{1}{\beta}+1\right) \text{ since } 1.0 < \gamma < 2.0,$$

so:

$$\frac{\rho_1}{p_1} v_1^2 > \frac{\gamma B_1^2}{\mu p_1} + \gamma > \frac{B_1^2}{\mu p_1} + \gamma,$$

and so:

$$v_1^2 > \frac{B_1^2}{\mu_1 \rho_1} + \gamma \frac{p_1}{\rho_1} = V_A^2 + c_s^2, \text{ consistent with shock formation.}$$

The above solution for a plasma shock is similar to that for *OFM*, except that we replace fluid variables with fluid and field variable properties, as:

Pressure:	p,	with:	$p + \dfrac{B^2}{2\mu}$,
Mach number:	$M = \dfrac{v_1}{c_s}$	with:	$\dfrac{v_1}{\left(c_s^2 + v_A^2\right)^{1/2}}$;

Energy: $\epsilon = \dfrac{P}{(\gamma - 1)\rho}$ with: $\epsilon + \dfrac{B^2}{2\mu p} \equiv \epsilon^*;$

Gamma: γ with: $\dfrac{\epsilon}{\epsilon^*}\gamma + 2\left(1 - \dfrac{\epsilon}{\epsilon^*}\right) \equiv \gamma^*;$

In order to determine the magnetic field after the shock: with $B_1 v_1 = B_y v_x$; we solve for v_x as above, and then determine B_y.

Shock Wave Structure

We had noted above that shock waves form when strong pressure differences drive collapsing compression waves until there is a balance with loss mechanisms. In ordinary gases, this balance is reached within a very thin layer the dimension of which is on the order of atomic processes. Analysis of the one-dimensional shock wave in fluids and plasmas allows simplifications not otherwise available to study the physics of atomic processes and transport phenomena in gases; experimental studies have provided important insights into the complexity of real gas shock behavior. The processes that occur in a shock wave, in the shock structure, relate to the intermolecular transport processes. Along with a central region that can demonstrate large changes in properties, the understanding of processes that take place in the transition regions of the leading and trailing boundaries of the shock wave have also proved illuminating in low-density and chemically active but nonionized flows. In the processes that occur in shock waves that involve ionization, electric and magnetic field effects are important and can become quite complex (Liberman and Velikovich, 1986).

In *OFM,* shocks steepen until there is a balance between driving pressure difference and viscous dissipation. In plasmas, joule heating due to electrical conduction transport can become an added dissipation, which adds to the complexity. To assist the analysis and the understanding of predicted physical behavior, for simplicity, we oftentimes assume full ionization. We will now consider some order of magnitude estimates to identify the influence of electrical current conduction.

The rate of dissipation of energy (ΔE) across a thickness, δ, characterized by particle speed, v_1, can be expressed:

$$\frac{\Delta E}{\Delta t} = \frac{\Delta E}{\delta / v_1} \sim \frac{J^2}{\sigma} = \frac{1}{\sigma}\frac{(\nabla \times B)^2}{\mu^2} = \frac{\Delta B^2}{\sigma \mu^2 \delta^2},$$

so:

$$\delta \sim \frac{\Delta B^2}{\sigma \mu^2 v_1 \Delta E}.$$

But if we take $\Delta E \approx \frac{1}{2}\rho_1 v_1^2$, then we can determine the order of the thickness of the shock wave:

$$\delta \sim \left(\frac{2\Delta B^2}{\sigma \mu^2 \rho_1 v_1^3}\right) \sim \Delta B^2, \quad \frac{1}{\sigma}, \quad \frac{1}{v^3}.$$

If we recall the magnetic Reynolds number parameter: $R_m = \mu\sigma LU \rightarrow \mu\sigma\delta v_1 \Rightarrow \delta \sim \frac{R_m}{\mu\sigma v_1}$.

So it can be seen that the inclusion of plasma properties such as conductivity and magnetic field effects can be expected to have a direct effect on shock waves in the plasma medium.

Extended Reviews of Plasma Shock Wave Physics

The occurrence and effects of shock waves in the plasma medium with electric and magnetic field configurations have important consequences for equilibrium, stability, energy balance, and deposition. Research on the subject of plasma shock waves has been important to a number of applications, and the results of studies, both theoretical and experimental, have appeared in the literature over an extended period of time. In particular, a number of research monographs that provide a framework for understanding the occurrence and complexity of plasma shock waves have been published. Most notable among these are by: Chu and Gross (1969), Liberman and Velikovich (1986), Tidman and Krall (1971), and Balogh and Treumann (2013). Primarily because shock waves in plasmas have an important role in terrestrial plasma experiments and space plasmas, a brief, general discussion of experimental evidence and theory on the topic will be presented here.

Directly related to the introduction of electromagnetic forces and occurrence of related transport dissipations, there are two distinct types of plasma shock wave behavior: collisional and collisionless. Where the dissipation is primarily due to interparticle collisions and transport, the collisional mean free path becomes the length scale of interest both with and without significant effects due to electric and magnetic fields in plasmas. Where the dominant dissipation in the plasma is related to microinstabilities, not viscosity, the shocks are called collisionless.

In both types of plasma shock waves there may be several distinct layers of interaction within the plasma shock thickness, as the dominance of different mechanisms can take place sequentially. Plasma shocks occur in media that are ionized in their quiescent state. Also, since strong shocks result in significant energy deposition in a gas, there may be self-ionizing effects as a part of the shock structure, and so plasma shock waves can manifest unique aspects. The consideration of the effects of ionization chemistry on plasma shock waves has been discussed by Gross (1965). The electron species and ion species in shock interactions can have different properties and unique fluid behaviors, including individual Mach numbers, and this can result in complex phenomena depending on the orders of magnitude of parameters.

Collisionless shocks are formed due to dissipation processes in plasmas that are not based on interparticle collisions, and so the thickness dimension can be much larger or smaller than the collisional mean free path. Collisionless shocks have been described by Chu and Gross (1969) as transition regions where plasma waves and wave energy density are important for the formation of the shocks. A concise, clear review of the current understanding of the physics of collisionless shock waves has been presented by Krall (1997). Conceptually, it is easiest to envision these shocks as being due to dissipation from instabilities generated by plasma (microscopic) waves. Different plasma waves have been associated with different types of collisionless shocks Krall (1997). Collisionless shocks have been described as laminar, turbulent and (quasi-laminar) mixed shocks. The Vlasov (collisionless Boltzmann) equation is considered the correct equation to describe the plasma (Tidman and Krall, 1971; Balogh and Treuman, 2013).

Theoretical evaluations of plasma shock wave phenomena have been carried out with various models, and that has assisted understanding. However, the most interesting results that allow physical understanding of collisional and collisionless plasma shocks are those that have been exhibited and identified in experiments. Based upon available analyses, extensive experimental research programs had been carried out in the 1960s and 1970s in the UK (Culham Laboratory) and at Columbia University in the United States. The presentation of data in the reported results is striking, and the experimental results were generally accompanied by a review of related theory, so qualitative and quantitative understanding is enhanced. Some examples of such studies that can help unify the basic physical phenomenology and theory with experimental data are now included as part of this basic review.

Most of the studies of plasma shocks have been reported for experiments where shocks were formed during linear, annular, or radial electromagnetic shock tube discharges. In those cases, the chamber is filled with gas, sometimes preionized and sometimes with applied magnetic field whose direction would produce flow along or normal to the magnetic field. The electromagnetic piston will drive a compressed region forward and will create plasma shocks with strong density and field gradients. Both collisional and collisionless plasma shocks have been created and studied in this fashion.

A range of plasma wave experiments was reported by Gross (1965) and Gross et al. (1966). The results shown in Figure 8.18 are for collisional shock waves, B_x is magnetic field normal to the shock wave and b_1 is upstream Alfven speed.

The data shown are for three waves: an ordinary gas dynamic shock, a switch-on shock, and an "extremal" shock, which is defined by the condition that the flow after the shock is at sound speed (this is a limiting Chapman–Jouget solution). It can be seen that, for the three cases, there are significant differences between the density ratios (and related properties)

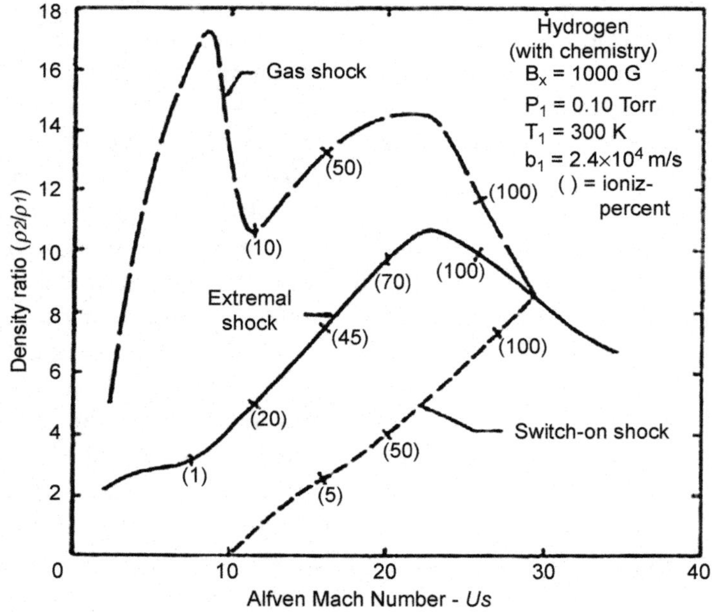

Figure 8.18 Density ratio across three types of shocks in plasma as a function of Alfven Mach number. *Gross et al. (1966), Figure 1, page 1034, with permission.*

across the shock waves. Specific attention should be given to the initiation of the switch-on shock at $M_A = 1$; the switch-on shocks are MHD shocks, not solely based on flow speed being equal to Alfven speed.

Another set of laboratory experiments with a cylindrical annular geometry also provided evidence of collisional switch-on shock wave behavior (Craig and Paul, 1973). With comprehensive diagnostic monitoring of the shock, particularly involving measurements of magnetic field components before and after the shock, ideal switch-on behavior was observed. The Alfven Mach number for the incident shock was altered by changing the magnitude of the applied (axial) magnetic field; results are shown in Figure 8.19. In order to reach agreement of theory with experimental results, enhanced (×20) collisional transport (viscosity) was introduced. With that adjustment, there was agreement of theory and experiment, thus providing evidence for the existence of enhanced (non-classical) transport in the plasma shock wave structures.

Collisionless shocks have been studied in a number of laboratory experiments. Early results were compiled and reviewed by Paul (1969), and he presented a discussion of theory on related microinstabilities. In a companion work, he addressed results of a specific experiment (Paul, 1970).

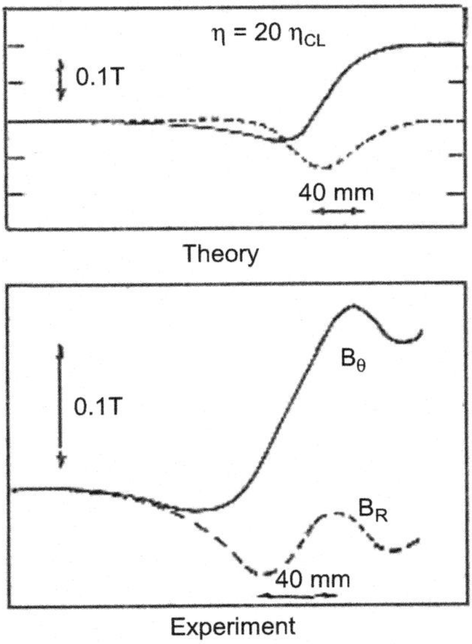

Figure 8.19 Comparison of experiment and computed fields for switch-on shock wave structure ($M_A = 1.3$). *Craig and Paul (1973), Figure 18, page 181, with permission.*

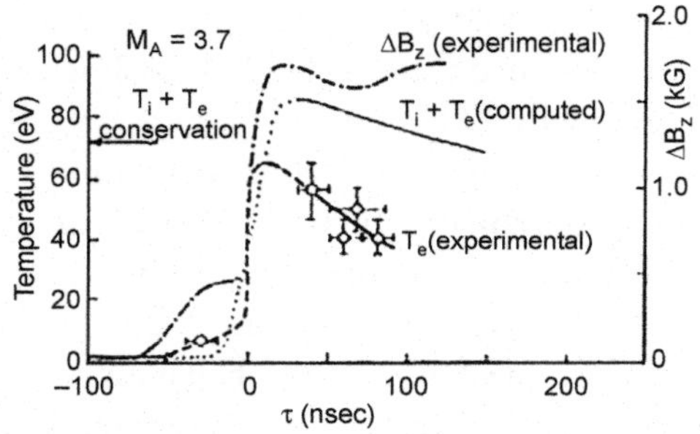

Figure 8.20 Measured and computed temperatures related to collisionless shocks at $M_A = 3.7$. *Paul (1970), Figure 17, with permission.*

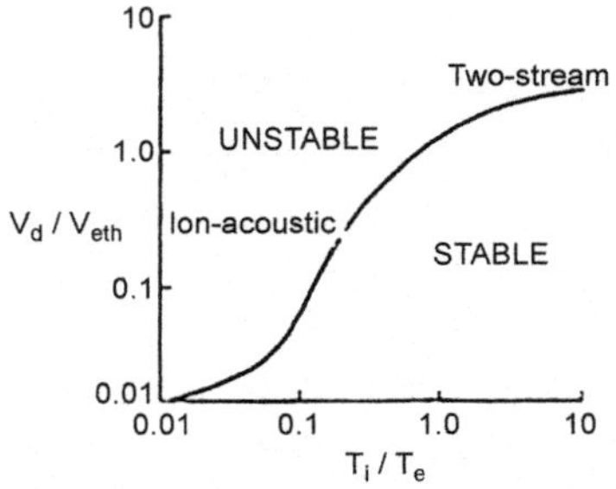

Figure 8.21 Species temperatures related to heating due to drift velocity instabilities. *Paul (1970), Figure 18, Stringer (1964), with permission.*

Along with measurements of shock jump parameters, that work presents data and analysis that supports the conclusion that the shocks produce enhanced heating of electrons due to instabilities, not collisional processes. Magnetic field and species temperature data and analytical evaluations are shown in Figures 8.20 and 8.21.

With reference to Figure 8.20, where $M = 3.7$, it can be seen that measurements of T_e are lower than energy conservation would predict, so that T_i is inferred to be significantly enhanced. The observed heating of particles is incompatible when only classical heating mechanisms were invoked to explain experimental data. Figure 8.21 shows the result of theory estimates that predict the onset of relevant instabilities (ion-acoustic and two-stream) for this experiment. As an interesting complement to this behavior, data for $M = 2.5$ did not show enhanced ion heating.

REFERENCES

Anderson, J.E., 1963. Magnetohydrodynamic Shock Waves. MIT, Cambridge, MA.

Balogh, A., Treumann, R.A., 2013. Physics of Collisionless Shocks. Springer, New York.

Boyd, T.J.M., Sanderson, J.J., 1969. Plasma Dynamics. Barnes & Noble, New York.

Braginskii, S.I., 1965. Transport processes in plasmas. In: Reviews of Plasma Physics, vol. 1. Consultants Bureau, New York.

Chandrasekhar, S., 1960. Plasma Physics. Univ. of Chicago, Chicago.

Chen, F.F., 1984. Introduction to Plasma Physics and Controlled Fusion, second ed. Plenum, New York.

Chu, C.K., Gross, R.A., 1969. Shock waves in plasma physics. In: Advances in Plasma Physics, Vol. 2. Wiley, New York.

Connor, J.W., Wilson, H.R., 1994. Survey of theories of anomalous transport. Plasma Phys. Control. Fusion 36, 719–795.
Courant, R., Friedrichs, K.O., 1948. Supersonic Flow and Shock Waves. Interscience, New York.
Cowling, T.G., 1957. Magnetohydrodynamics. Interscience, New York.
Craig, A.D., Paul, J.W.M., 1973. Observation of "switch-on" shocks in a magnetized plasma. J. Plasma Phys. 9 (Part 2), 161–186.
Gross, R.A., 1965. Strong ionizing shock waves. Rev. Mod. Phys. 37, 724–743.
Gross, R.A., et al., 1966. Ionizing Switch-on shock waves. Phys. Fl. 9, 1033–1035.
Jackson, J.D., 1963. Classical Electrodynamics. Wiley, New York.
Jeffrey, A., 1966. Magnetohydrodynamics. Interscience, New York.
Klages, R., Radons, G., Sokolov, M. (Eds.), 2008. Anomalous Transport: Foundations and Applications. Wiley, New York.
Karamcheti, K., 1966. Principles of Ideal-fluid Aerodynamics. Wiley, New York.
Krall, N.A., 1997. What do we really know about collisionless shocks? Adv. Space Res. 20, 715–724.
Liberman, M.A., Velikovich, A.L., 1986. Physics of Shock Waves in Gases and Plasmas. Springer, Berlin.
Liewer, P.C., 1985. Measurements of microturbulence in tokamaks and comparison with theories of turbulence and anomalous transport. Nucl. Fusion 25, 543.
Paul, J.W.M., 1969. Review of Experimental Studies of Collisionless Shocks Propagating Perpendicular to a Magnetic Field. UK-AEA Report CLM-P220. Culham Laboratory, UK.
Paul, J.W.M., 1970. Collisionless shock waves. In: Rye, B.J., Taylor, J.C. (Eds.), Physics of Hot Plasmas. Oliver & Boyd, Edinburgh.
Papadopoulos, K., 1985. Microinstabilities and anomalous transport in collisionless shocks. AGU Geophys. Monogr. 34, 59–90.
Shercliff, J.A., 1965. A Textbook of Magnetohydrodynamics. Pergamon, New York.
Spitzer, L., 1962. Physics of Fully Ionized Gases, second ed. Interscience, New York.
Stringer, T.E., 1964. Electrostatic instabilities in current-carrying and counter-streaming plasmas. J. Nucl. Energy C 6, 267.
Sutton, G.W., Sherman, A., 1965. Engineering Magnetohydrodynamics. McGraw-Hill, New York.
Tidman, D.A., Krall, N.A., 1971. Shock Waves in Collisionless Plasmas. Interscience, New York.
Wylie, C.R., 1966. Advanced Engineering Mathematics. McGraw-Hill, New York.

CHAPTER 9

Plasma Dynamics and Hydromagnetics: Reviews of Applications

INTRODUCTION

In the earlier chapters, concepts have been introduced and analyses have been developed that allow a better understanding of the unique behavior of ionized gases and their reaction under the influence of electric and magnetic fields. Ultimately, however, the usefulness of this information relates to its application to specific problems that are important because of the need to understand physical devices or phenomena. Such devices and phenomena span a wide spectrum because of the natural and artificial creation of ionized gases. High-temperature gases can occur in energy conversion, aerospace, and manufacturing devices; some of these exist and are operational, and some will be successfully developed when we know more about control of plasma processes. So it is in the context that theory and analysis become more understandable and useful when applied to specific situations that a number of applications of plasmas and plasma dynamics are now examined and reviewed in detail.

It is useful to stress here that while the mere presence of electric and magnetic fields can have an effect on the charged particles in a region where plasma is introduced, it is the boundary conditions and connection to external circuitry that are the operative source for a given device to function. If certain field configurations are consistent with current flow, there must be a possibility for providing that external current flow. For example, the proposal of infinite conductivity, while convenient to reach a theoretical result, can lead to curiosities rather than functional indications for experimental mechanisms. But key to the operation of a device is current flow in the plasma. It must be kept in mind that the basic momentum exchange mechanism is: $\vec{J} \times \vec{B}$; the basic energy addition term is: $\vec{J} \cdot \vec{E}$.

Introduction to Plasmas and Plasma Dynamics
ISBN 978-0-12-801661-9

Magnetic fields that are introduced to affect plasma interactions are normally generated external to the plasma region by standard coil arrangements; these are termed "applied fields." Whenever there is current flow in plasma, magnetic fields will be generated. Normally, fields generated by current flow in plasma are referred to as "self-field." Currents in plasma that are high enough to induce strong fields generally occur in pulsed or unsteady devices.

Plasma Acceleration and Energy Conversion

INTRODUCTION

In many proposed devices with configurations of electric and magnetic fields, the interactions are geometrically complex due to the three-dimensional character of the physics and the devices. We begin by examining a problem that allows the nature of the coupling of equations and ultimate result of the applied fields to be seen more transparently.

The flow of plasma in enclosed channels offers the possibility of controlled application of electric and magnetic fields for the purpose of acceleration or energy conversion. The principles of the interaction can be studied more easily in the one-dimensional (1D) and two-dimensional (2D) models of the flow. In fact, the effects of viscosity and other plasma interactions introduce considerable complexity, and operation of practical devices is much more problematic than simple models.

Early studies (Resler and Sears, 1958a,b; Culick, 1964) recognized the potential of electromagnetic influence on fluid flow. These works also identified the problem of electromagnetic acceleration of plasma from subsonic to supersonic flow regimes because of the coupled electromagnetic and fluid nature of the choking phenomenon (speed-limiting conditions in the flow) that became evident in the equations.

Channel Flow: Steady and 1D

In this case the physical problem is simple: in one dimension, (ionized) plasma flow is input to a rectangular channel and orthogonal electric and magnetic fields are applied during the flow process to achieve the desired effect on output flow or energy exchange. The examination of this configuration will allow a better understanding of the general, ideal behavior of plasma in accelerators and MHD (MagnetoHydroDynamic) generators. The equations are understandable, and the solutions are more transparent, however, the physical reality is much more complex; real, experimental devices are not modeled well by the simple analysis that follows.

We consider the geometry and interactions shown below (Figure 9.1); there is no area change.

We follow the developments of Resler and Sears (1958a). We consider the acceleration of flow in the x-direction with inlet ($x = 0$) value, u_0. The conductivity is assumed to be scalar, so, $\overrightarrow{J}, \overrightarrow{E}$ are in the same direction. There is a back electromotive force, $\overrightarrow{u} \times \overrightarrow{B}$. We assume the magnetic field

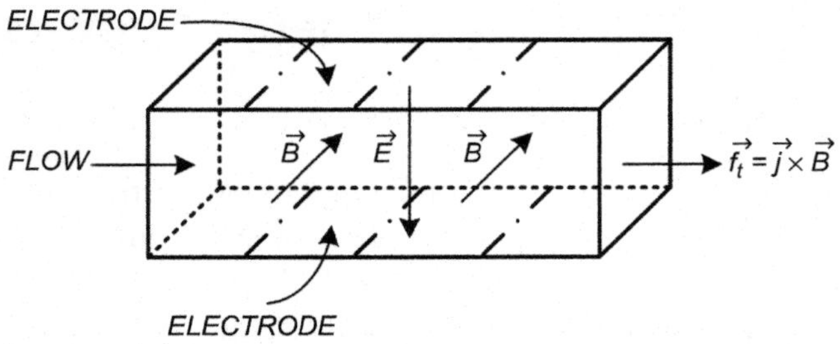

Figure 9.1 Geometry of fields and interactions in 1D channel flow.

generated by the current flow to be small compared to the applied field. Then the conservation equations and Ohm's law are:

$$\rho u = \text{constant},$$

$$\rho u \frac{du}{dx} = -\frac{dp}{dx} + jB,$$

$$\rho u \frac{d}{dx}\left(c_p T + \frac{u^2}{2}\right) = jE \quad (c_p \approx \text{constant}),$$

$$j = \sigma(E - uB),$$

$$p = \rho RT,$$

$$\sigma = \sigma(\rho, T).$$

There are six equations and eight unknowns. So, two variables must be specified in order to allow a solution. One can consider a restriction of the process during the flow; we can assume: (1) isentropic, (2) adiabatic, (3) isothermal, and (4) constant fields.

The isothermal assumption is consistent with the optimized case where all energy added is converted to flow kinetic energy. This is an ideal, limiting condition, and we apply it as:

$$\frac{dT}{dx} = 0, \quad \text{or} \quad T = 0, \quad \text{so:}$$

$$\rho u \left[\frac{d}{dx}\left(\frac{u^2}{2}\right)\right] = jE,$$

and we can use this form for energy to achieve analytic solutions (Jahn, 1969).

With isothermal flow, we can combine energy and momentum to get:

$$u = \frac{E}{B}\left(1 - \frac{RT}{u^2}\right),$$

where $RT = (c_s)^2$ with $c_s =$ sound speed. So, for a relevant solution: $u > c_s$. We can also see the coupling relationship of fluid and field properties; in particular, $u \sim \frac{E}{B}$.

To achieve a straightforward solution, we can take:

$$j \cdot E = \text{constant},$$

and so we get the solution:

$$u^*(x) = \sqrt{1 + 2\frac{jEL}{\rho u_0^3}x^*},$$

where:

$$u^* = \frac{u}{u_0}, \quad x^* = \frac{x}{L},$$

where $L =$ channel length.

For another analytic solution, substitute the $u \sim E/B$ relationship into energy to get:

$$\rho u^3 \frac{du}{dx} = \sigma E^2 c_s^2.$$

If we presume that $\sigma E^2 =$ constant along the channel, this can be integrated to give:

$$u^* = \left[1 + 4\frac{\sigma E^2 L}{\rho u_0^3}\left(\frac{c_s}{u_0}\right)^2 x^*\right]^{1/4},$$

where again the strong coupling of fluid and field properties is evident.

Analytic solutions have also been presented for σB^2 and σEB as constant (Jahn, 1969).

Solutions are presented in Figure 9.2 for the $j \cdot E$ and σB^2 analytic solutions.

(In the figure, magnetic interaction parameter: $I \equiv \frac{\textit{magnetic force}}{\textit{kinetic force}} = \frac{jB}{\rho \frac{u^2}{L}} = \frac{\sigma u B \cdot B}{\rho \frac{u^2}{L}} = \frac{\sigma B^2 L}{\rho u}$.)

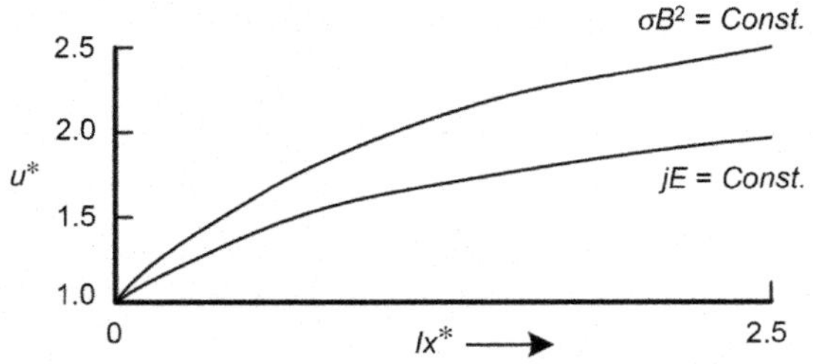

Figure 9.2 Velocity along channel for $j \cdot E$ or σB^2 held constant for isothermal supersonic flow. *Jahn, R.G., 1969. Physics of Electric Propulsion. McGraw-Hill, New York. Adapted from Figure 8-4, p. 205 with permission.*

For the jE case, an increase in velocity up to 2.0 times the inlet velocity is predicted when the interaction parameter has a value of 2.5. A second solution where the assumptions were for isothermal flow with σB^2 a constant along the channel is also shown; it can be seen that higher acceleration is predicted. In any case, the advantage of direct application of fields to the flow is evident, and this demonstrated the potential for electric and magnetic field influence on electrically conducting fluid flow.

The application of force interactions and the theoretical effects of applied electric and magnetic fields can be seen in the above approximate solutions. However, such theoretical gains have not been achieved with experimental conditions available in collisional plasmas. The most difficult problems encountered in channel flow experiments are related to Hall currents; devices with the ionization seed material needed to enhance conductivity and are prone to shorting due to Hall currents, even with segmented electrodes. There is also considerable flow complexity due to viscous boundary layer effects. This behavior will now be explored further, as it is representative of how fluid mechanics involved in plasma interactions can, in cases, overwhelm the basic electromagnetic effects.

Hydromagnetic Channel Flow with Viscous Interactions

The analysis of plasma flow in an enclosed channel when viscosity of the fluid is included is a more realistic model, but the analysis is inherently complex. The flow has three-dimensional character with secondary flow components. The analysis developed below will include simplifying assumptions to allow an examination of the primary mechanical aspects of viscous flow; we will not include the energy equation, considering dissipation and heat terms as effects to be added later. We consider interactions with a boundary in two dimensions; however, the analysis must include some three-dimensional terms (Shercliff, 1965; Sutton and Sherman, 1965; Boyd and Sanderson, 1969). Generally, we follow the calculation sequence outlined by Boyd and Sanderson (1969).

The basic equations for this analysis are as above, except that we include viscosity, as follows:

$$\frac{\partial \rho}{\partial t} + \nabla \cdot (\rho \vec{v}) = 0;$$

$$\rho \frac{D\vec{v}}{Dt} = -\nabla p + \vec{J} \times \vec{B} + \nabla \cdot \overleftrightarrow{\tau},$$

where the stress tensor, $\overleftrightarrow{\tau}$, can be expressed as:

$$\nabla \cdot \overleftrightarrow{\tau} = \frac{\overline{\mu}}{3} \nabla (\nabla \cdot \vec{v}) + \overline{\mu} \nabla^2 \vec{v},$$

and

$$\overline{\mu} = \text{viscosity};$$

$$\vec{J} = \sigma \left(\vec{E} + \vec{v} \times \vec{B} \right);$$

$$\nabla \times \vec{E} = -\frac{\partial \vec{B}}{\partial t};$$

$$\nabla \times \vec{B} = \mu \vec{J}.$$

We will consider incompressible flow to simplify the analysis; this type of channel flow is referred to as Hartmann flow. We take flow in the x-direction with applied magnetic field in the z-direction, as shown in Figure 9.3.

It is assumed that we know: dp/dx, σ, and d. We neglect end effects ($L \gg d$), neglect secondary flow ($w \gg d$), and take $\{f(y) = 0\}$. We take: $v(x)$, and $\vec{v} \times \vec{B} = \vec{E}_y$.

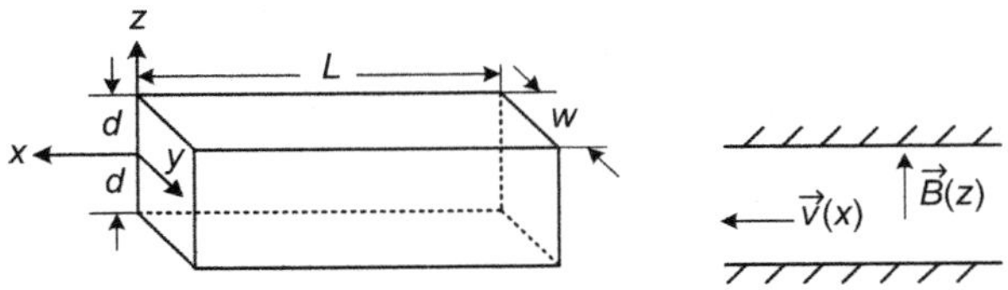

Figure 9.3 Schematic of geometry of channel flow with viscous interactions.

With the assumptions of steady, incompressible flow:
Continuity:

$$\nabla \cdot \vec{v} = 0; \quad v = f(z \text{ only}), \quad \frac{\partial v}{\partial x} = 0;$$

Momentum:

$$0 = -\nabla p + \vec{J} \times \vec{B} + \overline{\mu}\, \nabla^2 \vec{v};$$

and Momentum x-direction:

$$\frac{\partial p}{\partial x} = \left(\vec{J} \times \vec{B}\right)_x + \overline{\mu}\, \frac{\partial^2 v}{\partial z^2}.$$

From Faraday's law:

$$\vec{J} = \nabla \times \frac{\vec{B}}{\mu},$$

with:

$$J_x = \frac{1}{\mu}\frac{\partial B_y}{\partial z}, \quad J_y = \frac{1}{\mu}\frac{\partial B_x}{\partial z}, \quad J_z = 0.$$

Then, incorporating in directional momentum equations:

$$(x\text{-dir.}) \quad \frac{\partial p}{\partial x} = \frac{B_0}{\mu}\frac{\partial B_x}{\partial z} + \overline{\mu}\, \frac{d^2 v}{dz^2};$$

$$(y\text{-dir.}) \quad 0 = \left(\vec{J_x} \times \vec{B}\right)_y + 0,$$

but with no secondary flow:

$$\frac{dB_y}{dz} = 0, \qquad \text{so,} \qquad J_x = 0;$$

$$(z\text{-dir.}) \quad \frac{\partial p}{\partial z} = \left(\vec{J_y} \times \overrightarrow{B_x}\right)_z + 0,$$

and:

$$\frac{\partial p}{\partial z} = -\frac{B_x}{\mu}\frac{dB_x}{dz}.$$

Now with:

$$\overrightarrow{J} = \overrightarrow{J_y} = \frac{1}{\mu}\frac{\partial B_x}{\partial z}\overrightarrow{y},$$

and from Ohm's law:

$$\frac{\overrightarrow{J}}{\sigma} = \frac{\overrightarrow{J_y}}{\sigma} = \overrightarrow{E} + \overrightarrow{v_x} \times \overrightarrow{B_z},$$

therefore:

$$\overrightarrow{E_x} = \overrightarrow{E_z} = 0.$$

Since $\overrightarrow{v} \times \overrightarrow{B}$ has no y-component, we have:

$$\frac{J_y}{\sigma} = \frac{1}{\mu\sigma}\frac{\partial B_x}{\partial z} = E_y - v_{(x)}B_{0(z)}.$$

Furthermore, from Ampere's law:

$$\nabla \times \overrightarrow{E} = -\frac{\partial \overrightarrow{B}}{\partial t} = 0,$$

for steady flow. So:

$$\nabla \times \overrightarrow{E} = \left(\frac{\partial E_y}{\partial z}\right)_x = 0,$$

and:

$$E_y = \text{constant} \equiv E_0 \text{ (to be determined)}.$$

Now, taking x-momentum again:

$$\frac{\partial p}{\partial x} = \frac{B_0}{\mu}\frac{\partial B_x}{\partial z} + \overline{\mu}\frac{d^2 v}{dz^2},$$

and applying the operator $\frac{\partial}{\partial x}$ to the equation, we get:

$$\frac{\partial^2 p}{\partial x^2} = \frac{B_0}{\mu}\frac{\partial}{\partial z}\left(\frac{\partial B_x}{\partial x}\right) + \overline{\mu}\frac{d^2}{dz^2}\left(\frac{\partial v}{\partial x}\right) = 0;$$

integrating with respect to x:

$$\frac{\partial p}{\partial x} = \text{constant} \equiv -P_0 + f(z).$$

Now, with z-momentum, apply the operator $\frac{\partial}{\partial x}$ and interchange operation order:

$$\frac{\partial}{\partial z}\left(\frac{\partial p}{\partial x}\right) = -\frac{B_x}{4\pi}\frac{d}{dz}\left(\frac{\partial B_x}{\partial x}\right) = 0.$$

So:

$$f(z) = 0,$$

and

$$\frac{\partial p}{\partial x} \equiv -P_0.$$

We integrate again and get:

$$p = -P_0 x + P_1(z) + \text{constant},$$

then set:

$$\text{constant} = 0,$$

i.e.,

$$p(x = 0, z = 0) = 0.$$

Then in x-momentum:

$$\frac{\partial p}{\partial x} = -P_0 = \frac{B_0}{\mu}\frac{\partial B_x}{\partial z} + \overline{\mu}\,\frac{d^2 v}{dz^2},$$

but we know $\frac{\partial B_x}{\partial z}$ from Ohm's law, so we can substitute and express:

$$-P_0 = \frac{B_0}{\mu}\cdot \mu\sigma(E_0 - vB_0) + \overline{\mu}\,\frac{d^2 v}{dz^2},$$

or:

$$\frac{d^2 v}{dz^2} - \sigma\frac{B_0^2 v}{\overline{\mu}} = -\frac{P_0 + \sigma B_0 E_0}{\overline{\mu}},$$

which has the form:

$$\frac{d^2 v}{dz^2} - A^2 v = -B - C.$$

This is a second-order ordinary differential equation (ODE) with constant coefficients, which can be solved by standard methods to give:

$$v = \frac{B+C}{A^2} - \frac{(B+C)/A^2}{e^{Ad}+e^{-Ad}}\left(e^{Az}+e^{-Az}\right) = \frac{B+C}{A^2}\left(1 - \frac{e^{Az}+e^{-Az}}{e^{Ad}+e^{-Ad}}\right),$$

and substituting proper coefficients:

$$v = \frac{P_0+\sigma B_0 E_0}{\sigma B_0^2}\left(1 - \frac{e^{Az}+e^{-Az}}{e^{Ad}+e^{-Ad}}\right) = \frac{P_0+\sigma B_0 E_0}{\sigma B_0^2}\left[1 - \frac{\cosh(M_H z/d)}{\cosh M_H}\right],$$

where:

$$M_H = Ad = B_0 d\left(\frac{\sigma}{\mu}\right)^{1/2},$$

and M_H is referred to as the Hartmann number. Again, in this simplified model, $v = f$ (z only). So we have achieved the solution for the velocity in the channel as a function of z and known properties. Specifically, $\frac{\partial p}{\partial x} = -P_0$, and B_0, d, and σ are given, but E_0 is not given. The parameter, $\vec{E}_0$, is the electric field transverse to velocity and $J_y \sim E_y$, hence it depends upon physical constraints placed on the system that we are analyzing.

As an example of E_0 significance, we examine the channel configuration for an MHD (MagnetoHydroDynamic) generator. A schematic of the geometry, circuits, and field is shown in Figure 9.4.

In this application:

$$I = \left(\frac{\Delta V_L}{R_L}\right)_{ext} = (J\cdot A_{el})_{int},$$

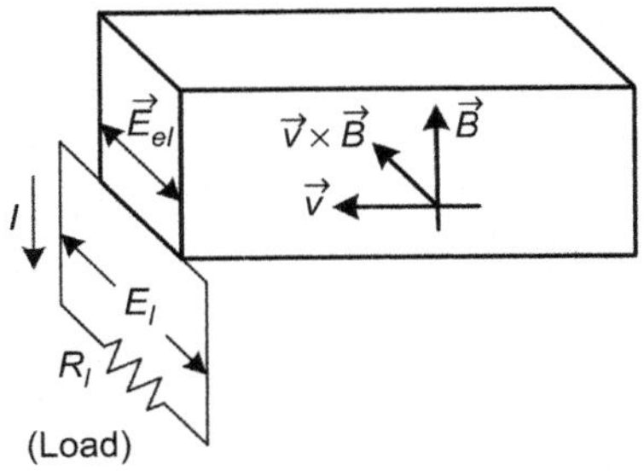

Figure 9.4 Flow and field geometry for MHD generator.

and:

$$\text{Power} \sim (I \cdot R_l)_{ext} = (J \cdot E_{el})_{int},$$

where:

$$\frac{\vec{J}}{\sigma} = \vec{E}_{el} + \vec{v} \times \vec{B},$$

and $\vec{E}_{el}$ is the field measured between electrodes and is related to the load, e.g.:

If $R_l \rightarrow 0$, $J \Rightarrow E_{el} \rightarrow 0$, no power delivered;

If $R_l \rightarrow \infty$, $J = 0 \Rightarrow E_{el} = -v \times B$, no power delivered.

We will consider other cases for MHD channel interaction below.

We continue the analysis and evaluate the average flow velocity, v_0, as a function of E_0. In standard fashion and using the results obtained above:

$$v_0 = \frac{1}{2d} \int_{-d}^{d} v dz = \frac{P_0 + \sigma B_0 E_0}{2d\sigma B_0^2} \int_{-d}^{d} \left[1 - \frac{\cosh(M_H z/d)}{\cosh M_H} \right] dz,$$

$$= \frac{P_0 + \sigma B_0 E_0}{2d\sigma B_0^2} \left[d - \frac{d}{M_H} \frac{\sinh\left(\frac{M_H d}{d}\right)}{\cosh M_H} + d + \frac{d}{M_H} \frac{\sinh\left(-\frac{M_H d}{d}\right)}{\cosh M_H} \right],$$

$$v_0 = \frac{P_0 + \sigma B_0 E_0}{2d\sigma B_0^2} \left(2d - \frac{2d}{M_H} \frac{\sinh M_H}{\cosh M_H} \right) = \frac{P_0 + \sigma B_0 E_0}{\sigma B_0^2} \frac{1}{M_H} (M_H - \tanh M_H).$$

Therefore, we can now solve for $E_0 = f(v_0)$, as:

$$\sigma B_0^2 M_H v_0 = (P_0 + \sigma B_0 E_0)(M_H - \tanh M_H),$$

and:

$$E_0 = \frac{\sigma B_0^2 M_H v_0 - P_0(M_H - \tanh M_H)}{\sigma B_0(M_H - \tanh M_H)}.$$

So the velocity profile, $f(z)$, can be determined as:

$$v(z) = \left[\frac{P_0}{\sigma B_0^2} + \frac{1}{B_0}\frac{\sigma B_0^2 M_H v_0 - P_0(M_H - \tan M_H)}{\sigma B_0(M_H - \tan M_H)}\right]\left[1 - \frac{\cosh(M_H z/d)}{\cosh M_H}\right],$$

$$= \frac{M_H v_0}{M_H - \tan M_H}\frac{1}{\cosh M_H}\left[\cosh M_H - \cosh\left(\frac{M_H z}{d}\right)\right],$$

so:

$$v(z) = M_H v_0 \frac{\cosh M_H - \cosh(\frac{M_H z}{d})}{M_H \cosh M_H - \sinh M_H}.$$

Also, the average current density J_0, is also related to average flow velocity, as:

$$J_0 = \frac{1}{2d}\int_{-d}^{d} J_y dz = \sigma(E_0 - v_0 B_0),$$

therefore:

$$v_0 \Rightarrow E_0, J_0 \quad \text{or} \quad E_0, J_0 \Rightarrow v_0.$$

Now as J, E are related to electrodes and loads $\Rightarrow$ with applied B_0: $v_0 = f(J_0, E_0, B_0)$.

In order to clarify the shape of the velocity profile, we examine, first, the functional form for small M_H by applying small perturbation analysis to get:

$$v(z) = \frac{\frac{3}{2}v_0\left(1 - \frac{z^2}{d^2}\right)}{\frac{M_H^2}{2!} - \frac{M_H^2}{3!}},$$

where v can be seen to decrease with increasing M_H. In the limit of: $M_H \to 0$, we recover the classic parabolic distribution of ordinary fluid mechanics:

$$(\lim_{M\to 0})v(z) = \frac{3}{2}v_0\left(1 - \frac{z^2}{d^2}\right).$$

The behavior of the velocity profiles for increasing M_H can be seen in Figure 9.5 ($M_{H_2} > M_{H_1}$).

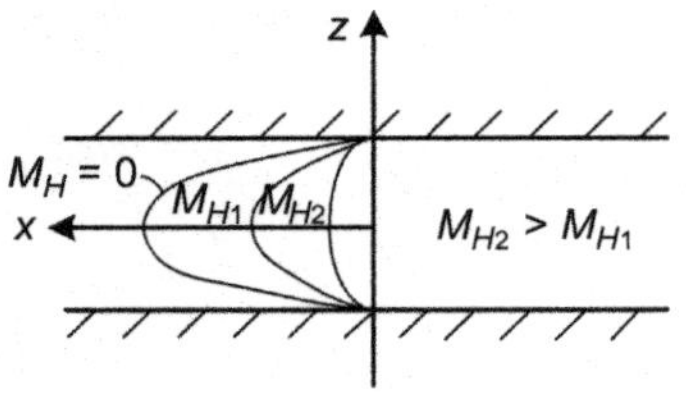

Figure 9.5 Velocity profiles for variable Hartmann number, $M_H \geq 0$.

We can also examine the functional form for large M_H, as:

$$v(z)_{M_H\ large} \approx v_0 \frac{\cosh M_H - \cosh\left(\frac{M_H z}{d}\right)}{\cosh M_H - \frac{\sinh M_H}{M}} \rightarrow v_0 \left[\frac{\cosh\left(\frac{M_H z}{d}\right)}{\cosh M_H}\right],$$

$$v(z)_{M_H\ large} \approx v_0 \left[1 - e^{-M_H\left(1-\frac{|z|}{d}\right)}\right] \approx \text{constant, except for } z \approx d.$$

The current density must be consistent with the flow behavior, as: $\vec{J} \times \vec{B} \rightarrow J_y \times B_z$, a force in the $+x$-direction. With $B_z \approx$ constant and $v = f(z) \Rightarrow v \times B_0 = f_1(z)$, then:

$$\frac{J_y}{\sigma} = E_0 - vB_0 = E_0 - v_0 B_0 \frac{\cosh M_H - \cosh\left(\frac{M_H z}{d}\right)}{\cosh M_H - \frac{\sinh M_H}{M}}.$$

Here, E_0 is the electric field in the y-direction, applied or related to current flow in the external circuit (E_0 is set constant). So the current density profile (Figure 9.6) is similar to the v profile.

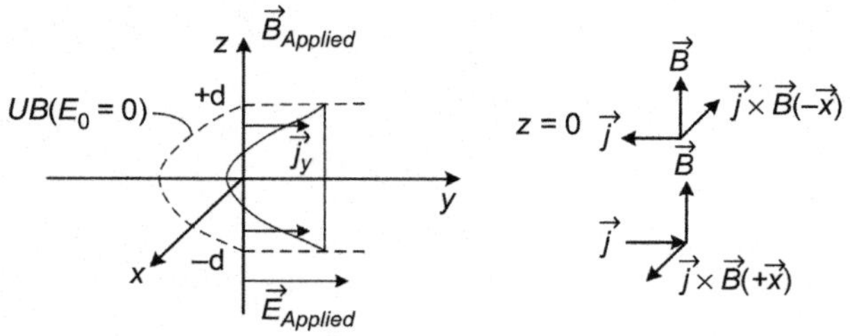

Figure 9.6 Induced current density in the channel geometry.

We consider now the effect of flow and current on induced magnetic field, B_x (B_z = constant, $B_y = 0$). From Maxwell's equations:

$$\frac{J_y}{\sigma} = \frac{1}{\mu\sigma}\frac{\partial B_x}{\partial z},$$

or:

$$dB_x(z) = \mu J_y(z)dz,$$

with:

$$\frac{J_y}{\sigma} = E_0 - v_0 B_0 \frac{\cosh M_H - \cosh\left(\frac{M_H z}{d}\right)}{\cosh M_H - \frac{\sinh M_H}{M_H}}.$$

In order to evaluate the field, the integration starts from the x–y plane, so:

$$B_x(z) - B_x(0) = \int_0^z dB_x(z) = \mu \int_0^z J_y(z)dz.$$

This integral can be evaluated exactly and in detail, but with complex functions. An approximation of linear function form can be made, $J_y \approx \mathrm{K}\, z$, and then we get:

$$B_x(z) \approx \mu\, \mathrm{K}\frac{z^2}{2}.$$

Assuming symmetry about the plane, the functional form for the field is shown in Figure 9.7.

Applications of Channel Flow Solutions

Based on the analysis presented above, various configurations of fields relate to devices that can perform different functions. The flow and field geometry orientation for channel flow are shown in Figure 9.8.

1. No applied E field ($E_0 = 0$) and "free" ($R_L = 0$) current flow.

   $$\text{So,}\quad J = -\sigma v B_0 = -J_y, \quad \text{and} \quad -\overrightarrow{J_y} \times \overrightarrow{B_z} = -\overrightarrow{f_x}.$$

   The channel acts as an *electromagnetic brake*.
2. No applied E field ($E_0 = 0$) and $J = 0$.
   Then:

   $$J \times B = 0 \quad \text{and} \quad E_0 = vB.$$

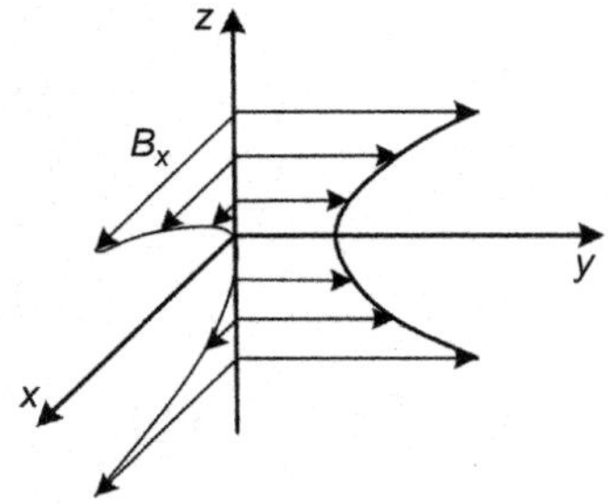

Figure 9.7 Variation of induced currents and magnetic fields.

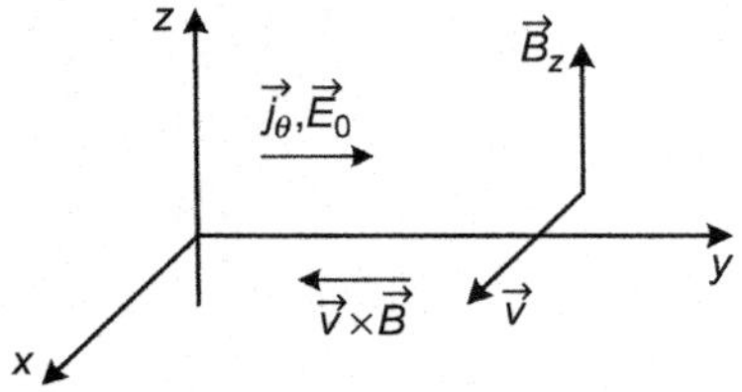

Figure 9.8 General model of geometry for fields and flow in a channel ($\frac{J}{\sigma} \approx E_0 - vB_0$).

Measure E_0 and B in order to determine v; the channel acts as an *electromagnetic flowmeter.*

3. No applied E, and allow current flow through an external load (R_L). We have:

$$0 < E < vB, \text{ and } \frac{J}{\sigma} = E - vB \Rightarrow I_{ext} \Rightarrow \text{Power} = I_{ext}^2 \cdot R_L.$$

This is an *MHD (power) generator.*

4. Applied E field such that $E_0 > vB_0$. Then:

$$\frac{J}{\sigma} = E_{0,x} - vB_0 \Rightarrow \overrightarrow{J_y} \times \overrightarrow{B_z} = +\overrightarrow{f_x}.$$

This is an electromagnetic *plasma accelerator* configuration.

Studies of MHD Power Generation

The basic analysis of MHD channel flow, as outlined above, served as foundation for research programs for MHD flow accelerators, and, most significantly, MHD power generation. Early (1938) work was done in the United States, and in the 1960s and the 1970s, experimental studies began in the United States and Russia; this was followed by studies in Japan, Australia, and Italy in the 1980s. The attraction of the technology is partly based on the use of combustion-fueled devices of relatively small size for a power plant.

MHD generators can be open cycle or closed cycle, the former being simpler. There are three general types of generator schemes and related geometries: Faraday, Hall, and Disc generators (Messerle, 1995). A closed-cycle MHD generator was developed at the National Aeronautics and Space Administration (NASA) (Sovie and Nichols, 1974). The MHD generator section is shown in Figure 9.9 to assist visualization of the components.

It can be seen that this is a Faraday channel (voltage generated laterally across channel) with segmented electrodes to diminish the effects of Hall currents. The gas is argon at 1.5 kg/s, 2.5 atm, 2000 K, Mach number 0.25–0.32, and $B = 0.1–1.8$ T. Cesium seed vapor was injected in the entrance region, at a rate about 0.03–0.10%. Under no-load condition the Faraday generating voltage was near ideal and Hall voltages were not excessive. The system performed as designed and was able to produce about 300 W and 0.036 W/cm^3. However, in such devices, Hall currents significantly reduce efficiency. Studies of comparable design and performance coal-fired open-cycle MHD generators were also reported (Galanga et al., 1982).

More recent results from the study of an MHD Hall configuration have been reported (Merkle et al., 2009). In that work, combustion gases of 2700 K and Mach 2 were seeded for ionization and injected into a cylindrical MHD generator chamber. The Hall configuration produced a driving potential difference between entry and exit and measurable current in an external load. Computational studies that were carried out produced results in agreement with experiment, as shown in Figure 9.10.

Power generated during the operation was 10.6 kW, and the analysis indicated bases for further improvements.

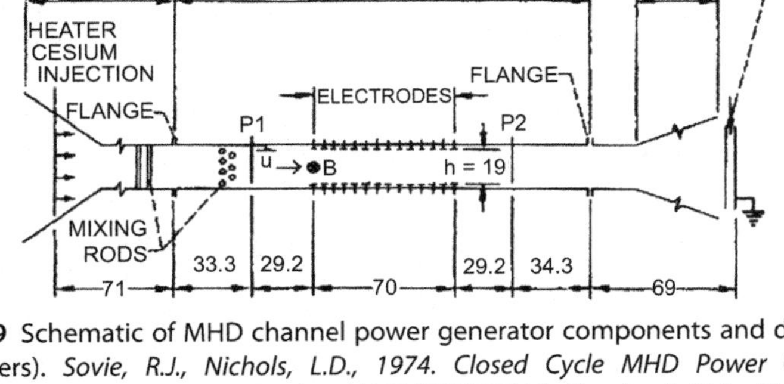

Figure 9.9 Schematic of MHD channel power generator components and dimensions (centimeters). *Sovie, R.J., Nichols, L.D., 1974. Closed Cycle MHD Power Generation Experiments in the NASA Lewis Facility, NASA TMX-71510, Cleveland, Ohio. Figure 2 with permission.*

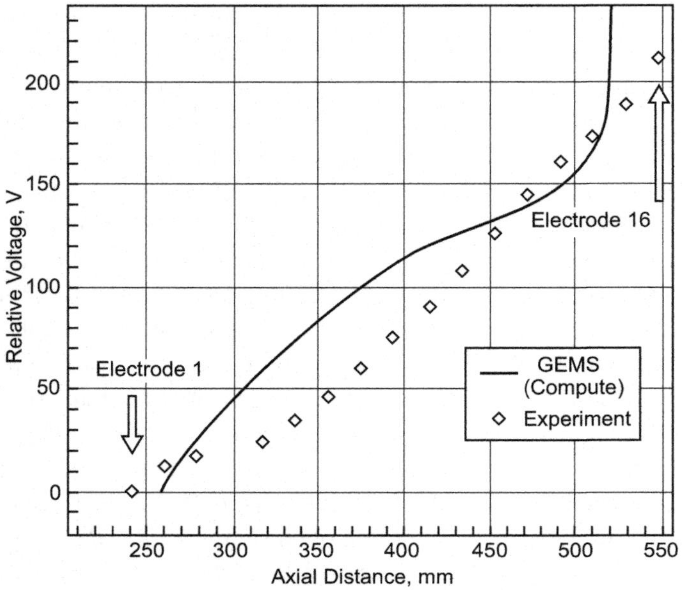

Figure 9.10 Comparison of computed and experimental voltage drop along MHD Hall generator channel. *Merkle, C.J., Moeller, T., Rhodes, R., Keefer, D., 2009. Computational simulations of power extraction in MHD channel. AIAA paper 2009-3827, 40th Plasmadynamics and Lasers Conference, San Antonio, Texas. Figure 17, p. 15 with permission.*

Channel Flow With Electromagnetic Acceleration of Gas to Supersonic Conditions

In the work presented above (Section Channel Flow: Steady and 1D) the effects of electromagnetic fields on ionized gases in steady, 1D, constant-area, inviscid flow were introduced. It was noted for that restricted ($T =$ constant) analysis that the equation: $u = \frac{E}{B}\left(1 - \frac{RT}{u^2}\right)$, displays a singular point, and we took: $u > c_s$ to enable solution. The topic of acceleration from static (or subsonic) conditions through sonic (M = 1) or "choking" plasma flow conditions was noted as significant, and it will now be examined further. The "sonic" transition is a unique and important consideration, and it continues to be a defining factor in the prediction of plasma flow behavior.

In the work of Resler and Sears (1958a) the plasma was assumed to behave as an ideal gas, and the passage through sonic conditions was found to be possible only through "tunnel" conditions where (the symbols of the authors were u_1, u_2, u_3) the condition to be satisfied by the fields was $u(\text{sonic}) = u_3 = E/H$. So it can be seen that for that unique and restrictive gas condition, the fields did control the flow behavior. This analysis did rely on primarily thermodynamic considerations for the flow behavior. Other work on accelerators and MHD generators proceeded similarly (Resler and Sears, 1958b; Culick, 1964; Sutton and Sherman, 1965), and velocity constraints on the field conditions in different combinations were identified. These analyses resulted in the definition of unique flow and field conditions necessary for the acceleration depending on the assumptions. There were few available data for comparison with theory or to allow guidance from experiment, which was an impediment to progress.

Later work incorporated equations of state that included appropriate terms for higher temperature gases as an improvement; the solutions became more relevant and allowed comparison with experimental data that was becoming available. In particular, data from MagnetoPlasmaDynamic (MPD) thruster experiments, in which gases were ionized by electrode discharges and accelerated to high speed in a nozzle expansion configuration, could be considered and evaluated with respect to channel flow theory. One such analysis (Lawless and Subramaniam, 1987) was based on an extension of the 1D equations presented above (Section Channel Flow: Steady and 1D) for mass, momentum, and Ohm's law, and added Ampere's law as follows:

(Note: Properties marked with* are at sonic conditions, $a^2 = \frac{5}{3}\frac{kT}{m_A}$):

Mass:

$$\rho u = F = \rho^* a^*;$$

Momentum:

$$p + Fu + \frac{B^2}{2\mu_0} = p^* + Fa^* + \frac{B^{*2}}{2\mu_0};$$

Ohm's and Ampere's laws:

$$J = \sigma(E - uB) = -\frac{1}{\mu_0}\frac{dB}{dx}.$$

Energy equation:

$$h + \frac{u^2}{2} + \frac{EB}{\mu_0 F} = h^* + \frac{a^{*2}}{2} + \frac{EB^*}{\mu_0 F}.$$

Equation of state (fully ionized plasma) was

$$h = h(p, \rho) = \frac{5\,k\,T}{m_A} + \frac{\epsilon_i}{m_A},$$

and this set allowed a relevant solution of flow acceleration in the channel geometry.

The velocity gradient expression was found to be singular at $M = 1$ as in ordinary fluid mechanics. In this model of the plasma flow, in order for the flow to be continuous, it is required that:

$$E = \rho^* a^* B^* \frac{\partial h}{\partial P}\Big|_{\rho^*} = \frac{5}{2} a^* B^*.$$

This is a choking condition that must be satisfied; it is distinctly different from that obtained for nonionized gases at the sonic condition in that it incorporates combined physical conditions on the gas reactions and magnetic field parameters.

An analytic solution was obtained for "frozen flow" (fully ionized), as:

$$v = a^* \left[-\frac{\zeta}{2} \pm \frac{\left(\zeta^2 - 4\xi\right)^{1/2}}{2} \right],$$

where, + is for supersonic and − is for subsonic flow, and:

$$\zeta = \frac{5}{8} S^* \left(\frac{B}{B^{*2}} - 1 \right), \quad \xi = \frac{5}{4} S^* \left(1 - \frac{B}{B^*} \right) + 1,$$

with:

$$S^* = \frac{B^{*2}}{\left(\mu_0 \rho^* a^{*2}\right)},$$

which is the ratio of magnetic to kinetic gas pressure. One result of this analysis was found to support experiment results. Data for thrusters were found to scale as: $(I^2/\dot{m})$, the discharge current squared divided by the mass flow rate through the thruster; the above analytic result demonstrates that

$$(I^2/\dot{m}) \approx \frac{(w/h)a^*}{\mu_0} S^*,$$

showing the same: $a^* S^*$ dependence.

A later work (Subramanium and Lawless, 1988) included the effects of finite rate ionization in a similar analysis; an analytic solution was not possible, but the solution was carried out using numerical computation methods. The analysis based on this model was compared to experimental data for axisymmetric accelerators at one set of operating conditions and showed reasonable agreement with trends that were observed.

The important point to note in this discussion of simplified analysis of 1D channel flow is that continuum equations that incorporate physically consistent modeling of gas, energy addition, and momentum expressions do predict trends that are observed in experiment. A second point is that the coupling of fluid and electromagnetic equations produces results that are unique in their physical behavior. An aspect worth noting is that attempts for more exact solution of problems such as this modeling of plasma thrusters, of necessity, become more complex in physics and require advanced mathematical and numerical techniques.

Flow Control Utilizing Plasma Interactions

Introduction

The inherent ability of an electric discharge in gases to introduce energy and body force in a local area holds considerable attraction for control of flow, particularly high-speed flow. Early studies (Mishin et al., 1981) addressed the possibility of mitigating or dispersing shock waves in weakly ionized plasmas. Understanding of gas discharges at higher pressures (0.1–1 atm) assisted the application to experimental study of discharges with lower temperature (Becker et al., 2005).

Shock Wave Dispersion

Studies of the dispersion of shock waves in plasma (Marcheret et al., 2001; Merriman et al., 2001) examined the energy deposition from the discharges and potential electromagnetic force application. Experimental and computational studies confirmed that shock attenuation, acceleration, and broadening are the result of thermal energy deposition.

Flow Control

The evidence of controlled energy deposition by gas discharges in high-speed flows has generated interest in the use of plasma discharges actuating control of the flow field. A study by Shin et al. (2007) examined the effects of diffuse and constricted local discharges on shock wave and boundary layer behavior; it was found that there was improved control with diffuse discharges. Specifically, the effectiveness of plasma discharges for flow control depends upon the dispersion of the discharge field for pressure enhancement.

Localized Flow Activation

A developing area with considerable potential for application is the utilization of localized sources of high-pressure plasma jets. High voltages generate localized discharge arcs, with the formation of "jets and bullets" within a convecting weakly ionized plasma flow. An examination of localized arc discharge actuators has been carried out, and the effects on shock wave and boundary layers have been studied (Narayanaswamy et al., 2010). The influence of localized arc filament plasma actuators has also been studied to define the possible effects on acoustic noise mitigation; operating conditions were defined where noise reduction was evident (Samimy et al., 2007).

REFERENCES

Becker, K.H., Kogelschatz, U., Schoenbach, K.H., Barker, R.J. (Eds.), 2005. Non-Equilibrium Air Plasmas at Atmospheric Pressure. IOP, London.

Boyd, T.J.M., Sanderson, J.J., 1969. Introduction to Plasma Dynamics. Barnes & Noble, New York.

Culick, F.E.C., 1964. Compressible magnetogasdynamic channel flow. ZAMP 15, 126–143.

Galanga, F.L., Lineberry, J.T., Wu, Y.C.L., et al., 1982. Experimental results of the UTSI coal-fired MHD generator. J. Energy 6, 179–186.

Jahn, R.G., 1969. Physics of Electric Propulsion. McGraw-Hill, New York.

Lawless, J.L., Subramaniam, V.V., 1987. Theory of onset in magnetoplasmadynamic thrusters. AIAA J. Propul. Power 3, 123–127.

Macheret, S.O., Ionikh, Y.Z., Chernysheva, N.V., Yalin, A.P., Miles, R.B., 2001. Shock wave propagation and dispersion in glow discharge plasmas. Phys. Fluids 13, 2693–2705.

Merkle, C.J., Moeller, T., Rhodes, R., Keefer, D., 2009. Computational simulations of power extraction in MHD channel. In: AIAA Paper 2009-3827, 40th Plasmadynamics and Lasers Conference. Texas, San Antonio.

Merriman, S., Ploenjes, E., Palm, P., Adamovich, I.V., 2001. Shock wave control by non-equilibrium plasmas in cold supersonic gas flows. AIAA J. 39, 1547–1552.

Messerle, H.K., 1995. Magnetohydrodynamic Power Generation. Wiley, New York.

Mishin, G.I., Bedin, A.P., Yushchenkova, N.I., Skvortsov, G.E., Ryazin, A.P., 1981. Anomalous relaxation and instability of shock waves in plasmas. Soviet Phys. Tech. Phys. 26, 1363.

Narayanaswamy, V., Clemens, N.T., Raja, L.L., 2010. Investigation of a pulsed-plasma jet for shock/boundary layer control. In: AIAA 2010-1089, 48th AIAA Aerospace Sciences Meeting, Orlando, FL.

Resler Jr., E.L., Sears, W.R., 1958a. The prospects for magneto-aerodynamics. J. Aeronaut. Sci. 25, 235–245.

Resler Jr., E.L., Sears, W.R., 1958b. Magneto-gasdynamic channel flow. Z. Angew. Math. Phys. (ZAMP) 9b, 509–518.

Samimy, M., Kim, J.H., Kastner, J., Adamovich, I., Utkin, Y., 2007. Active control of a Mach 0.9 jet for Noise Mitigation Using Plasma Actuators. AIAA J. 45, 890.

Shercliff, J.A., 1965. A Textbook of Magnetohydrodynamics. Pergamon, New York.

Shin, J., Narayanaswamy, V., Raja, L.L., Clemens, N.T., 2007. Characterization of a Direct-Current Glow Discharge Plasma Actuator in a Low-Pressure Supersonic Flow. AIAA J. 45, 1596–1605.

Sovie, R.J., Nichols, L.D., 1974. Closed Cycle MHD Power Generation Experiments in the NASA Lewis Facility. NASA TMX-71510, Cleveland, Ohio.

Subramaniam, V.V., Lawless, J.L., 1988. Onset in magnetoplasmadynamic thrusters with finite-rate ionization. AIAA J. Propul. Power 4, 526–532.

Sutton, G.W., Sherman, A., 1965. Engineering Magnetohydrodynamics. McGraw-Hill, New York.

Plasma Thrusters

INTRODUCTION

The development of devices for plasma propulsion in space has been of interest for more than 50 years (Choueiri, 2004). For a thruster in space, we can express the thrust as: $T = \dot{m} \cdot u_{ex}$, which is the product of propellant mass flow rate and exhaust velocity relative to the rocket.

The promise of high specific impulse ($I_{sp} = u_{ex}/g_0$, g_0 is Earth gravity) provides the possibility of deep space missions that are otherwise unattainable. This can most easily be seen in the mission equation:

$$\frac{m_f}{m_i} = e^{-\frac{\Delta v}{u_{ex}}},$$

where the final spacecraft mass (m_f) is strongly influenced by the propulsion exhaust velocity (u_{ex}) for a mission characteristic velocity (Δv).

There have been extensive, sustained research programs for a number of types of thrusters, most notably ion thrusters, arc thrusters, and pulsed plasma thrusters (PPTs), and more recently Hall thrusters. These devices have been the subject for research and development, as documented in journal articles, monographs, and textbooks (Jahn, 1969; Goebel and Katz, 2008). As the present work seeks to define the more general area of plasma interactions and flow phenomena, we will review a number of thruster topics conceptually, from the fundamental point of view, rather than to attempt to describe details of development of working thrusters, which has been well documented in other material. The goal of the present work, in this specific case and in general, is to provide fundamental understanding such that the reader is prepared and oriented to pursue further detailed work and study on advanced topics of interest.

ELECTROMAGNETIC TERMS AFFECTING PLASMA MOMENTUM AND ENERGY

For systems whose goal is the generation of thrust in space, the presence of plasma in electromagnetic fields will result in an alteration of plasma momentum and energy. Thruster configurations seek to apply these interactions to create a high specific impulse (exhaust velocity). In order to emphasize the terms and mechanisms that are operative and dominant, we repeat here, and examine, the relevant single fluid equations of plasma flow.

Momentum:

$$\rho\left(\frac{\partial \mathrm{v}}{\partial \mathrm{t}}+(\vec{\mathrm{v}}\cdot\nabla)\vec{\mathrm{v}}\right)=-\nabla \mathrm{p}+\nabla\cdot\overleftrightarrow{\tau}+\rho_{\mathrm{e}}\vec{\mathrm{E}}+\vec{\mathrm{J}}\times\vec{\mathrm{B}},$$

where $\overleftrightarrow{\tau}$ is a viscous stress with components: $\tau_{ij}=\overline{\mu}\left(\frac{\partial \mathrm{u}_i}{\partial \mathrm{x}_j}+\frac{\partial \mathrm{u}_j}{\partial \mathrm{x}_i}\right)-\frac{2}{3}\overline{\mu}(\nabla\cdot\mathrm{u})\delta_{ij}$.

Energy:

$$\rho\frac{\mathrm{D}}{\mathrm{Dt}}(\overline{e}_m)=-\nabla\cdot\mathrm{pv}+\nabla\cdot\vec{\mathrm{Q}}+\nabla\cdot(\vec{\mathrm{v}}\cdot\overleftrightarrow{\tau})+\vec{\mathrm{J}}\cdot\vec{\mathrm{E}},$$

with:

$$\overline{e}_m=e_{int}+\frac{\mathrm{v}^2}{2},$$

and for the caloric equation:

$$e_{int}=C_V T.$$

Ohm's law:

$$\vec{\mathrm{J}}=(\rho_{\mathrm{e}}\vec{\mathrm{v}}+\vec{\mathrm{J}}_{\mathrm{cond}}),$$

where:

$$\vec{\mathrm{J}}_{\mathrm{cond}}=\sigma(\vec{\mathrm{E}}+\vec{\mathrm{v}}\times\vec{\mathrm{B}}),$$

and $\sigma=\sigma_{\parallel},\sigma_{\perp}$ ($\parallel$, $\perp$ to $\vec{\mathrm{B}}$); a generalized Ohm's law can include Hall effect and ion slip.

So:

$$\vec{\mathrm{J}}\cdot\vec{\mathrm{E}}=\vec{\mathrm{v}}\cdot(\rho_{\mathrm{e}}\vec{\mathrm{E}}+\vec{\mathrm{J}}_{\mathrm{e}}\times\vec{\mathrm{B}})+\frac{\mathrm{j}_{\mathrm{e}}^2}{\sigma}=\vec{\mathrm{v}}\cdot(\vec{\mathrm{F}}_{\mathrm{EM}})+\frac{\mathrm{j}_{\mathrm{e}}^2}{\sigma},$$

or:

$$\text{Energy Addition}=\text{Work}+\text{Heat (joule)}.$$

PPTs (Electromagnetic: Pulsed, Unsteady)

PPTs (Pulsed Plasma Thrusters) are unique; they allow intense electromagnetic discharge, and they have been employed on spacecraft earlier than any other type of plasma thruster. Their high power (megawatt) is achieved by applying limited energy in short time periods (microseconds), and so they possess inherent high currents and $\vec{\mathrm{J}}\times\vec{\mathrm{B}}$.

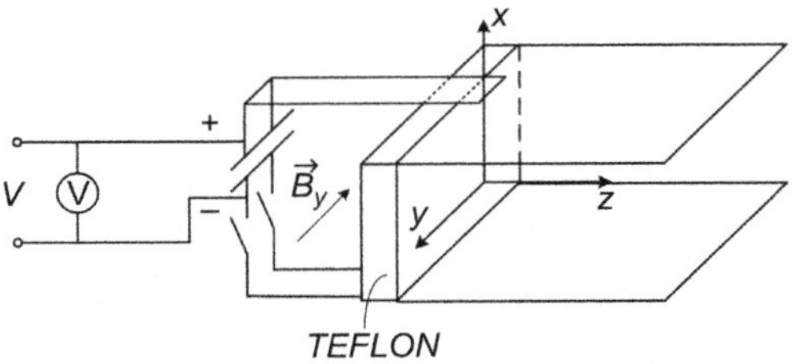

Figure 9.11 Schematic of rectangular PPT geometry.

The transient electromagnetic terms in PPT thrusters require that electrical terms and circuit equations receive careful attention. A schematic of a rectangular thruster is shown in Figure 9.11. Current flow (I) resulting from a capacitor discharge creates a current sheet (J_x) at the minimum inductance chamber boundary (Teflon) and a magnetic field (B_y) is formed in the circuit loop, thus creating $(\vec{J} \times \vec{B})_z$, which accelerates propellant mass along the electrodes (z).

An examination and analysis of the pulsed thruster electrical circuit equation has been presented by Jahn (1969), and these ideas will now be outlined. The voltages in the discharge circuit can be expressed:

$$\mathrm{V} = \mathrm{IR} + \dot{\varphi} = \mathrm{IR} + \frac{\mathrm{d}}{\mathrm{dt}}(\mathrm{LI}) = \mathrm{IR} + \mathrm{L}\dot{\mathrm{I}} + \mathrm{I}\dot{\mathrm{L}},$$

where V, I, and R are the circuit voltage, current, and resistance, respectively; L is the inductance; and φ is the circuit magnetic flux. The power balance for the capacitor discharge current flow is:

$$\mathrm{P} = \mathrm{IV} = \mathrm{I}^2\mathrm{R} + \mathrm{LI}\dot{\mathrm{I}} + \mathrm{I}^2\dot{\mathrm{L}},$$

or:

$$\mathrm{IV} = \mathrm{I}^2\mathrm{R} + \frac{\mathrm{d}}{\mathrm{dt}}\left(\frac{1}{2}\mathrm{LI}^2\right) + \frac{1}{2}\mathrm{I}^2\dot{\mathrm{L}}.$$

So, power is delivered to resistance dissipation, magnetic field energy and work done to drive propellant mass in acceleration by the current sheet. In such pulsed devices, the electromagnetic force component is of primary interest, as it can drive the mass to the desired high exhaust velocity (specific impulse). The resistance heating term will result in electrothermal acceleration of the heated particles, which has limited effectiveness as a thrust component. The magnetic field term can result in periodic discharges until ultimate dissipation, but later discharges are less intense and the acceleration is generally electrothermal in nature. The PPT can have various geometries (rectangular, cylindrical) in open or enclosed rail passages to exit.

For a space thruster system, both mass and energy of propellant are critical properties for a given mission. In the basic sense, energy supply can be considered as an independent term (by variation of pulse frequency), but most imperative for mission accomplishment is the need for maximum exhaust velocity of limited propellant mass. This means that the maximum electromagnetic acceleration is most desirable.

The optimization of performance of the PPT depends upon understanding the energy deposition terms for acceleration, heating, and field energy. We take the rectangular channel with dimensions: w (width), h (height), and $l_{c,t}$ (total length). With current, I, starting at $t = 0$, resulting in a current sheet (J_x in x, y plane) at an initial z location ($z = 0$), we can apply Maxwell's equation:

$$\frac{1}{\mu_0}\oint \vec{B}\cdot d\vec{l} = \int_A \vec{J}\cdot\vec{n}\, dA,$$

to determine that:

$$B_y = \frac{I\mu_0}{w}.$$

Applying that result:

$$\int_A \vec{B}\cdot\vec{n}dA = \frac{I\mu_0}{w}\int_{z_0}^{z} hdz = I\cdot\mu_0\frac{h}{w}z,$$

and as:

$$-\int_A \vec{B}\cdot\vec{n}dA = -LI$$

so:

$$L = \mu_0\frac{h}{w}z.$$

As the electric discharge initiates at the position of minimum inductance, it forms at $z = 0$. The magnetic field within the current loop acts to accelerate the current sheet with $\vec{J}\times\vec{B}$ and accelerates the plasma in the z-direction. The force interaction is

$$\vec{f}\left(\frac{force}{vol}\right) = \vec{J}\times\vec{B} = \vec{J}_x\times\vec{B}_y \sim \frac{I^2\mu_0}{w^2z},$$

with a total force:

$$F_z(z) = \int_{z=0}^{z} f_z(z)dVol. = \int_{0}^{z} \frac{I^2\mu_0}{w^2 z} \cdot whdz = I^2\mu_0 \frac{h}{w} \ln z.$$

The power addition in electromagnetic acceleration can be evaluated as:

$$P = \left(\frac{1}{2}\dot{L}I^2\right) = \frac{1}{2} I^2\mu_0 \frac{h}{w} \dot{z}.$$

Typical PPTs will utilize mass by ablation of solid propellant material at the back surface of the channel during the initial discharge, and so we can approximate that the mass being accelerated is constant (M), so:

$$P = F_z \cdot \dot{z} = (M\ddot{z})\dot{z} = \frac{1}{2} I^2\mu_0 \frac{h}{w} \dot{z},$$

and for constant current:

$$\dot{z} = \frac{1}{2} \frac{I^2}{M} \mu_0 \frac{h}{w} z,$$

which indicates exhaust velocity: $u_{ex} \sim I^2$ and $\sim z$, with thrust, $T \sim I^2$. Typically, a capacitor discharge drives the PPT acceleration and has a current form with $I(t) = I_0 \sin \omega t$. This allows a relatively straightforward analysis, with exhaust of the plasma within the half cycle (reversal) time of the current.

The optimization of PPT performance involves many factors in the plasma creation and acceleration processes. The general description, above, simply emphasizes components for energy deposition. Some details will be given further attention here to emphasize factors that can influence improved performance. The thermal energy deposition (heating) will be absorbed by ablation, dissociation, atomic excitation and ionization, and radiation. All these processes require analysis, but, in fact, are difficult to model in a continuum, nonequilibrium plasma. Most particularly, they are lumped into the IR term (j^2/σ, current density^2/conductivity). The initial ionization and subsequent current flow are strongly influenced by the atomic and plasma properties of the propellant gas. The energy deposition term for acceleration: $(\vec{J} \times \vec{B}) \cdot \vec{v}$, with local current flow and field interaction, ultimately dictates the efficiency and final ejection velocity.

Solid Teflon ablation has been almost universally used as a propellant material, but, while reliable in space, exhaust velocities were modest. Basic studies indicated that water could be an alternative to provide improved discharges and propellant acceleration. Typical discharge properties and

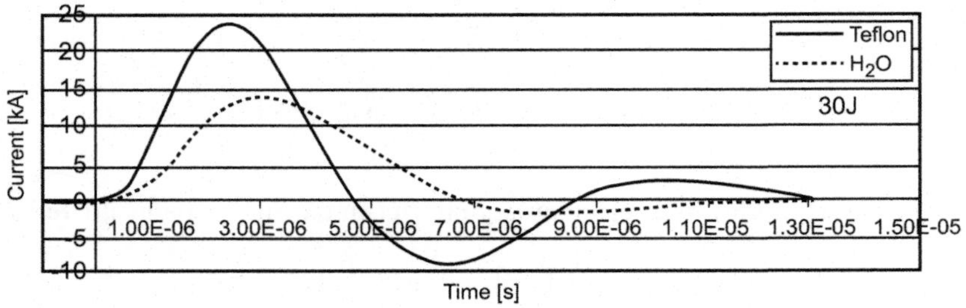

Figure 9.12 PPT discharge current history for Teflon and water propellant. *Scharlemann, C.A., York, T.M., 2003, Pulsed plasma thruster using water propellant, Part I: Investigation of thrust behavior and mechanism. AIAA-2003-5022. Figure 2, p. 3 with permission.*

experimental results for propellant acceleration of Teflon and water have been reported (Scharlemann and York, 2003). For a discharge circuit with $30\,J$ capacitor, the history of discharge current is shown in Figure 9.12.

A discharge forms at $z = 0$, and two significant half cycles are shown. The discharge history shows a large difference in circuit parameters due to propellant plasma properties, and this is indicative of significant differences in energy deposition. The ionization, conductivity, and transport properties of water are markedly different from those of Teflon. At 30 J and water metered at 1.1 μg/s and Teflon ablation at 36 μg/s, the response of a biased Langmuir probe for both propellants is shown in Figure 9.13 for probe locations of 2.5 and 5 cm from the exit plane.

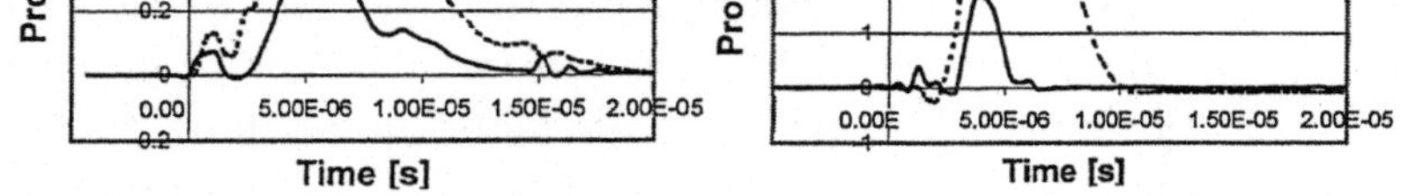

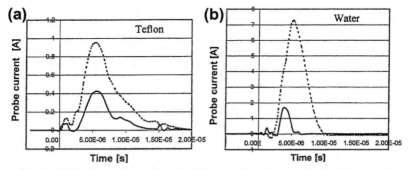

Figure 9.13 Comparison of constant bias Langmuir probe signals for Teflon and water propellants at 2.54 cm (dashed line) and 5.04 cm (solid line) downstream of the thruster exit. *Scharlemann, C.A., York, T.M., 2003, Pulsed plasma thruster using water propellant, Part I: Investigation of thrust behavior and mechanism. AIAA-2003-5022. Figure 5, p. 5 with permission.*

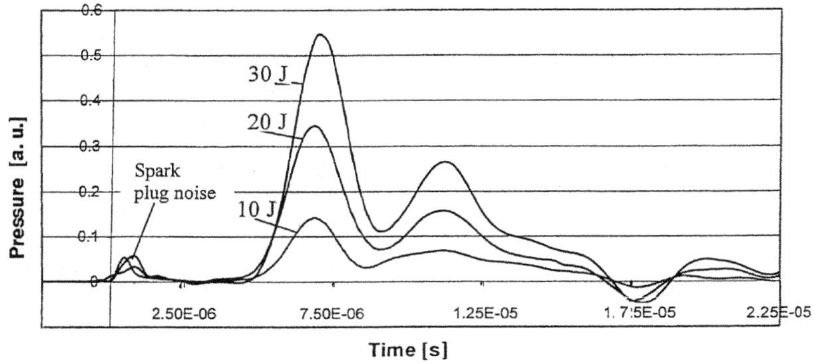

Figure 9.14 Signals of (flow momentum) impact pressure probe for different discharge energies at a 5.1 cm distance from thruster exit for Teflon propellant. *Scharlemann, C.A., York, T.M., 2003, Pulsed plasma thruster using water propellant, Part I: Investigation of thrust behavior and mechanism. AIAA-2003-5022. Figure 7, p. 7 with permission.*

It can be seen that the plasma signal rose much more sharply and arrived at an earlier time for water, clearly indicative of enhanced electromagnetic acceleration. Measurements of momentum flux (impact pressure) were made and are shown in Figure 9.14.

The momentum data verified the Langmuir probe electron density indications of plasma behavior; it also showed the significant differences in plasma mass flux history for different capacitor energies. Thrust performance of the PPT for the different propellants based on calibrated measurements is shown in Table 9.1.

The results presented include a thrust stand measurement (GRC, NASA Glenn Research Center) to compare with calibrated local measurements. The local probe measurements indicate that details of plasma interactions, such as current density geometry, ionization, and effective circuit elements in plasma devices are key to understanding and improving energy delivery and momentum deposition in the plasma.

MPD and Applied Field MPD Arcs (Electromagnetic: Steady or Quasisteady)

The MagnetoPlasmaDynamic (MPD) arc configuration was first reported in Ducati et al. (1964) as a high performance space thruster. The configuration involved the application of high currents (>1000 A) with low mass flow rates (0.1 g/s) in a relatively standard arcjet device, however, the reported data showed startling improvements in performance: very high exhaust velocities.

Table 9.1 Thruster Performance for the Water and Teflon Mode (Scharlemann and York, 2003)

Discharge Energy (J)	Impulse Bit (GRC) (μN-s)		Impulse Bit Pressure Probe (μN-s)		Mass Bit (μg/Discharge)		Specific Impulse (S)		Efficiency	
	Teflon	Water	Teflon	Water	Teflon	Water	Teflon	Water	Teflon	Water
10	122	–	124	47	11.9	~1.1	1060	4355	6.5	10
20	273	–	281	90	27.5	~1.1	1040	8340	7.2	18
30	440	–	440	128	35.3	~1.1	1270	11,860	9.1	24.8

(10^5 m/s) were attributed to electromagnetic effects. As the experimental performance showed exceptional potential for advancements in space propulsion, substantial research efforts were initiated. In fact, the high exhaust velocities have been verified, and the propellant acceleration has been identified with definable electromagnetic interactions (Jahn, 1969). A standard axisymmetric configuration with axial cathode and surrounding anode was associated with the formation of strong azimuthal magnetic "self-fields" (B_θ) which interact with "blowing" ($J_r \times B_\theta$) and "pumping" ($J_z \times B_\theta$) modes. While exceptionally high exhaust velocities were identified in the MPD arc, just as with normal arcjets, serious problems with electrode erosion and discharge stability were found to be inherent. Also, as currents on the order of kA result in power supply requirements on the order of MW, this presents a problem for potential space deployment and is even a difficulty for laboratory testing. This problem may be mitigated for laboratory testing and possible space application operation by energy storage in capacitors and using periodic discharges that run for milliseconds to as long as seconds.

Self-Field MPD

The mechanisms involved in the plasma–current interactions are of considerable interest and will be reviewed here. The processes are complex in the MHD representation, inherently involving models for transport (viscosity, conductivity and diffusion), and so, emphasis has generally been placed on identifying outcomes of the operation of the plasma discharge devices, such as thrust relationships, and to search for parametric relationships. A comprehensive discussion of these interactions and issues has been presented in the literature (Choueiri, 1998) and, related to that work, those issues will be reviewed here. A schematic of the self-field MPD is presented in Figure 9.15.

Again, the key interactions are: $J_r \times B_\theta$ and $J_z \times B_\theta$. An analytical evaluation of the thrust can be carried out by integration of $\vec{J} \times \vec{B}$ through the discharge volume as:

$$\mathrm{T} = \int \int_{r_c}^{r_a} f_z(\mathrm{z}, \mathrm{r})\mathrm{dVol} \approx \int_{r_c}^{r_a} \frac{\mathrm{I}}{2\pi\mathrm{r}\cdot\mathrm{L}} \cdot \frac{\mathrm{I}\mu_0}{4\mu\mathrm{r}} \cdot 2\pi\mathrm{rLdr} = \mathrm{I}^2\left(\frac{\mu_0}{4\pi}\right)\ln\left(\frac{r_A}{r_C}\right).$$

The fundamental functional form here is:

$$\mathrm{T} \approx (\mathrm{Const.})\mathrm{I} \cdot B_{self-field}.$$

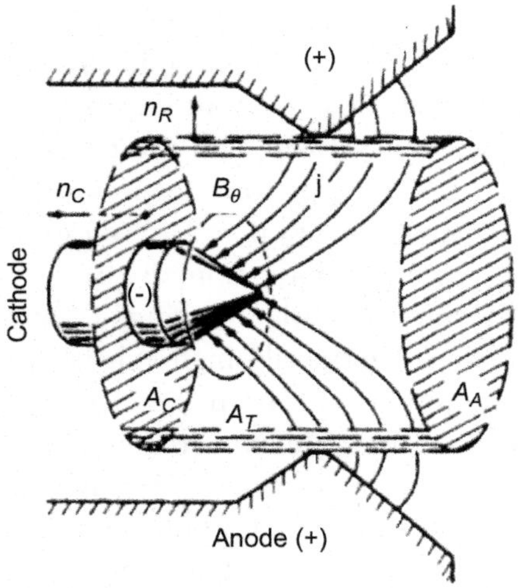

Figure 9.15 Geometry, currents and magnetic fields in the self-field MPD thruster. *Choueiri, E., 1998. Scaling of thrust in self-field magnetoplasmadynamic thrusters, AIAA J. Propul. Power 14, 744–753. Figure 5, p. 749 with permission.*

A formula incorporating both internal channel and exhaust regions takes the form:

$$T = \left(\frac{\mu_0}{4\pi}\right)\left[\ln\left(\frac{r_A}{r_C}\right) + \frac{3}{4}\right] \cdot I^2.$$

This was first derived by Maecker (1955) and also presented in Jahn (1969); an extension incorporating a solution of channel flow using single fluid equations was later reported by Tikhonov et al. (1993). The functional behavior is basically captured in the dimensionless Thrust Coefficient, as: $C_\tau = \left(\frac{4\pi}{\mu_0}\right)\left(\frac{T}{I^2}\right)$. Comparisons of the formulations with data are presented in Figure 9.16.

Most fundamentally, the simple formula (Jahn, 1969) for electromagnetic acceleration does capture the magnitude of the interactions at higher current ($I > 10{,}000$ A), where data display the dominance of electromagnetic effects. The experimental data presented are for what is termed a "full-size" MPD with $\dot{m} = 3.6$ g/s and $I > 2000$ A. Comprehensive evaluation of the thrust components was carried out (Choueiri, 1998), and the result is shown in (Figure 9.17).

The thrust at higher currents is generated predominantly by the (Back Plate) pinching (BP_p) and blowing (BP_b) effects expressed in terms noted above (A, Anode; C, Cathode; O, Outer; I, Inner; F, Face; T, Tip).

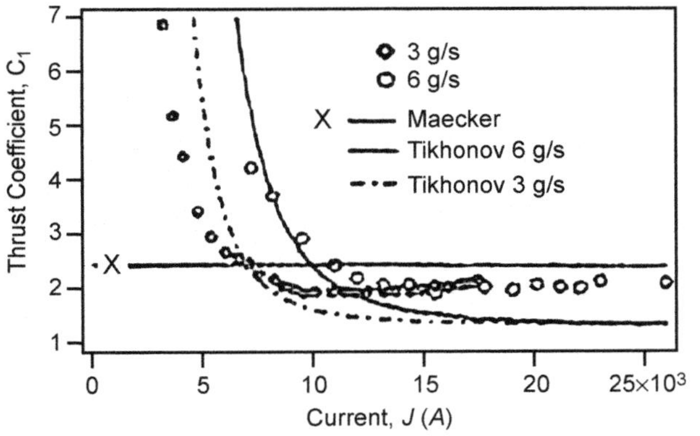

Figure 9.16 Comparison of self-field MPD thrust measurements with analytical formulas. *Choueiri, E., 1998. Scaling of thrust in self-field magnetoplasmadynamic thrusters. AIAA J. Propul. Power 14, 744–753. Figure 3, p. 746 with permission.*

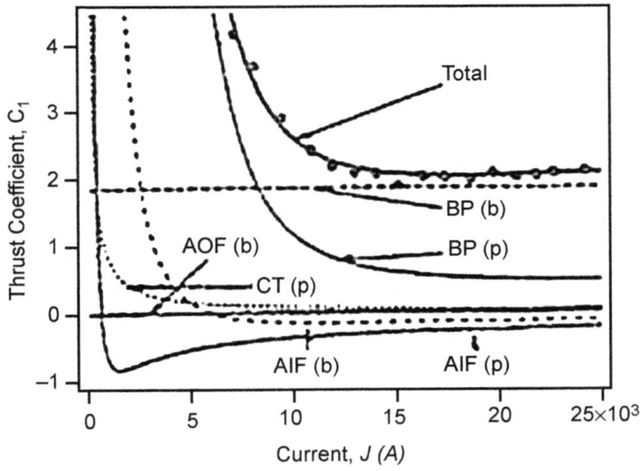

Figure 9.17 Comparison of calculated components of electromagnetic thrust with data for an MPD. *Choueiri, E., 1998. Scaling of thrust in self-field magnetoplasmadynamic thrusters. AIAA J. Propul. Power 14, 744–753. Figure 8, p. 750 with permission.*

As can be seen (Figure 9.16), the behavior of the MPD at currents below 10,000 A does not fit the simple electromagnetic acceleration model, and so the processes are decidedly more complex. In fact, it appears that there are two regimes of operation: (1) dominant electromagnetic mode, and (2) multiple interaction mode. Examination of the data shows that the effect of lower mass flow rate is to change this transition point; the ratio of

$\left(\frac{I^2}{\dot{m}}\right)$ does appear related. The behavior in the multiple interaction mode at lower current was (Choueiri, 1998) modeled utilizing relationships with dimensionless variables. One of the important variables is the Scaling Factor, expressed as:

$$\xi = \frac{I}{I_{Ci}},$$

where:

$$I_{Ci} = \left[\frac{\dot{m} u_{Ci}}{(\mu_0/4\pi) C_T}\right]^{1/2},$$

and where:

$$u_{Ci} = \left(\frac{2\varepsilon_i}{m_a}\right)^{1/2},$$

the Alfven critical speed, which is the flow speed where energy equals the ionization energy, ε_i, of an atom with mass, m_a. There is no universal analytical relationship of these variables to fit the lower current data, so empirical modeling was carried out to develop a relationship that could follow trends in the variations in data. An expression relating the thrust coefficient has the form:

$$C_T = \left(\frac{\nu}{\xi^4}\right)_{low\ \xi} + \left[\ln\left(\frac{r_a}{r_c} + \xi^2\right)\right],$$

with component terms for low ξ and high ξ, and $\nu = \frac{\dot{m}}{\dot{m}^*}$, where $\dot{m}^*$ is a parameter derived from thruster properties.

While considerable progress has been made in developing understanding of the self-field MPD acceleration processes, there are aspects of MPD behavior that require further study in order to achieve operational viability. Related to higher currents and lower mass flow rates (a regime termed "starvation") there is the "onset" of an instability in the discharge; again, the ratio. $\left(\frac{I^2}{\dot{m}}\right)$ does appear to be related to this event. This behavior has been typified by high frequency voltage oscillations, and one work has shown this to be related to the formation and collapse of anode spots or concentrations of current which disrupt the uniform acceleration process (Uribarri and Choueiri, 2008). The occurrence of microinstabilities has also been identified (Choueiri et al., 1987), and as these are current driven, they are problematic for stable operation as high currents are necessary as the acceleration source.

While much attention has been directed to the plasma formation and acceleration relationships, there is not a good understanding of the plasma termination of its interaction with magnetic fields in the guiding and expansion processes of the exhaust plasma and the ultimate ejection of the plasma from the thruster. This event does relate to the collisional nature of the expansion and/or plasma-field flux transport. Some reported work has identified basic mechanisms in the detachment, and that work provides analysis which can assist prediction of performance (Aheo and Merino, 2010; Hooper, 1993).

Applied Field MPD

Application of applied magnetic fields (AF) to the MPD configuration (Figure 9.18) was a rational step to attempt to alleviate limitations on behavior of self-field thrusters, such as the onset of instabilities in order to achieve more stable operation (Seikel et al., 1970).

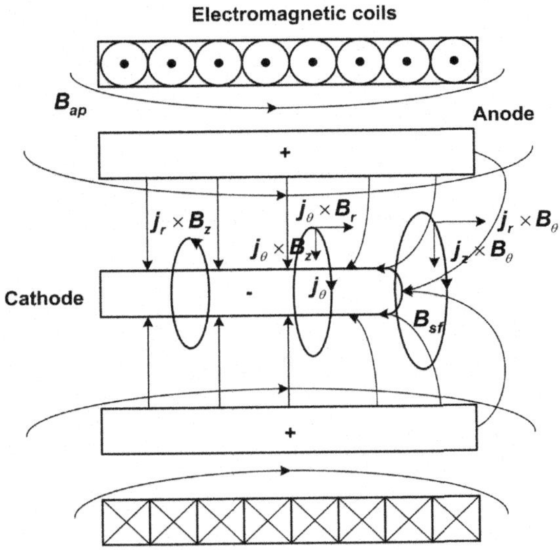

Figure 9.18 Geometry and magnetic field interactions in the AF MPD. *Tang, H.B., Cheng, J., Liu, C., York, T.M., 2012. Study of applied magnetic field magnetoplasma-dynamic thrusters with particle-in-cell and Monte Carlo collision. II. Investigation of acceleration mechanisms. Physics of Plasmas 19, 073108. (Figure 1(b) Applied-field MPD thruster) with permission.*

Applied fields from cylindrical coils surrounding the anode generate solenoidal axial magnetic fields in the body of the thruster and diverging fields with axial (B_z) and radial (B_r) fields at the exit in the expansion region. These fields are responsible for the creation of azimuthal currents (J_θ), which then interact with the fields to produce Hall force components ($\overrightarrow{J_\theta} \times \overrightarrow{B_z}$ and $\overrightarrow{J_\theta} \times \overrightarrow{B_r}$).

In early studies in conventional configurations, some improvements in performance were evident with the AF-MPD (Bishop et al., 1971); however, the force interactions are inherently complex and so were difficult to analyze. Experimental studies have been a primary basis for gaining insight. In some experiments at NASA with very high applied fields, significant enhancement of thrust above self-field values was noted (Michels and York, 1972). A pulsed, high-power MPD (3 g/s, 5–20 kA) with superconducting magnet with field strengths of 2.0 T was tested. Thrust was evaluated from local momentum flux (impact pressure) in the exhaust; thrust was approximately linear ($T \sim B$) at lower fields, but at higher thrust was proportional to a higher power of B. The radial profiles of impact pressure showed definable effects of plasma being guided along the diverging field lines, but thrust was not directly measured. Accordingly, as with the self-field MPD, it is probable that the identification of parametric behavior for operation at a specific AF-MPD power density or field regime may not result in performance characteristics that are universal. Because of the burden of power supply for an MPD, research studies have been directed to lower power levels and lower applied magnetic fields.

Following earlier reported results from Russian laboratories, Fradkin et al. (1970) investigated a lower power lithium propellant device (25 kW) and modest applied fields; what was exceptional in the experiments were the very high exhaust velocities and efficiencies. A proposed model of interactions in the thruster included azimuthal (v_θ) swirling motion induced by $J_r \times B_z$ (AF); this rotation was assumed to be solid body and was to be completely converted to an axial thrust component. Using this model, the thrust relationship was predicted to be linear with discharge current and applied field strength, as:

$$\mathrm{T_{AF}} = \mathrm{f}(\mathrm{r}, \dot{\mathrm{m}}) \cdot (\mathrm{I} \cdot \mathrm{B_{AF}}).$$

The measured thrust in these experiments did show a linear increase with applied magnetic field strength at one current. This was consistent with earlier reported results by others, which showed a linear variation of thrust with the product ($I \cdot B_{AF}$). However, power deposition due to sheath and plasma voltage drops in the experimental data did not scale; the relationship to thruster geometry (anode and cathode radii) was also not compatible with available models. This relationship of thrust to discharge current and magnetic field is similar to that in the self-field MPD, except that applied field dominates self-field effects. So, while improvements with applied fields had been identified, the physical and functional relationships remained to be identified.

Efforts to understand geometric and power scaling of the MPD were pursued at NASA centers to advance the development of operational thrusters. A study of geometric changes in the thruster configuration for an AF-MPD in a lower power range (20–130 kW) as they relate to acceleration mechanisms through measurement of performance parameters was carried out (Myers, 1995). Taking a typical configuration (Figure 9.18) the dimensions of cathode radius, r_c; anode radius, r_a; cathode length, L_c; and anode length, L_a were methodically varied and performance parameters (power, thrust, efficiencies) were evaluated. Discharge voltage was found to increase linearly with applied magnetic field, and input power increased along with thrust. Efficiency and exhaust velocity were also found to increase linearly with AF strength. One result from data evaluation showed that power deposition to the anode was a dominant loss factor, absorbing 50–80% of the input power. Power to the anode could be expressed as:

$$P_a = I_{se}\left(V_a + \frac{5kT_e}{2e} + \o_e\right),$$

where, V_a is the voltage drop at the anode, the second term is the thermal energy delivered from the electrons, and $\o_e$ is the anode work function. Regarding the influence of geometry, the electrode radii were found to have significant effect on the efficiency and specific impulse; this is different from the behavior in the self-field MPD. Due to the applied magnetic field effects, the voltage drop in the anode fall region was seen to increase, an evident reaction to the altered current flow (conduction behavior) in the electrode region.

The effects of applied magnetic fields are significant in two regions and processes: (1) inside the thruster discharge chamber and (2) outside in the diverging fields in the expansion region. In almost all AF-MPD studies, magnetic fields that are generated by a solenoidal coil aligned with the thruster axis penetrate the discharge chamber and near the coil end, and near the thruster exit they create an expanding magnetic field geometry. Any measurement of overall device performance, in effect, combines the results of these interactions, and this will obscure the identification of component contributions. The diverging fields outside the exit have a geometry similar to, and potential function of, a "magnetic nozzle."

In order to separate and examine the effects of thrust chamber and magnetic expansion, a series of investigations was carried out and reported. The first experiment (York et al., 1993) evaluated the behavior of an MPD configuration with no AF in the discharge chamber but with magnetic fields from a solenoidal coil in the exit/exhaust region. This effectively allowed comparison of self-field MPD with an MPD with applied magnetic nozzle in the exhaust region. The study was conducted with a lower power, $^1/_4$-scale standard thruster configuration, with smaller size and lower currents (~1000 A and 100 kW) power. The size scaling attempted to maintain plasma force density to "full-scale" interactions by maintaining $\dot{m}/A$ and $\vec{J} \times \vec{B}$. The force per unit volume in the chamber was:

$$\vec{J} \times \vec{B} \propto \frac{I^2}{r^2 \cdot Z} \propto I^2 z.$$

The exhaust velocity associated with the electromagnetic thrust component, was:

$$U_{em} \propto \frac{I^2}{\dot{m}} \propto \frac{J^2}{(\dot{m}/A_{cs})} \cdot z^2 \propto \frac{JB}{(\dot{m}/A_{cs})} \cdot z.$$

So, this indicated that any reduction of geometric scale does influence the force interaction and exhaust velocity in a definable way. With an anode of 2.5 cm internal diameter and a cathode 0.5 cm outer diameter and 1.25 cm long, a solenoidal coil surrounding the anode extended 5 cm out beyond the face of the anode and the thrust chamber exit. The MPD arc and AF circuits were discharged separately. This was done in order that the applied solenoidal field did not penetrate the arc chamber during the test time but did form a guiding and expanding configuration outside the thruster for the plasma exiting the MPD. The current–voltage data of the thruster showed little difference for the MPD with no AF in the chamber, without or with an applied external magnetic nozzle. However, the plasma flow magnitude and radial profile in the exhaust flow from the thruster were changed significantly by the applied magnetic fields in the expansion flow region. The thrust determined from impact pressure measurements demonstrated an increase as: $T_{ext\ AF} = 1.6 \cdot T_{self\ F}$, due to the applied magnetic fields in the exhaust. So, a distinct improvement in thrust was measured with the addition of applied magnetic nozzle fields in the expansion region in this experiment.

A second experiment (York and Kamhawi, 1993) examined the effects of an extended magnetic field configuration on an AF-MPD. This thruster was operated in the conventional AF-MPD sense with solenoidal magnetic fields in the chamber, but with two different magnetic field geometries in the exhaust region. Configuration (A) had a solenoidal coil that extended 5 cm beyond the thruster exit plane and allowed typical AF magnetic expansion of the plasma. Configuration (B) had a solenoidal coil that extended 10 cm beyond the exit plane and maintained magnetic field magnitudes for 5 cm longer in the axial direction before the magnetic expansion in the exhaust. From records of current–voltage for the thruster, the power deposition was about the same for both configurations. Thrust magnitudes were again calculated from integrated momentum flux (impact pressure) measurements in the exhaust; the thrust evaluation showed: $T_{AF-B} = 1.8 \cdot T_{AF-A}$; so, there was significantly higher thrust with the extended magnetic nozzle (B), while there was a slightly higher power input. With calculation of the values of thrust components, the thrust due to swirl was found to be largest, and there was a significant electrothermal component. It was proposed that the extended magnetic (nozzle) fields in the exhaust allowed for increased thrust, however, the detailed plasma processes and related acceleration mechanisms that brought this about remain to be clarified.

In the continued study of the MPD thruster by NASA, computational modeling of AF-MPD processes using MHD representation of the AF force interactions and plasma behavior was carried out (Mikellides et al., 2000). Continuum plasma physics understanding with transport coefficients that included relevant anomalous effects (resistivity due to microinstabilities) was incorporated. The experiment that was modeled was again a low-power, 100-kW AF-MPD (NASA, argon, 0–0.12 T). The results of the analysis indicated strong effects of plasma rotation due to azimuthal currents and, more importantly in this MHD continuum model, conversion of this energy into thermal thrust due to viscous dissipation. The rotational speed was seen to be proportional to $(I \cdot B)$ and was identified as reaching a limiting value due to viscous dissipation balancing. The discharge voltage was seen to be linear with current (constant B), and thrust was predicted to be linear with applied magnetic field (constant I), which is in agreement with the limited thruster data. The modeling of electrode sheath effects was given careful consideration, but it was defined as problematic for the solution. This type of MHD model of necessity relates physical behavior to definable macroscopic transport properties. However, recent calculations of an AF-MPD using particle-in-cell computations have offered alternative interpretations of processes that may occur (Tang et al., 2012).

A following MHD computational study (Mikellides and Turchi, 2000) developed general analytical scaling relationships for AF-MPD thruster voltage and thrust. Thrust was proposed to behave as:

$$\mathrm{T} \approx R, (\dot{\mathrm{m}}\mathrm{IB})^{1/2},$$

where $R = \mathrm{r_a}/\mathrm{r_c}$, $\dot{\mathrm{m}}$ is the mass flow rate, I is the discharge current, and B is the applied magnetic field strength at the cathode tip. While this is a different functional form than proposed earlier, in the range of data considered, it was compatible. So comparisons of the scaling with data for the 100-kW NASA thruster, including effects of geometry and fields, showed general agreement in this limited performance envelope.

In summary, the AF-MPD is a plasma space thruster that has unique performance and efficiency characteristics that could prove effective to satisfy mission and power requirements in the future. However, in order to achieve that goal, it is evident that there is a need to understand and control the physics of detailed particle interactions in the discharge region and also in the region external to the thruster in the magnetic expansion fields.

Ion Thruster

Introduction

An ion thruster is an electrostatic acceleration propulsion system; it is unique in that extremely high particle exhaust velocities can be generated, as the effective exhaust velocity is not limited by the energy released from chemical reactions. In the ion thruster, the propellant is ionized by an electrical discharge with a unique cathode geometry, the ions are electrostatically extracted from a body of plasma and accelerated to a high velocity by an applied potential difference to produce the thrust. This means that to achieve a thrust increment, the required propellant quantity is relatively small; for a spacecraft, this can be translated into an extended lifetime, a larger payload, a lower cost, or a combination of these factors (Wilbur et al., 1998). The required electrical power is not large, so it can be obtained from solar, nuclear, and other sources of energy. Based on the advantages of high efficiency, high specific impulse, and long lifetime, ion thrusters have been widely used in spacecraft propulsion systems (Anderson et al., 2000; Soulas and Patterson, 2007).

Thrust Components, Geometry, and Potential Arrangements

The most widely used ion thruster concept generates ions by the electron bombardment method. A typical (Kaufman-type) ion thruster is shown in Figure 9.19; it includes a gas feed system, a hollow cathode, a discharge chamber (usually used as an anode), permanent magnets, an ion optics system, and a neutralizer.

In an operating ion thruster, electrons are generated from the hollow cathode. These electrons enter the discharge chamber and produce ions through electron bombardment ionization collisions with propellant neutral atoms. Permanent magnets are used to increase the electron path length and enhance the collision rate to produce more ions. Ions are then extracted and accelerated by the ion optics system to reach a very high velocity and escape from the thruster. Finally, ions are neutralized with the electrons from the neutralizer (Rovey and Gallimore, 2008).

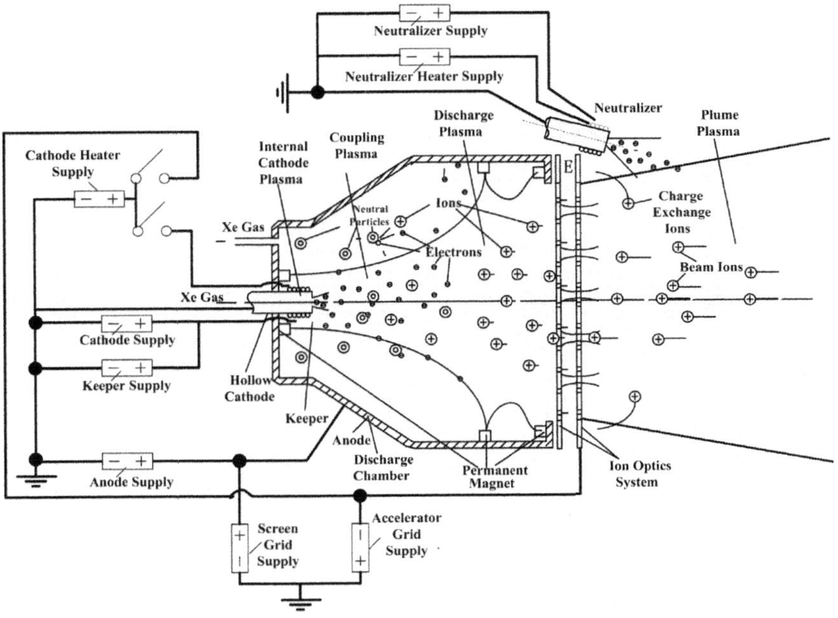

Figure 9.19 Schematic of Kaufman-type ion thruster.

The electrical potentials of components of the thruster have typical values as follows:

1.	Cathode	+1500 V
2.	Anode/discharge chamber	+1490 V
3.	Inner grid	+1000 V
4.	Outer grid	−250 V

The plasmas in an operating thruster can be categorized into four types (Monterde et al., 1997; Milligan and Gabriel, 1999, 2009):

1. The internal cathode plasma, which is present inside the hollow cathode.
2. The coupling plasma, which is immediately downstream of the cathode before the main discharge plasma.
3. The discharge plasma, which is mainly located in the discharge chamber.
4. The plume plasma, which is present outside the ion optics system.

The ion thruster has been the subject of intensive research efforts over many years, and these devices have been successfully flight tested. A detailed discussion of the physics and engineering of such devices is available in the literature (Goebel and Katz, 2008), and so current issues will not be treated further here. This discussion of ion thrusters will present a basic description of the operating principles and characteristics of the components of typical devices.

Discharge Chamber

The ionization takes place in the discharge chamber. Neutral propellant gas (e.g., Xe) is fed into the discharge chamber through a hollow cathode, and the neutral gas atoms collide with the primary electrons emitted from the hollow cathode. After collisions, some atoms are ionized, while the primary electrons become thermalized and are eventually collected by the anode. The ionization process is enhanced by the presence of a magnetic field; this induces cyclic motion about the field lines and prolongs the residence time of the primary electrons and so increases the probability of an ionization collision (Herman and Gallimore, 2004).

The effectiveness of ion production inside the discharge chamber is measured in terms of plasma ion production cost, ε_P, and the beam ion production cost, ε_B. The plasma ion production cost is given by:

$$\varepsilon_P = \frac{(I_D - I_P)\,V_D}{I_P},$$

where I_P represents the ion production current inside the discharge chamber and I_D and V_D represent the discharge current and the discharge voltage, respectively.

The beam ion production cost is computed as:

$$\varepsilon_B = \frac{(I_D - I_B)V_D}{I_B},$$

where I_B represents the ion beam current.

The beam ion production cost is related to the plasma ion production cost with the following expression:

$$\varepsilon_B = \frac{\varepsilon_P}{f_B} + \frac{1 - f_B}{f_B} V_D,$$

where $f_B = \frac{I_B}{I_P}$ is the fraction of ions extracted from the chamber by the grids.

Another important parameter that is generally used to measure the ion engine discharge chamber performance is the discharge chamber propellant utilization efficiency, η_P, as:

$$\eta_P = \frac{I_B}{I_n},$$

where I_n is the total neutral flow rate supplied into the discharge chamber measured in Amperes equivalent.

The beam ion production cost can be evaluated by the following expression (Brophy and Wilbur, 1985):

$$\varepsilon_B = \frac{\varepsilon_P^*}{f_B[1 - \exp(-C_0\dot{m}(1 - \eta_P))]} + C,$$

where ε_P^* is the "baseline ion production cost," which is a function of a particular thruster geometry, magnetic field strength, and propellant (i.e., ionization and excitation energy); C_0 is a parameter that depends on propellant characteristics such as the temperature and ionization cross-section; and C is a function of the discharge chamber operating parameters—mainly the discharge voltage V_D. This analytical expression is related to the typical graph (Figure 9.20) of the experimental relationship that is found between ε_B and η_P for constant propellant flow rate.

A characteristic "knee" is evident on the curve; this behavior indicates that attempts to increase the propellant utilization efficiency beyond

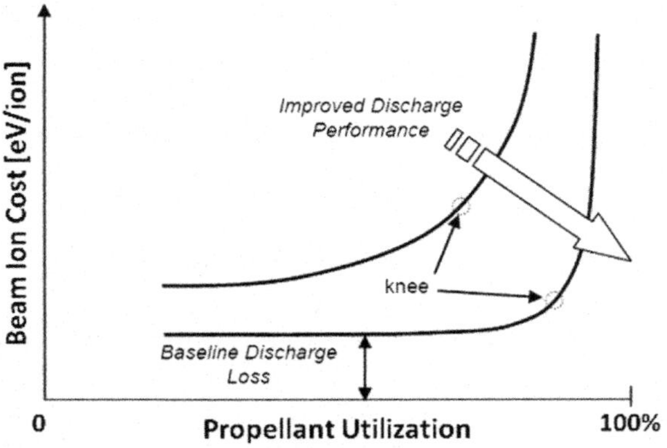

Figure 9.20 Typical relationship between beam ion cost and propellant utilization efficiency. *Wirz, R.E., 2005, Discharge Plasma Processes of Ring-Cusp Ion Thrusters, PhD Thesis, California Institute of Technology, Pg. 16, Figure 2.2-1.*

this "break" point, leads to excessive discharge power requirements. For conventional ion thrusters, ε_B is typically near 200 eV/ion and $\eta_P > 80\%$.

Ion Optics

Ion optics is inherent in the geometry and physics of the ion thruster grids operation. The grids are located at the exit side of the discharge chamber and accelerate ions from the discharge chamber to produce thrust. In Figure 9.19, two grids are shown; however, three or more grids can be used depending on the application of the ion source. The grid closest to the discharge chamber plasma is called the screen grid, and the downstream grid is called the accelerator grid. Commonly the screen grid is set at high positive potential (1000 V above spacecraft common), and the acceleration grid is set at negative potential (100 V below spacecraft common).

In the ion optics, ions are accelerated to high exhaust velocity by electrostatic force. The thrust relationship can be written as $T \approx \dot{m}_i v_i$, where $\dot{m}_i$ represents the ion mass flow rate and v_i is the ion velocity. The ion exhaust velocity can be calculated by conservation of energy as:

$$qV_B = \frac{1}{2}Mv_i^2,$$

or:

$$v_i = \sqrt{\frac{2qV_B}{M}},$$

where V_B represents the net voltage through which the ions are accelerated, q is the charge of the ions, and M is the ion mass. The mass flow rate of ions is related to the ion beam current, I_B, by:

$$\dot{m}_i = \frac{I_B M}{q}.$$

From these equations, the thrust for a singly charged propellant ($q = e$) can be calculated as:

$$T = \sqrt{\frac{2M}{e}} I_B \sqrt{V_B},$$

which defines that thrust is proportional to the beam current and the square root of the acceleration voltage.

The primary function of the acceleration grid is to prevent electrons within the beam plasma from traveling or backstreaming into the discharge chamber. This function is realized by means of setting the acceleration grid at a negative potential relative to the downstream beam plasma potential. Charge exchange or charge transfer takes place between the ions in the beam and the neutral atoms flowing through and downstream of the grids. Charge exchange ions created in the beam can be accelerated into the screen and grids and as they are the source of erosion, they are the subject of considerable interest (DeBoer, 1997).

SPACE CHARGE LIMITED CURRENT

While the ion thruster is capable of generating high-speed particles and high specific impulse, it has a fundamental limitation on the current that can be processed, and so there is a limitation on the magnitude of the thrust that can be generated. This limitation is related to the population of particles that can exist and be accelerated in the gap between the grids. Fundamentally, charged particles create and destroy electric field, so there is a limit in the number of ions in a volume of electrode gap space before the electric field is neutralized. A basic analysis can be carried out to quantify this limit.

A sketch of the acceleration grid region is shown in Figure 9.21. We take the grid adjacent to the discharge region as being at V_B, and for simplicity take the acceleration gradient at $V = 0$.

Recall from Maxwell's equations that:

$$\nabla^2 V = \frac{\rho_e}{\varepsilon_0}.$$

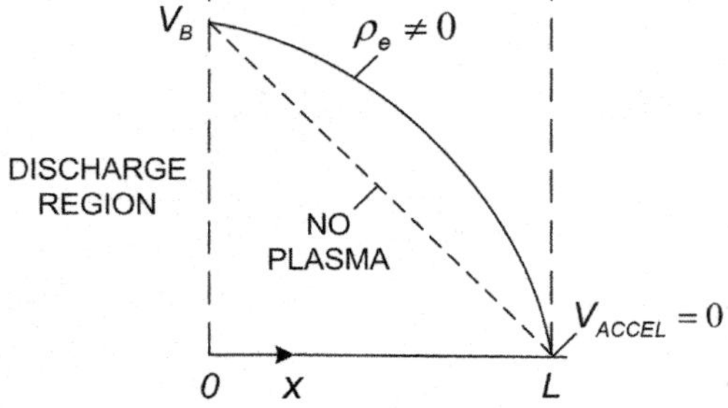

Figure 9.21 Schematic of the ion accelerating grid region.

If:

$$\rho_e = 0, \quad \nabla^2 V = 0,$$

so:

$$\nabla V = \frac{dV}{dx} = \text{Const.}$$

If $V_{\text{ACCEL}} = 0$, then the acceleration electric field is

$$E = \frac{V_B}{L} = \text{Constant.}$$

If there are ions in the grid region, they will be accelerated to velocity, v_+, as:

$$\frac{d^2 V}{dx^2} = \frac{\rho_e}{\varepsilon_0} = \frac{n_+ e_+}{\varepsilon_0}$$

and the current density is:

$$J_+ = \frac{I_B}{A_B} = v_+ n_+ e_+,$$

where A_B is the cross-sectional area of the ion beam. We can relate the velocity and the potential using the energy equation:

$$e_+ V_B = e_+ V(x) + \frac{m_+ v_+^2(x)}{2}$$

and:

$$v_+^2(x) = \frac{2e_+(V_B - V(x))}{m_+}.$$

So we can solve for $V(x)$ and $v_+(x)$ with appropriate boundary conditions.

The limiting beam current (J_{max}) through the grids occurs when $\frac{dV}{dx} = 0$ at $x = 0$. Then:

$$V_B - V(x) = \left[\frac{9}{4} \cdot \frac{J_{max}}{\varepsilon_0 (2e_+/m_+)^{1/2}}\right]^{2/3} x^{4/3}.$$

Furthermore, taking $V = 0$ at $x = L$ gives us:

$$J_{max} = \frac{4}{9} \varepsilon_0 \left(\frac{2e_+}{m_+}\right)^{1/2} \frac{V_B^{3/2}}{L^2},$$

which defines the charge-limited current or maximum ion flux through the grids in a two-grid system. So we can write:

$$\frac{T}{A} = \frac{m_+}{e_+} \cdot J_+ u_+ = \frac{m_+}{e_+} \left(\frac{2e_+}{m_+} V_B\right)^{1/2} J_+,$$

and:

$$\left(\frac{T}{A}\right)_{max} = \frac{8}{9} \varepsilon_0 \left(\frac{V_B}{L}\right)^2.$$

Exit Beam Neutralizer

The ion thruster expels positive ions from the exit plane at a significant rate, so an equal amount of electrons must be expelled to avoid charge imbalance with the spacecraft. A neutralizer is usually located at a radius outside of the discharge chamber. In the hollow cathode neutralizer, additional gas is passed over a thermionic emitter, such as lanthanum hexaboride (LaB_6) or porous tungsten impregnated with oxides. When the thermionic emitter is heated, it emits electrons that are extracted through a small orifice using a positively biased electrode called a keeper. This has proven to be an effective technique to maintain charge long term neutrality of the thruster.

Hall Thruster

Introduction

The Hall thruster is a coaxial device in which orthogonal electric and magnetic fields ($\vec{E} \times \vec{B}$) are employed to ionize propellant gases, such as

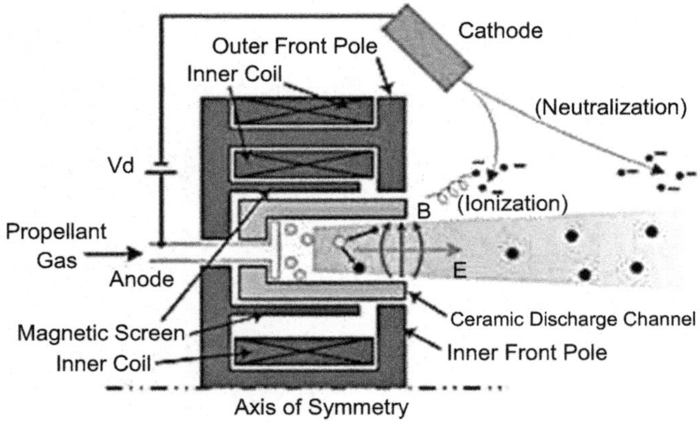

Figure 9.22 Schematic diagram of the Hall thruster. *http://ieeexplore.ieee.org/stamp/stamp.jsp?arnumber=6164235. Liu, C., Gu, Z., Xie, K., Sun, Y.K., Tang, H.B., June 2012. Influence of the magnetic field topology on Hall thruster discharge channel wall erosion. IEEE Trans. Appl. Supercond., 22(No. 3), (Figure 2 Schematic diagram of a Hall thruster.) with permission.*

xenon, and accelerate the resulting ions to produce thrust (Goebel and Katz, 2008). The plasma that is generated develops a cross-field discharge due to the Hall effect on plasma particle motion. A schematic diagram with thruster components and mechanisms displayed is shown in Figure 9.22.

In the annular discharge chamber, orthogonal electric field and magnetic fields are effective to create and accelerate the plasma. The applied electric field, which is generated between the anode and the cathode, is mainly in the axial direction. The electrons, which are ejected by the cathode, enter the discharge chamber at the channel exit. The applied magnetic field generated by the ferromagnetic core is mainly in the radial direction. The electrons are influenced by the magnetic field and have a cyclotron motion, which results in a closed-drift Hall current in the azimuthal direction. Electrons then collide with neutral atoms that are injected from the propellant gas feeder in the anode region and cause ionization. The ions are unaffected by the magnetic field due to their heavier mass and are accelerated by the axial electric field and are ejected to produce thrust.

Particle Interactions

As a consequence of the crossed electric field and magnetic field, the azimuthal motion of the electrons will cause a Hall current. The drift velocity can be calculated by the electric field strength and magnetic field strength as:

$$\vec{V}_{E\times B} = \frac{\vec{E} \times \vec{B}}{B^2},$$

and the Hall current density can be calculated as:

$$J_{Hall} = n_i q V_{E\times B}.$$

By integrating the Hall current density over the channel cross-section, we can obtain the total Hall current.

In the discharge channel, the balance of the electron charge and the ion charge means that the plasma is electrically neutral, so the ion current density is not space charge limited as in an ion thruster. The operation of a Hall thruster requires the electrons to be impeded and the ions not to be affected by the magnetic field, so the electron cyclotron radius must be much smaller and the ion cyclotron radius must be much larger than the thruster geometry dimension, as:

$$R_{ce} \ll L \ll R_{ci},$$

where L represents a thruster dimension, e.g., length of the channel; R_{ce} is the electron cyclotron radius; and R_{ci} is the ion cyclotron radius.

The Hall parameter is defined as:

$$\Omega = \frac{\omega_e}{\upsilon_t},$$

where:

$$\omega_e = \frac{eB}{m_e},$$

and υ_t is the total electron collision frequency. In the case of: $\Omega \gg 1$, electrons complete many cycles of their cyclotron motion before undergoing a collision; in the case of $\Omega \ll 1$, the electrons complete few cycles. Analysis of the plasma flow in the acceleration channel has been the subject of considerable attention, and these results have been well documented in the literature (Ahedo et al., 2001; Keidar et al., 2001) An efficient Hall thruster operation can only be realized over a limited range of electron Hall parameters (Hofer et al., 2006; Hofer and Gallimore, 2006).

Thruster Design and Performance

Compared with electric propulsion devices such as ion thrusters, the advantages of the Hall thruster include a simpler structure and, more importantly, no space charge limitation. The Hall thruster has other significant advantages such as: long life (10,000 h), high power density (0.4–1.3 kW/kg), and advantageous specific impulse range (1000–2000 s). The efficiency and specific impulse of a Hall thruster are somewhat less than those of the ion thruster, but, very importantly, the thrust is higher at a given power; also, it requires fewer power supplies than ion thrusters.

There are two types of Hall thrusters that are in use: the stationary plasma thruster (SPT) and the thruster with anode layers (TAL). The SPT Hall thruster, which is referred to as a one-stage thruster, uses a dielectric discharge chamber. In comparison, the TAL Hall thruster is a two-stage thruster and uses a metallic discharge chamber, which is divided into an ionization region and an acceleration region by an intermediate electrode. The SPT thruster has been more thoroughly researched (Morozov and Savelyev, 2000; Shagayda and Gorshkov, 2013) and has been utilized in space more often than the TAL thruster. The SPT-100 Hall thruster (which was developed in Russia) is shown in Figure 9.23.

Hall Thruster Components

As displayed in the schematic (Figure 9.22), a Hall thruster is composed of four main components: the discharge chamber, the magnetic circuit, the anode (and gas feeder), and the hollow cathode. Although both racetrack

Figure 9.23 Photograph of SPT-100 Hall thruster (Stetchkin/Fakel, Russia). *Image from Encyclopedia Astronautica, SPT-100 Image, downloaded from: http://www.astronautix.com/engines/spt100.htm, 8/26/2014, with permission.*

and linear geometries have been investigated, the typical hall thruster has a coaxial cylindrical discharge chamber.

Discharge Chamber

The discharge chamber is also an accelerating channel, which is composed of two dielectric cylinders of different radii fixed together on the same axis. The wall of the channel is typically manufactured from ceramic materials such as boron nitride (BN) or Borosil ($BN\text{-}SiO_2$). Other materials such as alumina (Al_2O_3) and silicon carbide (SiC) have also been used. The ratio of the discharge chamber axial length and radial length is usually greater than 1.

Magnetic Circuit

Although the magnetic circuit in a Hall thruster is complex, its basic configuration is the C-core magnetic circuit. When a coil winding is wrapped around a ferromagnetic core (magnetic pole), a magnetic field will occur between the gap. In a Hall thruster, the radial magnetic field, B_r, is generated to be applied throughout the volume of the annular channel. After the geometry of the magnetic pole is fixed, the topology of the magnetic field can be adjusted by changing the current and the number of turns of the solenoid coil. Figure 9.24 shows a typical distribution of magnetic field strength along the discharge channel axis.

The radial magnetic field strength can be seen to increase from the channel rear wall to a maximum magnitude and then decreases to the channel exit.

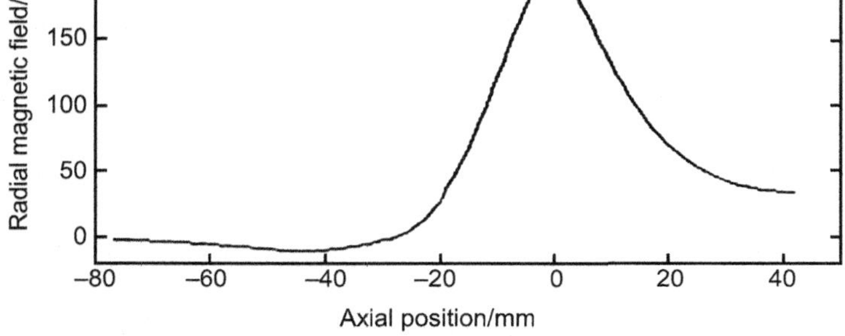

Figure 9.24 Typical radial magnetic field variation along the axis of the discharge channel. *Peng, E., Wu, Z.W., Han, K., Yu, D.R., 2009. On the role of magnetic field intensity effects on the discharge characteristics of Hall thrusters. Acta Phys. Sin. 58(4), 2535–2542. Figure 3 with permission.*

Anode

The anode is located at the rear wall of the discharge channel. The anode has two functions: acting as a positive electrode to create an electric field in the channel and allowing passage of neutral gas to the discharge channel. The potential of the anode is required to be on the order of a 100 V to facilitate ionization. A series of equally spaced holes is required to allow the uniform azimuthal distribution of the neutral gas. The neutral gas propellant is typically xenon, based upon ionization energy and atom mass; krypton has also been investigated for high-power Hall thrusters.

Hollow Cathode

The hollow cathode is usually located at a radius outside of the discharge chamber (Figure 9.21). In the hollow cathode, additional gas is passed over a thermionic emitter, such as lanthanum hexaboride (LaB_6) or porous tungsten impregnated with oxides. When the thermionic emitter is heated, it emits electrons that are extracted through a small orifice using a positively biased electrode called a keeper. The electrons emitted by the hollow cathode divide into two parts. One part of the electrons enters the discharge chamber to ionize the atoms, and the remaining electrons move downstream of the thruster to neutralize the ions ejected from the discharge chamber.

REFERENCES

Plasma Thrusters

Ahedo, E., Merino, M., 2010. Two-dimensional supersonic plasma acceleration in a magnetic nozzle. Phys. Plasmas 17, 073501.

Bishop, A.R., Connolly, D.J., Seikel, G.R., 1971. Test of permanent magnet and superconducting magnet MPD thrusters. AIAA Pap. 71–696.

Choueiri, E.Y., Kelly, A.J., Jahn, R.G., 1987. MPD thruster plasma instability Studies. In: AIAA-87-1067.

Choueiri, E., 1998. Scaling of thrust in self-field magnetoplasmadynamic Thrusters. AIAA J. Propul. Power 14, 744–753.

Choueiri, E.Y., 2004. A critical history of electric propulsion: The first 50 years (1906–1956). AIAA J. Propul. Power 20, 193–205.

Ducati, A.C., Giannini, G.M., Muehlberger, E., 1964. Experimental results in high specific impulse thermionic acceleration. AIAA J. 2, 1452–1454.

Fradkin, D.B., et al., 1970. Experiments using a 25-kW hollow cathode lithium vapor MPD arcjet. AIAA J. 8, 886–894.

Goebel, D.M., Katz, I., 2008. Fundamentals of Electric Propulsion: Ion and Hall Thrusters. Wiley, New York.

Hooper, E.B., 1993. Plasma detachment from a magnetic nozzle. AIAA J. Propul. Power 9, 757–764.

Jahn, R.G., 1969. Physics of Electric Propulsion. McGraw-Hill, New York.

Maecker, H., 1955. Plasma jets in arcs in a process of self-induced magnetic compression. Z. Phys. 141 (1), 198–216.
Michels, C.J., York, T.M., 1972. Exhaust flow and propulsion characteristics of a pulsed MPD arc thruster. In: AIAA-72-500.
Mikellides, P.G., Turchi, P.J., Roderick, N.F., 2000. Applied-field magnetoplasmadynamic thrusters, Part 1: Numerical simulation using the MACH2 code. AIAA J. Propul. Power 16, 887–893.
Mikellides, P.G., Turchi, P.J., 2000. Applied-field magnetoplasmadynamic thrusters, Part 2: Analytic expressions for thrust and voltage. AIAA J. Propul. Power 16, 894–901.
Myers, R.M., 1995. Geometric scaling of applied-field magnetoplasmadynamic thrusters. AIAA J. Propul. Power 11, 343–350.
Scharlemann, C.A., York, T.M., 2003. Pulsed plasma thruster using water propellant, Part I: Investigation of thrust behavior and mechanism. In: AIAA-2003-5022.
Seikel, G.R., et al., 1970. Plasma Physics of Electric Rockets – Plasmas and Magnetic Fields in Propulsion and Power Research. NASA SP-226.
Tang, H.B., Cheng, J., Liu, C., York, T.M., 2012. Study of applied magnetic field magnetoplasmadynamic thrusters with particle-in-cell code with Monte-Carlo collisions: II. Investigation of acceleration mechanisms. Phys. Plasmas 19, 073108.
Tikhonov, V.B., et al., 1993. Research of plasma processes in self-field and applied magnetic field thrusters. In: IEPC, 93-076.
Uribarri, L., Choueiri, E.Y., 2008. Relationship between anode spots and onset voltage hash in magnetoplasmadynamic thrusters. AIAA J. Propul. Power 24, 571–582.
Wirz, R.E., 2005. Discharge Plasma Processes of Ring-Cusp Ion Thrusters. Dissertation (PhD), California Institute of Technology.
York, T.M., Zakrzewski, C., Soulas, G., 1993. Diagnostics and performance of a low-power MPD thruster with applied magnetic nozzle. AIAA J. Propul. Power 9, 553–560.
York, T.M., Kamhawi, H., 1993. Plasma expansion in a low-power MPD thruster with variable magnetic NOZZLE. In: AIAA-93-138.

Ion Thrusters

Anderson, J.R., et al., 2000. Performance characteristics of the NSTAR ion thruster during an on-going long duration ground test. In: IEEE Aerospace Conference Proceedings, vol. 4, pp. 99–122.
Brophy, J.R., Wilbur, P.J., 1985. Simple performance model for ring and line cusp ion thrusters. AIAA J. 23, 1731–1736.
De Boer, P.C.T., 1997. Measurements of electron density in the charge exchange plasma of an ion thruster. AIAA J. Propul. Power 13, 783–788.
Herman, D.A., Gallimore, A.D., 2004. Discharge chamber plasma structure of a 30-cm NSTAR-type ion Engine. In: AIAA-2004-3794.
Milligan, D.J., Gabriel, S.B., 1999. Investigation of the baffle Annulus region of the UK25 ion thruster. In: AIAA 99-2440.
Milligan, D.J., Gabriel, S.B., 2009. Generation of experimental plasma parameter maps around the baffle apeeture of a Kauffman (UK-25) ion thruster. Acta Astronaut. 64, 952–968.
Monterde, M.P., Haines, M.G., Dangor, A.E., Malik, A.K., Fearn, D.G., 1997. Kaufman-type xenon ion thruster coupling plasma: Langmuir probe measurements. J. Phys. D Appl. Phys. 30, 842–855.
Rovey, J.L., Gallimore, A.D., 2008. Dormant cathode erosion in a multiple-cathode gridded ion thruster. AIAA J. Propul. Power 24, 1361–1368.
Soulas, G.C., Patterson, M.J., 2007. NEXT ion performance dispersion analyses. In: AIAA 2007-5213.
Wilbur, P.J., Rawlin, V.K., Beattie, J.R., 1998. Ion thruster development trends and status in the United States. AIAA J. Propul. Power 14, 708–715.

Hall Thrusters

Ahedo, E., Martinez-Cerezo, P., Martinez-Sanchez, M., 2001. One-dimensional model of the plasma flow in a Hall thruster. Phys. Plasmas 8, 3058.

Encyclopedia Astronautica, SPT-100 Image, downloaded from: http://www.astronautix.com/engines/spt100.htm, 8/26/2014.

Hofer, R.R., Jankovsky, R.S., Gallimore, A.D., 2006. High-specific impulse hall thrusters, Part 1: influence of current density and magnetic field. J. Propul. Power 22, 721.

Hofer, R.R., Gallimore, A.D., 2006. High impulse hall thrusters, Part 2: efficiency analysis. J. Propul. Power 22, 732.

Keidar, M., Boyd, I.D., Beilis, I.I., 2001. Plasma flow and plasma-wall transition in Hall thruster channel. Phys. Plasmas 8, 5315.

Liu, C., Gu, Z., Xie, K., Sun, Y.K., Tang, H.B., 2012. Influence of the magnetic field topology on hall thruster discharge channel wall erosion. IEEE Trans. Appl. Supercond. 22, 4904105.

Morozov, A.I., Savelyev, V.V., 2000. Fundamentals of stationary plasma thrusters. In: Reviews of Plasma Physics, vol. 21. Springer, New York.

Peng, E., Wu, Z.W., Han, K., Yu, D.R., 2009. On the role of magnetic field intensity effects on the discharge characteristics of Hall thrusters. Acta Phys. Sin. 58 (4), 2535–2542.

Shagayda, A.A., Gorshkov, O.A., 2013. Hall-thruster scaling laws. J. Propul. Power 29, 466.

Magnetic Compression and Heating

INTRODUCTION

The application of pulsed magnetic fields to compress and heat gases has been the subject of research for more than 60 years (IAEA, 1961). Because the effects of pulsed electromagnetic discharges have proved to be useful in inducing atomic and nuclear processes, they continue to be the subject of research for nuclear energy, lasers, and space thrusters. While a number of geometries and electrical discharge configurations have been of interest, the present work will examine one configuration with unique characteristics: the linear theta pinch. This physical geometry first generated near thermonuclear plasmas in the laboratory, and it has received considerable experimental and theoretical evaluation (Quinn and Siemon, 1981). Such pulsed devices apply very high power for short periods of time, and that time is sufficient for important atomic interactions to occur and to be studied in the laboratory.

Dynamic (Theta) Pinch

The geometry and electrical circuit of a linear theta pinch device is shown in end view in the following diagram (Figure 9.25); the device has length, *L*.

When the switch is closed the current will flow in the coil with a periodic discharge, as:

$$I = I_{max} \sin \omega t,$$

where:

$$I_{max} = \omega C V_{max} = \omega Q.$$

The azimuthal current in the coil will generate an axial magnetic field:

$$B_z = \frac{\mu_0 I}{l},$$

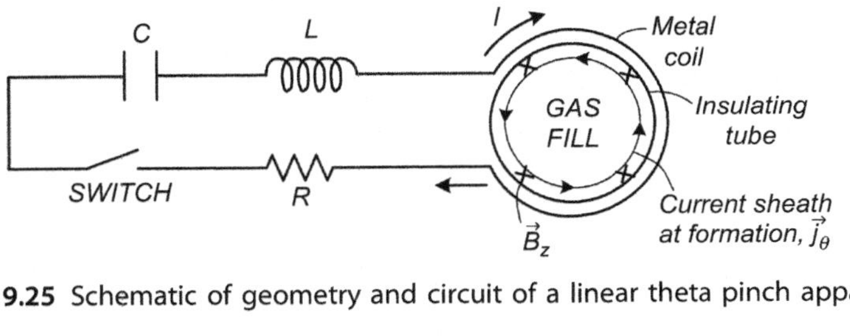

Figure 9.25 Schematic of geometry and circuit of a linear theta pinch apparatus.

and the changing field ($\dot{B}_z$) will generate an azimuthal field (E_θ) inside the gas-filled insulating tube. With proper circuit parameter design, this field will be large enough to break down the gas and form a discharge in the gas-filled tube at the outer radius, R_t. The electric field can be calculated from Maxwell's equations, as:

$$\nabla \times \vec{E} = -\frac{\partial \vec{B}}{\partial t},$$

and in cylindrical coordinates:

$$\vec{\nabla} \times \vec{E} = \left(\frac{1}{r}\frac{\partial E_z}{\partial \theta} - \frac{\partial E_\theta}{\partial z}\right)\vec{r} + \left(\frac{\partial E_r}{\partial z} - \frac{\partial E_z}{\partial r}\right)\vec{\theta} + \left[\frac{1}{r}\frac{\partial}{\partial r}(rE_z) - \frac{1}{r}\frac{\partial E_r}{\partial \theta}\right]\vec{z},$$

and over the area:

$$\int_A (\nabla \times \vec{E}) \cdot d\vec{A} = -\frac{1}{\partial t}\int_A \vec{B} \cdot d\vec{A},$$

or, applying the Kelvin-Strokes theorem:

$$\oint \vec{E} \cdot d\vec{l} = -\frac{1}{\partial t}\int_A \vec{B} \cdot d\vec{A},$$

then:

$$E_\theta(r) \cdot 2\pi r = \frac{1}{\partial t}\int_{A(r)} \vec{B} \cdot d\vec{A} = \frac{1}{\partial t}\int_0^r \frac{\mu_0 I}{l} 2\pi r dr = \frac{\mu_0}{l}\frac{\partial I}{\partial t}\pi r^2,$$

and:

$$E_\theta(r) = \frac{\mu_0}{l}\frac{r}{2}\frac{\partial I}{\partial t},$$

where the maximum E_θ occurs at the outer radius of the discharge tube and forms J_θ. The highly conductive plasma in the current sheet will exclude the magnetic field from the coil current in the region (radii) inside the current sheet. Across the current sheet we can examine the interactions, as:

$$\nabla \times \frac{\vec{B}}{\mu} = \vec{J},$$

and in cylindrical coordinates as:

$$\left(\frac{1}{r}\frac{\partial B_z}{\partial \theta} - \frac{\partial B_\theta}{\partial z}\right)\vec{r} + \left(\frac{\partial B_r}{\partial z} - \frac{\partial B_z}{\partial r}\right)\vec{\theta} + \left[\frac{1}{r}\frac{\partial}{\partial r}(rB_z) - \frac{1}{r}\frac{\partial B_r}{\partial \theta}\right]\vec{z} = \mu\vec{J_\theta},$$

so:

$$-\frac{\partial B_z}{\partial r} = \mu J_\theta,$$

and we can write:

$$\frac{\partial B_z}{\partial r} = \frac{\partial B_z}{\partial t}\left(\frac{dt}{dr}\right) = \frac{\dot{B}_z}{V_s},$$

where V_s is the current sheet velocity.

So we can calculate

$$\frac{F}{V}(\text{Force/Volume})_r = \mu J_\theta \times B_z = -\frac{B_z}{\mu}\mu\frac{\partial B_z}{\partial r} = \frac{d}{dr}\left(\frac{B_z^2}{2\mu}\right),$$

and $(\text{Force/area})_r$ on the current sheet is:

$$p_B = \int \frac{F}{V}dr = \frac{B_z^2}{2\mu}.$$

Current Sheet Implosion

There are several models that have been developed to describe the behavior of the current sheet and gas in the discharge tube as the current sheet implodes. One model assumes that the current sheet collects all the mass that it encounters, ionizes it, and carries it toward the center; it is referred to as the snowplow model (Rosenbluth, 1954). A schematic of the geometry is shown in Figure 9.26.

The force/mass relationship is:

$$F_{piston} = \frac{d}{dt}(momentum,\ sheet) = \frac{d}{dt}\left[M(R)\dot{R}\right],$$

where:

$$M(R) = \pi\rho_0\left(R_t^2 - R^2\right)l,$$

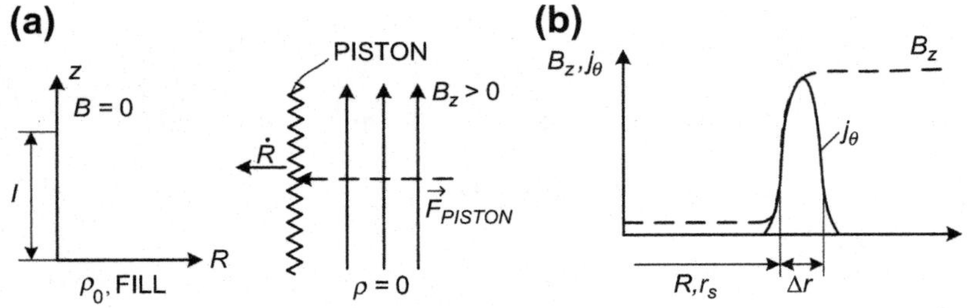

Figure 9.26 Current sheet implosion: (a) geometry and dynamics and (b) radial variation of current density.

with R_t being the tube radius, l being the length, and R being the sheet radius; then:

$$F_p = p \cdot A = -\frac{B_z^2}{2\mu} \cdot 2\pi R \cdot l = \frac{d}{dt}\left[\pi\rho_0\left(R_t^2 - R^2\right) l\dot{R}\right].$$

If we take: $\ddot{R} = 0$, no acceleration, we get an approximate value for sheet velocity, as:

$$-\frac{B_z^2}{2\mu} \cdot 2\pi R \cdot l = -\pi\rho_0 l\dot{R}\frac{dR^2}{dt} = -2\pi\rho_0 lR\dot{R}\frac{dR}{dt},$$

so:

$$\frac{B_z^2}{2\mu} = \rho_0\left(\frac{dR}{dt}\right)^2 = \rho_0 V_s^2.$$

Taking:

$$B_z = B_{max},$$

which can be calculated from current data, will allow $V_s(B_{max})$ to be determined. Then pinch time ($R \approx 0$, $t = t_p$) can be calculated from:

$$t_p \cong \frac{R_t}{V_s}.$$

Also, with circuit discharge data from experiment:

$$B \approx B_{max} \sin \omega t,$$

so:

$$\frac{dB}{dt} = B_{max}\omega \cos \omega t = B_{max}\omega$$

at $t = 0$, we can take:

$$B_z = \dot{B}t \approx B_{max}\omega t,$$

and calculate:

$$[V_s(t)]^2 \approx \frac{(B_{max}\omega t)^2}{2\mu\rho_0}.$$

An analytical derivation for pinch time has also been carried out (Artsimovich, 1964), and it will be outlined here. The implosion model assumes that current flows in a (thin) sheet of thickness Δr, as shown in (b) above. The total force acting on the sheet (piston) is:

$$F_{\mathrm{R}}(r) = -\frac{2\pi r_s l}{\mathrm{c}} \int_{r_s}^{r_s+\Delta r} J_\theta(r) \cdot B_z(r) dr.$$

As above, expressing momentum balance for the sheet implosion:

$$F_{\mathrm{R}} = \frac{d}{dt}[\mathrm{M}(r) V_s],$$

where:

$$\mathrm{M} = \pi(R_t^2 - r_s^2)\rho_0 l,$$

and:

$$\rho_0 = n_0 m_i, \qquad V_s = \frac{dr_s}{dt},$$

(where m_i is the ion mass) so:

$$2\pi r_s l \frac{B_z^2(r)}{8\pi} = \frac{d}{dt}\left[\left\{\pi(R_t^2 - r_s^2)\rho_0\right\}\frac{dr}{dt} \cdot l\right],$$

and with:

$$B_z \approx B_{max}\omega t,$$

$$\frac{r_s}{4} B_{max}^2 \omega^2 t^2 = -\frac{d}{dt}\left[\left\{(R_t^2 - r_s^2)\rho_0\right\}\frac{dr_s}{dt}\right].$$

If we define:

$$X \equiv \frac{r_s}{R_t}, \quad \text{and} \quad \tau = t\left[\frac{B_{max}^2\omega^2}{4R_t^2\rho_0}\right]^{1/4},$$

the equation can be written as:

$$-\tau^2 X = \frac{d}{dt}\left[\{1 - X^2\}\frac{dX}{dt}\right],$$

which has the approximate solution:

$$X(\tau) = \left(1 - \frac{\tau^2}{\sqrt{12}}\right).$$

This provides the functional form:

$$\frac{r_s}{R_t} \sim \left(1 - \frac{t^2 \cdot \text{const.}}{\sqrt{12}}\right),$$

and:

$$V_s = \frac{dr_s}{dt} \sim t.$$

So at: $r = 0$, we get:

$$\tau = (12)^{1/4};$$

and:

$$t_p = 1.86\left[\frac{4R_t^2\rho_0}{B_{max}^2\omega^2}\right]^{1/4}.$$

This result from analysis is similar to the function derived from the simple approach of calculating a pinch time based on an average sheet velocity, as:

$$t_p \approx \frac{R_t}{V_s(avg)} = \frac{R_t}{\frac{V_s(B_p) - V_s(0)}{2}} = \frac{R_t}{\frac{1}{2}\left[\frac{B_{max}^2\omega^2 t_p^2}{2\mu\rho_0}\right]^{1/2}},$$

so:

$$t_p = \left[\frac{32\pi\rho_0 R_t^2}{B_{max}^2\omega^2}\right]^{1/4} = 2.24\left[\frac{4R_t^2\rho_0}{B_{max}^2\omega^2}\right]^{1/4}.$$

Now, continuing, the temperature of the plasma *at pinch* can be calculated from:

$$\frac{1}{2}MV_s^2 = \left(\frac{3}{2}kT_i + \frac{3}{2}kT_e\right)n_0\left(\pi R_t^2 l\right),$$

$$\frac{1}{2}n_0\left(\pi R_t^2 l\right)m_i V_s^2 = \frac{3}{2}k(T_i + T_e)n_0\left(\pi R_t^2 l\right),$$

and:

$$m_i V_s^2 = 3k(T_i + T_e).$$

If $T_i = T_e = T$:

$$m_i V_s^2 = 6kT_{pinch}.$$

The number density *at pinch* can be calculated from pressure balance as:

$$\frac{B^2}{2\mu}\Big|_{pinch} = n_i k T_i + n_e k T_e,$$

and if:

$$T_i = T_e = T, \quad n_i = n_e = n,$$

then:

$$\frac{B^2}{2\mu}\Big|_{pinch} = 2n_p k T_{pinch}.$$

The radius of the plasma column *at pinch* can be calculated from:

$$M = m_i \cdot n_0\left(\pi R_t^2 l\right),$$

with:

$$N = n_0\left(\pi R_t^2 l\right),$$

so:

$$n_p = \frac{N}{volume} = \frac{N}{\pi R_c^2 l},$$

and:

$$R_c^2 = \frac{N}{\pi n_p l}.$$

While the above model is basic, the results show the mechanics and interactions that take place in magnetic compression and heating. The evaluations have assumed no energy loss in the compression and heating process and no end effects.

A conceptual model that includes the effect of energy loss through electron transport and energy loss at the ends during collapse was used for analysis of loss time, and it will be outlined here (York, 1977). The total number of particles in the confined plasma column of cross-sectional area, A_p, and length, l_p, is:

$$N(t) = nA_p \cdot l_p,$$

so:

$$\frac{dN}{dt} = -2(A_{ex} n <v>),$$

where A_{ex} is the area at the end where the flow speed is $<v>$.

If $T =$ constant, then $<v> =$ constant, and:

$$N(t) = N_0 e^{-t/\tau},$$

where:

$$\tau = \frac{A_p \cdot l_p}{2A_{ex} <v>} \sim l_p.$$

Now, loss speed has been related to sound (random thermal) speed:

$$<v> \approx \left(\frac{2\gamma k_B T_i}{m_i}\right)^{1/2}.$$

So:

$$\tau = \frac{l}{2} \cdot \frac{1}{<v>},$$

is the $1/e$ loss time $(N = \frac{N_0}{e})$ with no thermal effect.

However, if $T = f(t)$, due to simultaneous thermal energy loss, primarily due to electron transport to the ends, then the loss rate and loss time (τ) will change. Taking N_0 = constant, then the primary mechanism for thermal loss can be approximated as follows: For long machines, T_e approaches T_i and the loss time is taken as the electron conduction time, as:

$$t_{th-loss}(long) = t_{el-cond}.$$

For short machines, where the transport is still electron loss but the ion–electron equilibration time becomes the limiting mechanism, the thermal loss time is:

$$t_{th-loss}(short) = t_{i-e\ eq},$$

and:

$$T_i = T_{io} e^{-\frac{t}{t_{th-loss}}}.$$

Then the time for (1/e) particle loss with simultaneous loss of thermal energy can be expressed as:

$$t_e = 2t_{th-loss}\, e^{\frac{1}{1-\tau/2t_{th-loss}}}.$$

For particle loss time less than thermal loss time: $t_e \rightarrow \tau$. As a simple model, we take:

$$T_i \approx T_{io} - K_t t;$$

then:

$$t_e = \frac{T_{io}}{K_t}\left\{1 - \left[1 - \frac{3}{2}\frac{K_t}{T_{io}}\tau\right]^{2/3}\right\},$$

which expresses the extension of particle loss time due to simultaneous thermal energy loss.

In general, the particle loss time for a given linear theta pinch experiment is given as:

$$\tau = \left\{\left(\frac{l}{2}\right)\cdot\left(m_+/2k_B T_i\right)^{1/2}\right\}\cdot\chi,$$

where χ is the end loss factor, and (>1) due to thermal energy loss (McKenna and York, 1977). In a comprehensive evaluation of experiments and theories of loss physics in linear devices (Stover et al., 1978), the overall

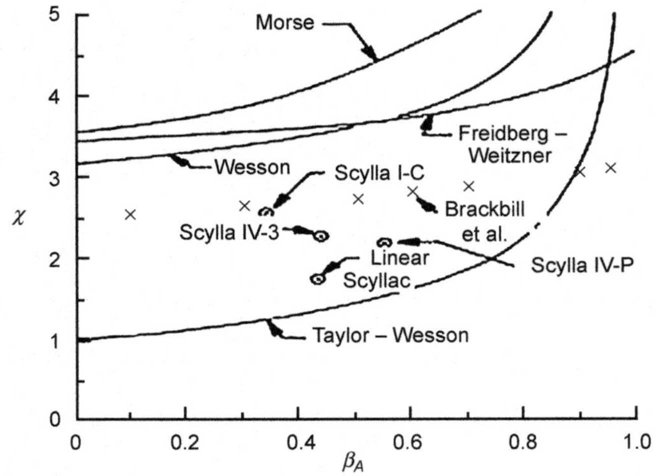

Figure 9.27 Comparison of theoretical models of end loss with linear theta pinch experimental results. *Stover, E.K., Klevans, E.H., York, T.M., 1978. Computer modeling of linear theta pinch machines. Phys. Fluids 21, 2090–2102. Figure 2, p. 2097 with permission.*

loss behavior of a number of devices was compared and the results are shown in Figure 9.27; β_A is the ratio of plasma kinetic pressure divided by magnetic pressure with on-axis values.

The experiments evaluated were conducted over a range of conditions, including higher temperatures, and with longer devices. Various available theories to predict plasma loss from these experiments are shown in Figure 9.27. As can be seen, there was not good agreement of theory with experiment, and this was the impetus for improved modeling. The work reported by Stover et al. (1978) presented results of a numerical solution of an MHD model with multiple species temperatures and multispecies transport. The model resulted in good agreement with the experiments, in details of local conditions, and in loss behavior. The following phenomenology was evidenced by the results of numerical modeling: (1) energy loss by electron conduction was important, (2) classical ion thermal conduction is inaccurate in modeling collisionless experiments, and (3) magnetic field diffusion effects were found to be important in collisional flow but not so important in flow of collisionless plasmas.

Plasma Flow within Magnetic Field Lines: Collisional

The geometry of the linear theta pinch generates axial magnetic fields (B_z) along the length of the coil. After pinch, with the plasma confined on the

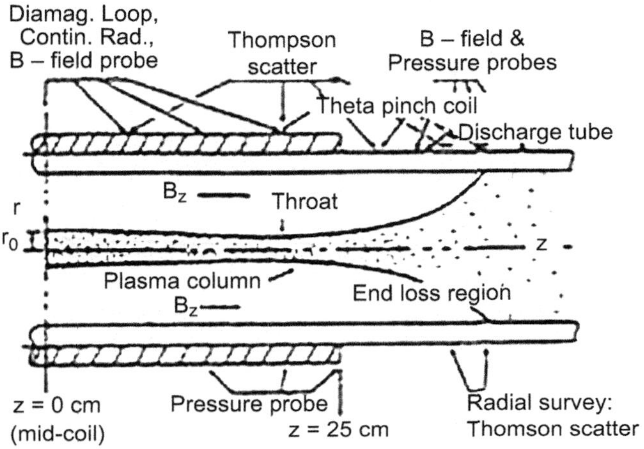

Figure 9.28 Schematic of the end region of a linear theta pinch experiment to study loss. *York, T.M., Jacoby, B.A., Mikellides, P., 1992. Flow processes within magnetic nozzle configurations. AIAA J. Propul. Power 8, 1023–1030. Figure 2, p. 1024 with permission.*

axis, the radial force equilibrium and end losses require an understanding of plasma flow and energy loss along confining axisymmetric (B_z) magnetic field lines: this is inherently an occurrence of plasma flow in a "magnetic nozzle." Some of the fundamental aspects of the physics of this behavior will now be discussed.

A schematic of the geometry and diagnostics in the end region in one experiment directed toward understanding the plasma flow processes (York et al., 1992) is shown in Figure 9.28.

The diagnostics applied were extensive in order to allow data overlap and so identify mechanisms and characteristic values of important parameters. In Figure 9.29, experimental parameters (a) and plasma radius as a function of axial position at different times in the experiment are shown (b).

The plasma properties indicate a collisional plasma. The physical equilibrium includes radial pressure balance between an external magnetic field (B_{ex}) confining a plasma with internal field and plasma, as:

$$\frac{B_{ex}^2}{8\pi} = \frac{B_i^2(r)}{8\pi} + p(r).$$

The experimental values of the throat (minimum) plasma radius agreed with a theoretical result predicted for the unsteady event (Wesson, 1966; Freidberg and Weitzner, 1975), as:

$$r_{th}(theory) = r_{1/e}\left(\frac{1-\beta}{\gamma}\right)^{1/6},$$

(a)

Implosion parameters (Experiment)

Implosion velocity	$7.17 \pm 0.43 \times 10^4$ cm/s
Implosion time	0.80 ± 0.03 μs
Electron temperature	20.70 ± 3.3 eV
Electron density	$4.34 \pm 0.61 \times 10^{13}$/cm^3
Plasma radius	1.97 ± 0.38 cm

(b)

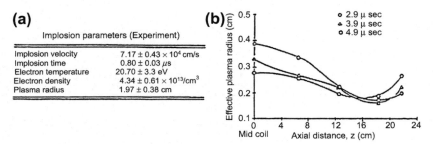

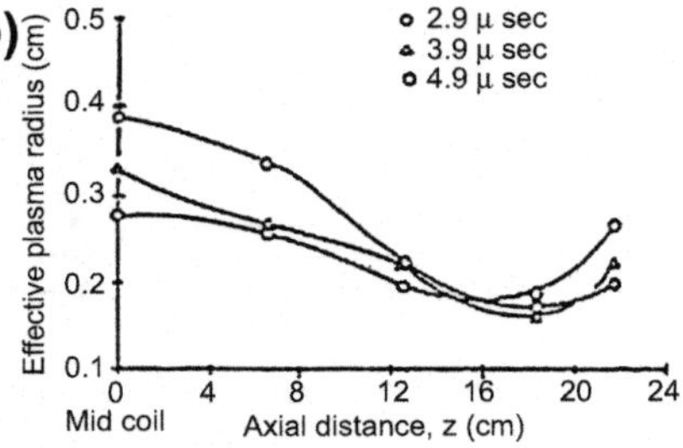

Figure 9.29 Plasma properties (a) and (b) variation of plasma radius with axial position. *York, T.M., Jacoby, B.A., Mikellides, P., 1992. Flow processes within magnetic nozzle configurations. AIAA J. Propul. Power 8, 1023–1030. Table 1 and Figure 4, p. 1025 with permission.*

where γ is the ratio of specific heats and:

$$r_{1/e}\,(expt.) = \int_0^{\infty} \left\{ 1 - \left[1 - \frac{n(r)}{n_0}\beta_0 \right]^{1/2} \right\} 2\pi r dr,$$

and n_0, β_0 are the values on-axis. The flow velocity at the throat is an important determination, as it is expected to be a limiting value on mass loss rate. The characteristic velocities involved in analyses of plasma flow considerations are: Alfven velocity, cusp velocity, and sonic velocity, as:

$$V_A = \frac{B}{\sqrt{4\pi\rho}}, \qquad V_c = \frac{V_A C_s}{\sqrt{V_A^2 + C_s^2}},$$

and:

$$C_s = \left(\frac{\gamma p}{\rho} \right)^{1/2}.$$

The cusp velocity was defined as the appropriate (magnetoacoustic) throat velocity for the magnetic field orientation in this geometry (Weitzner, 1977); sound speed is the established limiting speed for non-plasma flow.

Experimental results for flow velocity at the throat are compared to the computed flow speeds in Figure 9.30.

The cusp speed is seen to be closest to the experimentally determined speeds, but based on flow properties, at later times the values of sound speed were also close to experimental values. In principle, the throat speed is most

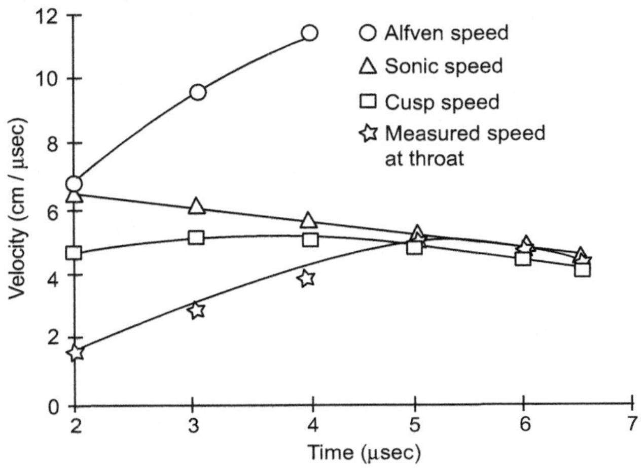

Figure 9.30 Values of sonic, Alfven, and cusp speeds determined from experiment compared to measured flow speeds at the throat as a function of time in the pinch discharge. *York, T.M., Jacoby, B.A., Mikellides, P., 1992. Flow processes within magnetic nozzle configurations. AIAA J. Propul. Power 8, 1023–1030. Figure 8, p. 1027 with permission.*

appropriately determined by the magnetosonic (cusp) speed, and this is evident in the results.

The expansion of the plasma in the diverging magnetic field lines was also examined in the experiment. There was radial diffusion of plasma into the confining field region and diffusion of field into the plasma near the axis region. The plasma in the expansion was collision dominated; the particles did not freely spiral about the magnetic field lines. Rather, the data were compatible with radial balance between magnetic pressure external to the plasma and the sum of fluid plus magnetic field pressure, as (for B, subscript ex denotes external; i denotes internal):

$$\frac{B_{ex}^2}{8\pi}(r > r_{plasma}) > nk(T_i + T_e) + \frac{B_i^2(r = 0)}{8\pi}.$$

To assist understanding of the physical processes, a numerical model was developed to compare with experiment (York et al., 1992). A model for two-temperature plasma with classical viscosity, resistivity, and thermal conduction functional forms was appropriate for the density and temperature in the experiment. The Hall parameter was $\omega\tau \approx 10^{-3}$ and 10^{-4} in the coil and expansion regions, respectively. It was found that

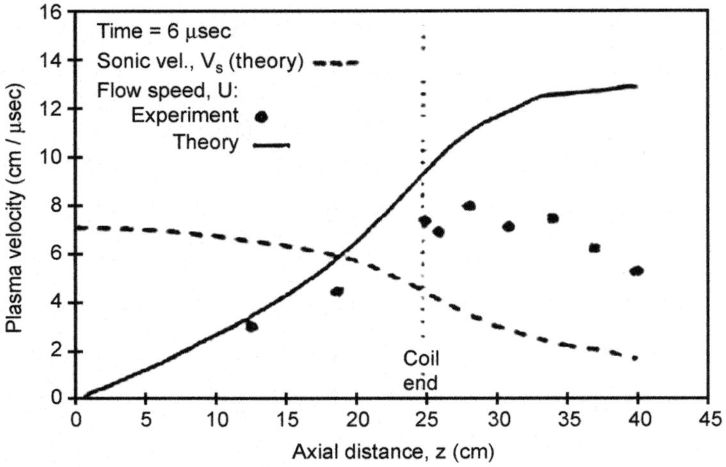

Figure 9.31 Values of (experimental) flow speed, computed flow speed, and sonic speed as a function of axial location calculated from experimental data. *York, T.M., Jacoby, B.A., Mikellides, P., 1992. Flow processes within magnetic nozzle configurations. AIAA J. Propul. Power 8, 1023–1030. Figure 14, p. 1029 with permission.*

energy loss by electron thermal conduction in the experiment was about 10 times greater than that calculated for classical conduction at early times, but at later times, thermal loss did fit classical behavior. At given times, the experimental flow velocity was measured at different axial locations; the results of the experimental flow speed are compared with computed flow speed and sonic speed calculated from data at different axial locations inside and outside the throat location at 6.0 μs in Figure 9.31.

It can be seen that the (collisional MHD model) theory did match the measured flow speeds up to the throat, which reached the magnetosonic condition, but the expansion region flow did not follow the computed expansion related to magnetic field lines. Again, there was a balance between external field magnetic pressure and the sum of fluid and internal magnetic pressure. It appears that the axial fields did assist the plasma flow axial alignment. It can be concluded for this relatively well-defined collisional plasma in a simple field–plasma configuration that plasma guiding did occur to some extent, but, beyond the magnetosonic point, detachment of the plasma from field lines was a more complex process.

REFERENCES

Artsimovich, L.A., 1964. Controlled Thermonuclear Reactions. Gordon and Breach, New York.

Freidberg, J.P., Weitzner, H., 1975. End loss from a linear theta pinch. Nucl. Fusion 15, 217–223.

(IAEA) International Atomic Energy Agency, 1961. In: Plasma Physics and Controlled Nuclear Fusion Research (Conference Proceedings, September 4–9), Salzburg; Also Nuclear Fusion Supplement (1962).

McKenna, K.F., York, T.M., 1977. End loss from a collision dominated theta pinch plasma. Phys. Fluids 20, 1556–1565.

Quinn, W.E., Siemon, R.E., 1981. Linear magnetic fusion systems. In: Teller, E. (Ed.), Fusion, vol. 1, B. Academic, New York.

Rosenbluth, M., 1954. Infinite Conductivity Theory of the Pinch. LA-1850. Los Alamos Scientific Laboratory Report.

Stover, E.K., Klevans, E.H., York, T.M., 1978. Computer modeling of linear theta pinch machines. Phys. Fluids 21, 2090–2102.

Wesson, J.A., 1966. Plasma Flow in a Theta Pinch. Plasma Physics and Controlled thermonuclear Research 1, 233.

Weitzner, H., 1977. End Loss from a Theta Pinch. Phys. Fluids 20, 384–389.

York, T.M., 1977. Scaling of End Loss Times in Linear Theta Pinches. USDOE Report C00-2040-1. Pennsylvania State University.

York, T.M., Jacoby, B.A., Mikellides, P., 1992. Flow processes within magnetic nozzle configurations. AIAA J. Propul. Power 8, 1023–1030.

Wave Heating of Plasmas

INTRODUCTION

Several mechanisms for heating plasmas have been noted and evaluated in the topics discussed earlier, such as Ohmic heating, magnetic compression heating, and shock wave heating. The addition of energy to plasma is a critical aspect for achievement of successful operation of plasma devices, and heating techniques have been tailored to the unique ranges of properties in different devices.

One of the most significant physics and engineering challenges has been the achievement of controlled magnetic fusion. Experiments with controlled fusion devices such as tokamaks, mirror machines, as well as *Z*- and theta pinch machines have required heating techniques such as electron beams, laser heating, and of direct interest here, plasma wave heating (e.g., ion cyclotron resonance heating (ICRH), electron cyclotron resonance heating (ECRH), Alfven/magnetosonic wave heating). These techniques have been adapted and applied to linear and toroidal geometries (Cairns, 1991). Controlled fusion devices utilize hydrogen-like gases, and so these have been of primary interest, but other wave heating experiments have been carried out with higher molecular weight gases.

Heating by Plasma Waves

A number of reviews of the subject of heating by plasma waves have been published (Cairns, 1991; Koch, 2006, 2008). While this is a topic involving detailed physics and engineering aspects, because of its importance in a number of applications, some basic considerations will be reviewed here. Specific utilization of plasma wave heating on several devices will be considered in the following sections.

The general procedure for heating with plasma waves involves several component steps:

1. Coupling of electromagnetic waves from a source to generate plasma waves
2. Transfer of plasma wave energy (absorption) to particle energy
3. Development of a definable distribution of the particle energy in the absorbing species
4. Transfer of particle energy by collisions to all plasma species

This sequence of events is presented in schematic form in Figure 9.32.

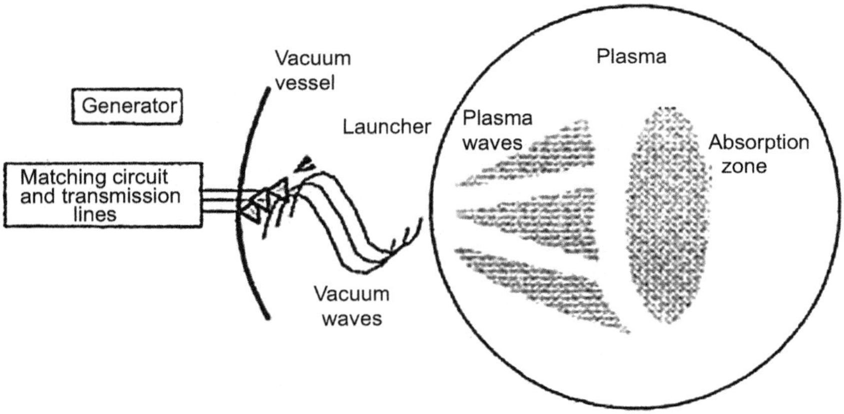

Figure 9.32 Schematic of plasma wave heating process. *Koch, R., 2006. The coupling of electromagnetic power to plasmas. Trans. Fusion Sci. Tech. 49, 177–186. Figure 1, p. 177 with permission.*

Of primary interest here is the plasma wave/energy transfer process. The absorption occurs by a resonance process of some (small fraction) particles with the waves. Interparticle collisions distribute this energy among the resonant species particles and also to the other species. For gases such as hydrogen (deuterium), different plasma waves can be used to heat the plasma in different ranges of properties, and these waves will be initiated at different frequencies of the source. A general listing of waves generated at different frequencies for specific plasma is presented in Table 9.2.

As the plasma wave mechanism generally involves magnetic field interactions, the waves are direction sensitive. Many plasmas are contained in vessels that are cylindrical with axes that may be linear or toroidal. A 2D slab model of the source–plasma local interface is shown in Figure 9.33.

The waves are launched in the x-direction, whereas the plasma is taken as uniform in the y-(poloidal) and z-(toroidal) directions.

Table 9.2 Plasma Waves and Wave Properties for Heating

Plasma Wave (Type)	**Source Freq. (MHz)**	**Plasma Medium (5 keV, D, 10^{19} m^3, 3 T)**
Alfven wave	1.0	$V_A = 7 \times 10^6$ m/s
ICRF (Ion cyc. range freq.)	10	$f_{ci} = 20$ MHz
Lower hybrid wave	10^3	$f_{LH} = 3$ GHz
ECRF (Elec. cyc. range freq.)	$>30 \times 10^3$	$f_{ce} = 80$ GHz

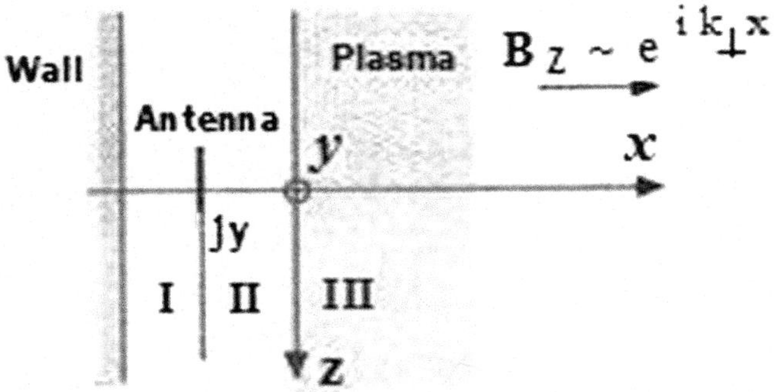

Figure 9.33 Plasma–wave coupling model. *Koch, R., 2006. The coupling of electromagnetic power to plasmas. Trans. Fusion Sci. Tech. 49, 177–186. Figure 2, p. 179 with permission.*

Collisionless Heating: Landau Damping

The process of energy transfer between particles takes place most commonly with interparticle collisions. The (random thermal) collision frequency can be expressed as

$$\nu = 2.9 \times 10^{-12}\, n \cdot \ln \lambda \cdot T^{-3/2}\ (\text{mks, eV}),$$

indicating that lower densities and high temperatures shift the process toward collisionless behavior. However, for wave oscillation accelerations of frequency, ω, it is the ratio, ω/ν, that is characteristic of the role of dissipation (and energy transfer) due to collisions. In the study of plasmas, it was recognized early that collisions are not exclusive in transferring energy from waves to particles. The wave dissipation due to collisions is negligible for frequencies in the megahertz range (Sagdeyev and Shafranov, 1958). Furthermore, plasma waves that have been excited can accelerate particles and transfer energy without collisions; it is when there are wave resonances that absorption can be significant. In early experimental studies of plasma waves, it was recognized by Stix and Palladino (1958) that strong absorption and plasma heating occurred when the wave frequency was equal to the ion cyclotron frequency, $\omega = \Omega_i$.

The collisionless transfer of energy from waves to plasmas was discovered in theoretical studies by Landau (1946); it is termed Landau damping. This behavior has been termed "the single most famous mystery of classical plasma physics" by Mouhot and Villani (2011), who extended the original linear analysis to a nonlinear domain. The concept and physics of this transfer of energy has considerable significance, particularly related to heating, in a number of applications areas in plasmas and plasma dynamics. Landau damping can occur with electron and ion waves.

The appropriate description of collisionless plasmas begins with the Boltzmann equation (Landau, 1946; Chen, 1983). The distribution function is assumed to be the sum of equilibrium and perturbation (1) terms, as:

$$f(\vec{r}, \vec{v}, t) = f_0(\vec{v}) + f_1(\vec{r}, \vec{v}, t).$$

For simplicity, assuming background electromagnetic fields, $\vec{B}_0$, $\vec{E}_0 \equiv 0$, the linearized Vlasov (collisionless Boltzmann) equation becomes:

$$\frac{\partial f_1}{\partial t} + \vec{v} \cdot \nabla f_1 - \frac{e}{m}\vec{E}_1 \cdot \frac{\partial f_0}{\partial \vec{v}} = 0,$$

where $\vec{E_1}$ is a perturbation electric field; this has a solution for particles of mass, m and charge Z:

$$f_1 = \frac{iZeE_1}{m(\omega - kv)} \cdot \frac{\partial f_0}{\partial \vec{v}}.$$

In terms of the distribution function, the power dissipated can be expressed as (Koch, 2006):

$$P_{DISS} = \frac{1}{2} Re[E_1 J_1]$$

where:

$$J_1 = Ze \int_{-\infty}^{+\infty} v f_1(v)\, dv$$

and:

$$P_{DISS} = \frac{Z^2 e^2 |E|^2}{2m} \int_{-\infty}^{+\infty} Im \frac{v}{\omega - kv} \frac{\partial f_0}{\partial \vec{v}}\, dv.$$

This is the change of power of the electromagnetic wave, and when negative, it is the power transferred to the particles. The evaluation of the integral is difficult because of the singularity at $\omega = kv$, where v is the particle velocity. Landau (1946) evaluated this integral properly, and for an analysis of electron oscillations, found that the wave energy dissipation is proportional to an exponential factor. Specifically, his results showed that the value of the perturbation wave electric field intensity was (x being the direction of propagation):

$$E(x) = \frac{E_0}{\varepsilon}\left[1 - \exp\left\{\frac{i}{a}\left(\frac{\varepsilon}{3}\right)^{\frac{1}{2}}x - \frac{3}{2a}\left(\frac{\pi}{2\varepsilon}\right)\exp^{\frac{-3x}{2\varepsilon}}\right\}\right],$$

where ε is the dielectric constant of the plasma, a is the Debye length, and E_0 is the magnitude of the initial wave electric field intensity. The third term in the above expression identifies the wave damping (energy absorption by particles).

In a more physical description of the analysis for electron plasma oscillations, Chen (1983) provided extensive discussion of the mathematics issues and physics concepts of the Landau damping effect. The results were more simply stated; wave energy was expressed as:

$$W_w = \frac{\varepsilon_0 E_0^2}{2},$$

and:

$$\frac{dW_w}{dt} = 2[Im\ \omega]W_w,$$

where:

$$[Im\ \omega] = -(\pi)^{1/2}\omega_p\left(\frac{\omega_p}{kv_{th}}\right)^3 \exp\left(\frac{-1}{2k^2\lambda_D^2}\right)\exp\left(\frac{-3}{2}\right),$$

and ω_p is the plasma frequency, v_{th} is the electron thermal speed, and λ_D is the Debye length. From this expression, it can be seen that the absorption becomes significant for larger values of $k\lambda_D$.

A conceptual description of the process of power absorption by collisionless heating from plasma waves is related to behavior of particle distribution in force fields. The absorption of energy from the wave is related to particles in the distribution function that have a velocity near the phase velocity of the plasma wave. This small group of particles more or less travels along with the wave. The energy transfer occurs with the particles near the high-energy side of the Maxwellian distribution; for particles near the wave phase velocity there are more particles with less velocity than there are with a higher velocity, so a net transfer is effected by the wave motion. Once energy is transferred to some particles, it is transferred to other particles and randomized in the species energy distribution. This heated species will subsequently transfer energy to the other plasma species. Landau damping has been effectively incorporated into heating schemes that launched high-power, high-frequency plasma waves, for example, near the lower hybrid frequency (Cairns, 1991; Koch, 2008).

As noted above, the ion cyclotron wave resonance is effective in transferring energy to plasmas; at large phase velocity and weak damping, the dispersion relationship for ICRF was determined to be (Hooke and Rothman, 1964; Chen, 1983):

$$\frac{k^2c^2}{\omega^2} = \frac{\omega_{pi}^2}{\Omega_i^2 - \omega^2},$$

where $\Omega_i =$ ion cyclotron frequency.

This relationship is singular at $\omega = \Omega_i$ and is indicative of strong absorption for that wave. Aspects of the physics of this absorption have been related to Landau damping, as well as other mechanisms (Koch, 2008).

Plasma Wave Heating for Space Propulsion

With an understanding of wave heating of collisionless plasmas, devices that demonstrated relevant plasma confinement could then be heated further to achieve improved operating conditions. The plasmas that were created in mirror machine fusion experiments displayed well-defined collisionless particle behavior and were in appropriate regimes for wave heating. The confinement characteristics, heating, plasma trapping at the ends, and expansion of the plasma in the loss regions of tandem mirror machines had been investigated and reported over a number of years (Simonen, 1981). There is also much current interest in the axisymmetric magnetic confinement geometry for high-energy particle generation (Simonen, 2008). This behavior is related to the inherent feature of particle loss from the linear configuration with open ends and the unique physics properties of the escaping plasma. A second potentially important application of this escaping high-velocity plasma stream has been studied for utilization in thrust generation in space (Chang-Diaz et al., 2004). As the confined plasma in mirror configurations has temperatures in the range of kilo-electronvolts, expansion of the escaping particles generates velocities of 10^5 m/s; mass flow rates are adequate to generate significant thrust values, thus providing an attractive concept for a deep space propulsion system powered by a fusion reactor system. Basic aspects of a mirror machine thrust system, along with the inherent plasma wave heating theory and the results of experiments that have been reported, will now be examined.

Variable Specific Impulse Magnetoplasmadynamic Rocket Concept

Utilizing the understanding of axisymmetric tandem mirror machine theoretical and experimental physics that was achieved in earlier magnetic fusion research (Simonen, 1981), the concept of an open-ended mirror machine thruster was developed (Chang-Diaz et al., 1988, 2004). A schematic of the thruster device (acronym, VASIMR) is shown in Figure 9.34.

The source plasma was created with (RF, Radio Frequency waves) ECRH (Cairns, 1991; Erckmann and Gasparino, 1994), and the plasma was heated further in magnetic confinement with ICRH (Koch, 2008); each of these had a power of 30 kW. Based upon previous reported results, plasmas with densities of about 10^{+13} cm^{-3} and temperatures of about 1 keV were anticipated in the experiments.

Numerical modeling of the plasma processes was carried out to predict experimental behavior in the VX-25 machine (Ilin et al., 2005). Propellant was assumed to be deuterium with plasma produced by a helicon antenna with up to 20 kW power. This was followed by use of ICRF plasma wave heater with powers from 1.5 to 10 kW. The predicted results of the calculations are shown in Figure 9.35; there was reasonable agreement with experiment for plasma loading (Ohms) for the ICRF.

Further experimental studies (VX-50 device) at higher (50 kW) total power and with deuterium, neon, and argon propellants were carried out.

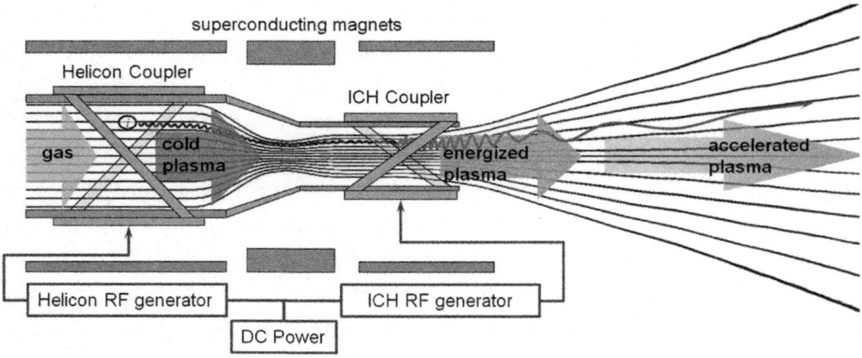

Figure 9.34 Schematic representation of the VASIMR® (Variable Specific Impulse Magnetoplasma Rocket) engine. *Downloaded from web site: www.adastrarocket.com, 6/20/2015. With permission of Ad Astra Rocket Company.*

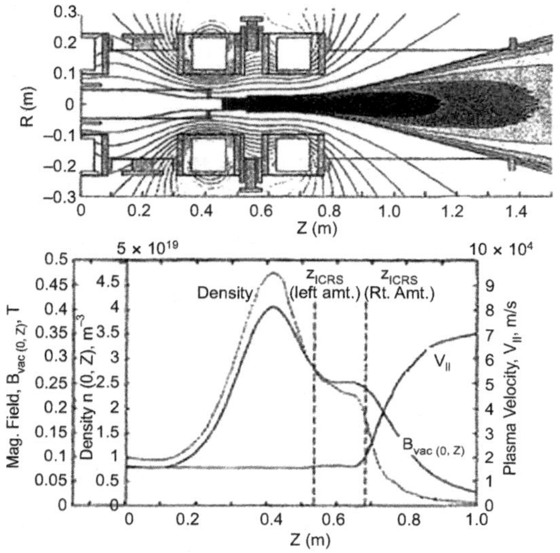

Figure 9.35 Calculated values (2D, 1D) of particle density, velocity, and magnetic field in VX-25. *Ilin, A.V., et al., 2005. Plasma heating simulation in the VASIMIR system. AIAA-2005-0949. Figure 2, p. 4; Figure 4, p. 6 with permission.*

With a helicon source (ECRF) of 25 kW RF at 13.56 MHz and an ICRF transmitter with 25 kW at 3.6 MHz, the ICRF lasted for 300 ms during the pulsed experiment (Squire et al., 2007). With deuterium propellant, ICRF power of 20 kW having a coupling efficiency over 90% at 3.6 MHz, bulk flow velocity of 100 km/s, and ion energies of up to 300 eV were measured. The indicated specific impulse was over 10^4 s, and the momentum flux was about 0.2 N. Experiments with argon and neon achieved exit velocities of 30 km/s and 40 km/s, respectively; this demonstrated the ability to operate efficiently at lower values (4000 s) of specific impulse, which is important in some mission planning profiles.

Plasma End Loss in a Collisionless Magnetic Expansion

The collisional flow of plasma in magnetic field lines has been discussed above, and the results of analysis using MHD equations have been reviewed. However, the lower density and higher temperature plasmas that are not controlled by collision processes behave differently, and they must be analyzed using appropriate methods. Furthermore, for either regime, the important question of plasma detachment from the guiding magnetic field lines and ultimate ejection of the plasma from boundaries of the region controlled by the thruster will have aspects of collisionless plasma behavior. A discussion of this problem and an analysis of the detachment processes has been presented in the literature (Arefiev and Breitzman, 2005; Breitzman et al., 2008) and will be reviewed here.

The governing equations of the collisionless flow are the Vlasov and Maxwell equations. In Breitzman et al. (2008), to model the flow in the expansion, a supersonic stream with cold electrons and an absence of rotation (no $\vec{E} \times \vec{B}$ drift) was assumed. The steady-state Vlasov equation for the distribution function (ion) is written as:

$$m_i \cdot v_i \cdot \nabla f_i + \frac{q_i}{c}\left[\vec{v}_i \times \vec{B}\right] \cdot \nabla_v f_i = 0,$$

and with cold electrons the equation is:

$$\frac{q_e}{c}\left[\vec{v}_e \times \vec{B}\right] \cdot \nabla_{ve} f_e = 0.$$

With a gyro-averaged distribution function (f_i) specified, the magnetic field profiles can be determined. The momentum flux was accounted for within the calculation framework so as to provide the plasma-field description. Magnetic field outside the plasma is a controlling influence on the flow, and a solution for the fields from an external coil configuration can be carried out.

A reference solution for the method described above was presented (Breitzman et al., 2008); this was for the case of super-Alfvenic cold plasma in a conical magnetic nozzle for which an analytic solution had been reported. The results were shown to be in good agreement with the model, and the details of the flow processes were consistent.

A numerical solution for plasma flow from a cylindrical magnetic coil nozzle was presented. The flow transition from sub-to super-Alfvenic flow, for cases with and without conversion of ion gyroenergy from perpendicular to parallel (direction along the field lines), were examined. The calculated results for the case of allowed gyroenergy transfer are shown in Figure 9.36.

The further use of this computation method to simulate flow in an experimental device that was designed to study detachment of the plasma from the end-expansion magnetic field lines (Detachment Demonstration Experiment) was also reported. The plasma was argon with a source density of 10^{30} m^{-3} and ion energy $\varepsilon_i = 2$ eV and source magnetic field of 0.1 T. The computation results provided details of the plasma acceleration, guiding, and detachment. In Figure 9.37, the thin solid lines are the magnetic field lines in the absence of plasma, while the insets show the

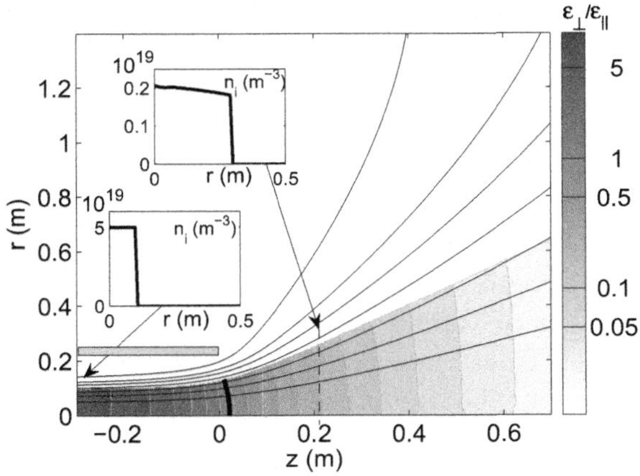

Figure 9.36 Numerical prediction of collisionless magnetic nozzle flow with gyroenergy transfer. *Breizman, B.N., Tushentsov, M.R., Arefiev, A.V., 2008. Magnetic nozzle and plasma detachment model for a steady-state flow. Phys. Plasmas 15, 057103. Figure 5, p. 057103-8 with permission.*

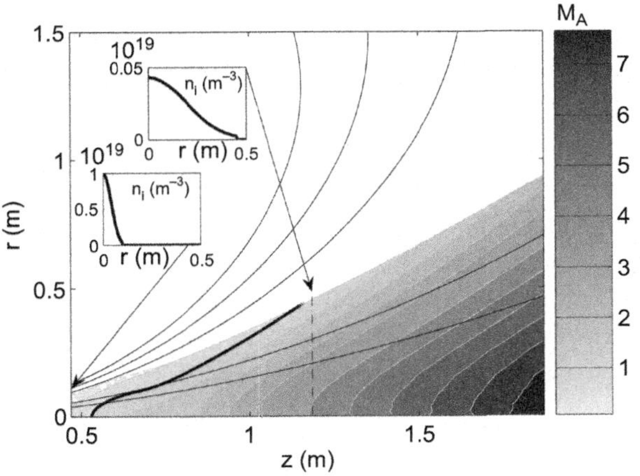

Figure 9.37 Comparison of simulation and density data of the plasma plume in the detachment demonstration experiment. *Breizman, B.N., Tushentsov, M.R., Arefiev, A.V., 2008. Magnetic nozzle and plasma detachment model for a steady-state flow. Phys. Plasmas 15, 057103. Figure 8, p. 057103-9 with permission.*

radial density profiles at inlet and at $z = 1.17$ m; the heavy solid line separates the sub- and super-Alfvenic flow speed regions.

The experimental density data agreed with the simulation radial profiles. The density data showed that the plasma flow divergence was significantly less than that of the magnetic field lines and that detachment was demonstrated.

Laser Heating of Plasmas (Theta Pinch)

In the laboratory, pulsed lasers have been used as a unique diagnostic tool for plasma diagnostics, such as Thomson scattering. High-power lasers are also effective for heating gases. The absorption of laser energy in gases occurs at line wavelengths and also by continuum processes in the electron population. As lasers of various powers are readily available in different wavelength ranges, laser interactions with plasmas have become an important application. The use of lasers for supplemental heating of confined plasma will be examined here in order to provide a framework for understanding the physical principles involved in the laser heating processes, the magnitude of heating effect, and the timescales of the process. The application to a linear theta pinch experiment, the physics of which was broadly presented in the section, Magnetic Compression and Heating, will be considered.

In order to achieve higher temperature plasmas in a magnetic confinement fusion application, high-power laser heating of a (theta) pinch plasma column was investigated. This technique was considered promising, as near-fusion conditions had been achieved in early theta pinch experiments, and the supplemental heating could improve performance. The Scylla I-C experiment at Los Alamos National Laboratory (LANL) (York and McKenna, 1975) was used for laser heating studies. An analysis was carried out in order to estimate the ultimate result of CO_2 laser heating of magnetically confined plasma following dynamic pinch compression. The plasma properties at pinch equilibrium prior to laser heating following magnetic compression and heating of deuterium gas at 100 mT fill pressure were:

p_0(mTorr)	R_f(cm)	n_f (cm^{-3})	$T_{ef} = T_{if}$(eV)	E_f(J)
100	0.249	3.7×10^{17}	41	143

where the subscript f defines the final stable pinch properties. The incident laser beam had a beam diameter of 8 cm for a pulse energy of 80 ± 20 J in a period of 60 ns, resulting in a peak power of 2×10^9 W; the focused beam diameter was 2 mm and the impinging power was 3×10^{10} W/cm^2.

The dominant absorption mechanism in this plasma was inverse bremsstrahlung, by which electrons are accelerated by the laser fields, and electrons heat the ions by collision. The absorption of the flux density of the beam, $I\left(\frac{W}{m^2}\right)$, as it passes through plasma along a path, s, can be expressed as:

$$\frac{dI}{ds} = -KI, \qquad \text{so,} \qquad I(s) = I(0)e^{-\tau(s)},$$

with:

$$\tau(s) = -\int_0^s \overrightarrow{K} \cdot d\overrightarrow{s},$$

where K is the inverse bremsstrahlung absorption coefficient. The absorption coefficient defines the rate of change of flux density, and along with the absorption length, (l_{ab}), is given by:

$$l_{ab} = \frac{1}{K} = 1.15 \times 10^{29} \frac{(k_B T_{ef})^{3/2}}{\lambda^2 Z n_f \ln \Lambda} \left(1 - \frac{\lambda^2}{\lambda_p^2}\right)^{1/2},$$

where λ is the laser light wavelength, λ_p is the wavelength of radiation at plasma frequency, Z is the electronic charge number, and $\ln \Lambda$ is the Coulomb logarithm. With heating of the electrons, the temperature and absorption coefficient will vary. Assuming uniform plasma density along the column, the temperature along the beam would vary as:

$$\frac{T_e(x)}{T_{ef}} = \left[1 + Q(0)e^{-Kx}\right]^{2/5},$$

where x is the distance along the beam in the plasma and Q(0) is the ratio of entry flux to the thermal energy in one absorption length; the result incorporates electron thermalization, electron–ion equilibration, and plasma column pressure equilibration after heating.

Laser light scattering instabilities can result in anomalous backscatter and anomalous absorption effects. Scattering of the incident laser light is possible due to backward scattering with either a low-frequency ion wave (stimulated Brillouin scattering instability) or an electron plasma wave (stimulated Raman scattering instability). These depend on the power level and wavelength through the electron velocity,

$$v_0 = 25\lambda_0\sqrt{P}\ (\text{cm/s}),$$

where λ_0 is in micrometers and P is in Watts per square centimeter.

The Brillouin instability occurs when

$$\omega_0 = \omega_s + \left[c^2(k - k_0)^2 + \omega_{pe}^2\right]^{1/2},$$

where

$$\omega_s = k_B\left(\frac{T_e}{m_i}\right)^{1/2}$$

is the ion wave frequency, k is the wave number of the ion wave, k_0 is the wave number of the incident light of frequency ω_0, and ω_{pe} is the plasma frequency. When $T_e \approx T_i$, the instability growth rate is

$$\gamma_0 = \frac{k}{4}\frac{v_0\omega_{pi}}{(\omega_s\omega_0)^{1/2}}\ \left(\mathrm{s}^{-1}\right).$$

For Raman scattering, the balance of instability growth with damping occurs when:

$$\omega_0 = \left[\omega_{pe}^2 + 3k^2v_e^2\right]^{1/2} + \left[c^2(k - k_0)^2 + \omega_{pe}^2\right]^{1/2},$$

then:

$$k = 2k_0 - k_0\left(\frac{\omega_{pe}}{\omega_0}\right)\left(\mathrm{cm}^{-1}\right),$$

where k relates to the electron plasma wave.

The instability growth rate is:

$$\gamma_0 = \frac{kv_e}{4}\frac{\omega_{pe}}{\left[\omega_k(\omega_0 - \omega_k)\right]^{1/2}}\ \left(\mathrm{s}^{-1}\right),$$

where:

$$\omega_k^2 = \omega_{pe}^2 + 3k^2v_e^2,$$

and the threshold power for the instabilities can be determined.

With the laser beam directed along a cylindrical plasma column and positive radial gradients of density, the beam would be diffracted toward regions of lower density and higher index of refraction (Steinhauer and Ahlstrom, 1971). This effect results in channeling of the laser beam and self-focusing. Experimental evidence of this type of "drilling" laser-generated density gradient behavior had been observed in experiment. For the Scylla I-C experiment, laser absorption parameters that resulted from pinched plasma properties for a fill pressure of 100 mT were predicted to be as follows:

Absorption length:

$$l_{ab} = 31 \text{ cm};$$

Brillouin instability onset:

$$p = 6.1 \times 10^{9} \text{ W/cm}^2;$$

Raman instability onset:

$$p = 2.1 \times 10^{10} \text{ W/cm}^2;$$

Laser-induced gradient:

$$\frac{dn_e}{dr} = 9.6 \times 10^{17} \left(\text{cm}^{-4}\right).$$

So, for this experiment, the instability thresholds would not be reached; the predicted heating at time of laser pulse would be $T_e(x = 0) = 221$ eV, $T_e(x = 1 \text{ m}) = 178$ eV, and after thermal and radial force equilibration, a period of $\Delta t = 100$ ns: $T_{e,i} = 56$ eV and $R_{f+\Delta t} = 0.274$ (cm).

REFERENCES

Arefiev, A.V., Breizman, B.N., 2005. Magnetohydrodynamic scenario of plasma detachment in a magnetic nozzle. Phys. Plasmas 12, 043504.

Boyd, T.J.M., Sanderson, J.J., 1969. Introduction to Plasma Dynamics. Barnes & Noble, New York.

Breizman, B.N., Tushentsov, M.R., Arefiev, A.V., 2008. Magnetic nozzle and plasma detachment model for a steady-state flow. Phys. Plasmas 15, 057103.

Cairns, R.A., 1991. Radiofrequency Heating of Plasmas. IOP Publishing Ltd, New York.

Chang-Diaz, F.R., et al., 1988. A tandem mirror hybrid plume plasma propulsion facility. In: International Electric Prop. Conf., IEPC-88-126.

Chang-Diaz, F., Squire, J., Shebalin, J., 2004. Plasma production and heating in VASIMIR-A plasma engine for space exploration. In: EPS Conference on Plasma Physics, ECA, 28G, P-1.025.

Chen, F.F., 1983. Introduction to Plasma Physics and Controlled Fusion, second ed. Plenum, New York.

Erckmann, V., Gasparino, U., 1994. Electron cyclotron heating and current drive in a toroidal fusion plasma. Plasma Phys. Controlled Fusion 36, 1869–1962.

Hooke, W.M., Rothman, M.A., 1964. A survey of experiments on ion cyclotron resonance in plasmas. Nucl. Fusion 4, 33.

Ilin, A.V., et al., 2005. Plasma heating simulation in the VASIMIR system. In: AIAA-2005-0949.
Koch, R., 2006. The coupling of electromagnetic power to plasmas. Trans. Fusion Sci. Tech. 49, 177–186.
Koch, R., 2008. The ion cyclotron, lower hybrid, and Alfven wave heating methods. Trans. Fusion Sci. Tech. 53, 194–201.
Landau, L., 1946. On the vibration of the electronic plasma. J. Phys. USSR 10, 25. English translation in JETP, 16, 574.
Mouhot, C., Villani, C., 2011. On landau damping. Acta Math. 207, 29–201.
Sagdeyev, R.S., Shafranov, V.D., 1958. Absorption of high-frequency electromagnetic energy in a high-temperature plasma. In: Proc. of 2nd Int. Conf., IAEA, Geneva.
Simonen, T.C., 1981. Experimental progress in magnetic-mirror research. Proc. IEEE 69, 935–957.
Simonen, T.C., 2008. The Axisymmetric Tandem Mirror: A Magnetic Mirror Game Changer. Lawrence Livermore Nat. Laboratory. TR-408176.
Squire, J.P., et al., 2007. High power VASIMIR experiments using deuterium, neon and argon. In: Int'l Electric Propulsion Conf., IEPC 2007-181.
Steinhauer, L.C., Ahlstrom, H.G., 1971. Propagation of coherent radiation in a cylindrical plasma column. Phys. Fluids 14, 1109.
Stix, T., Palladino, R.W., 1958. Ion cyclotron resonance. In: Proc. of 2nd Int'l. Conf. on Peaceful Uses of Atomic Energy, vol. 31. U.N., Geneva, p. 282.
York, T.M., McKenna, K.F., 1975. Laser-Plasma Interactions in the Scylla I-c Experiment-Preliminary Analysis. Los Alamos National Laboratory. Report LA-5957-MS.

Magnetic Fusion Plasmas

INTRODUCTION

In general, we know that the advancement of technology is intimately related to the acquisition and control of energy. The energy we now use is largely derived from fossil fuels whose source was energy deposition from the radiant energy of the Sun. The one new source of energy available for development with current technology is nuclear fusion, the same source of energy resident in the Sun, and it is the goal of controlled fusion research to harness that energy source (Glasstone, 1980). When two hydrogen atoms fuse together to form helium, the binding energy of the helium is significantly greater than that of the hydrogen, and that energy increment is released into the surroundings. The energy is released suddenly, and control of the rate and energy content of the particles has not yet been achieved. Current physics studies seek to control the fusion process in two primary ways: inertial confinement fusion (ICF) and magnetic confinement fusion; a hybrid of these is magnetized target fusion (MTF). The most accessible path for controlled fusion uses the deuterium (D) and tritium (T) isotopes of hydrogen. The conditions required to reach ignition were first presented in 1957 and are identified as the Lawson Criterion (1957). This condition defines that plasma which reaches ignition will release sufficient energy density to maintain the temperature and compensate for losses. The ignition plasma is specified by plasma density (n_e), temperature (T), and confinement time (τ); the confinement time can be expressed as:

$$\tau_E = \frac{E\ (plasma\ energy)}{P_{loss}\ (rate\ of\ energy\ loss)}.$$

The energy release is volume dependent and the energy loss is surface area dependent: a low limit on ignition conditions with D–T is $n_e\tau_e > 1.5 \times 10^{20}$ s/m^3 at 25 keV. In terms of threshold for magnetic fusion, a plasma at $T = 25$ keV and $n_e = 1.5 \times 10^{20}$ m^{-3} for 1 s can achieve ignition. Initial efforts to use magnetic fields to confine, compress, and heat ionized gases began in the 1950s, and efforts have progressed to the present with the construction of a prototype fusion reactor experiment, the International Thermonuclear Experimental Reactor (ITER) (Green and ITER Team, 2003).

Several generations of fusion devices have moved through conception and evaluation of fusion physics experiments. A number of research monographs on fusion plasma topics and plasma physics textbooks have been written and are available for detailed study (Freidberg, 2007; Miyamoto, 1989; Chen, 1984). In the present introductory work, the goal is to identify the development of physics understanding related to experiments and devices that have evolved with improved configurations to achieve controlled magnetic fusion. As magnetic topology is unique to these devices, there will be an attempt to introduce the conductor geometries and currents and their relationships to basic laws.

Z-pinch: Plasma Parameters and Device Development

The concept that magnetic fields created by currents flowing in a conducting medium can result in a radial compression "pinch" effect has a relatively long history. The formalism of the physics for the plasma pinch was established by the Bennett relationships published in the 1930s (Bennett, 1934), and experiments on Z-pinches were devised in the late 1940s. Physically, in a cylindrical coordinate system (z,r,θ), current in the z-direction induces magnetic field in the azimuthal (θ) direction in relationship to current enclosed within a radius (r); this produces a radially inward $\vec{J} \times \vec{B}$ force. The current conduction in the gas produces ionization, plasma compression, and heating.

The initiation of voltage and induced currents along the z-direction can result in a dynamic sequence involving the following: current sheet formation at minimum inductance (maximum radius); dynamic implosion that collects swept mass (snowplow); as well as compression on-axis and pinched column dynamics (due to instabilities). Aspects of the dynamic collapse and pinch equilibrium on axis geometries are presented in Figure 9.38.

Pinch equilibrium: The physics of charged particle streaming and an equilibrium state was first derived by Bennett (1934). Appealing to radial

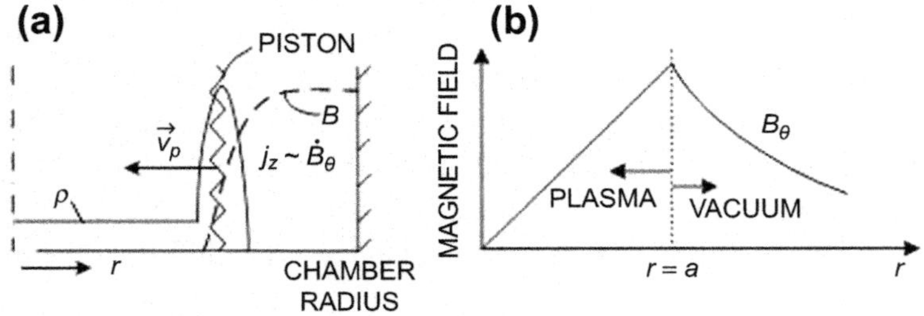

Figure 9.38 (a) Pinch collapse dynamics. (b) Equilibrium plasma column magnetic fields.

plasma force balance, the relationship of density, temperature, and current within radius a is:

$$8\pi N k_B (ZT_e + T_i) = \mu_0 I_{\mathrm{ENC}}^2(a),$$

where $N(r)$ is the number density per unit length and Z is the ion charge number; this behavior is shown in Figure 9.38(b). These concepts form the basis of magnetic compression in pinch devices.

Heat loss to electrodes: Initial experiments with the pinch were in a simple geometry and involved axial (z) current flow between electrodes at the ends. Several models of heat transport to anode and cathode were identified and equating power input and loss results in the general functional form (Haines, 2011):

$$I = \frac{\pi a^2}{Z_0} \frac{5}{2} \frac{k_B}{e} \cdot \alpha \cdot T^{5/2},$$

where Z_0 is the axial length, α is a constant from conductivity, $\sigma \sim \alpha T^{5/2}$, and a is the column radius.

Magnetic fusion conditions: If the plasma was assumed to be stable, and the energy loss was balanced by power addition, the energy confinement time can be expressed as:

$$\tau_E = \frac{\frac{3}{2} N k_B (T_e + T_i) Z_0}{I \cdot V},$$

where $I \cdot V$ is the input power to the plasma. To achieve fusion conditions, the plasma should satisfy:

$$n\tau_E = 1.71 \times 10^{-11} N T^{3/2}.$$

Typical orders of magnitude for this equilibrium would be: $V = 6.5 \times 10^4$ V, $I = 10^6$ A, $Z_0 = 1.0$ cm, $n = 10^{20}$ cm^{-3}, $N = 5.6 \times 10^{18}$ m^{-1}, $n\tau_E = 10^{20}$ s m^{-3}, and $T = 10^8$ K (10 keV).

Toroidal Z-pinch configuration: To avoid the loss of energy and particles from the ends of the pinched column, the notion of a toroidal column geometry to mitigate loss effects was an early concept. A patent proposal for a toroidal Z-pinch concept to achieve controlled fusion was submitted in the United Kingdom (Thomson and Blackman, 1947). The proposed device would have: geometry $R/a = 1.3/0.3$ m, current $I_p = 0.5$ MA, with confinement time, $\tau = 65$ s; $T_i = 500$ keV, using deuterium, D–D as fuel. Experiments to apply this concept were first carried out with the ZETA Z-pinch; a schematic and summary of the Z-pinch device are shown in Figure 9.39.

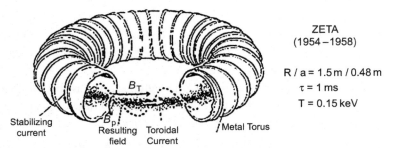

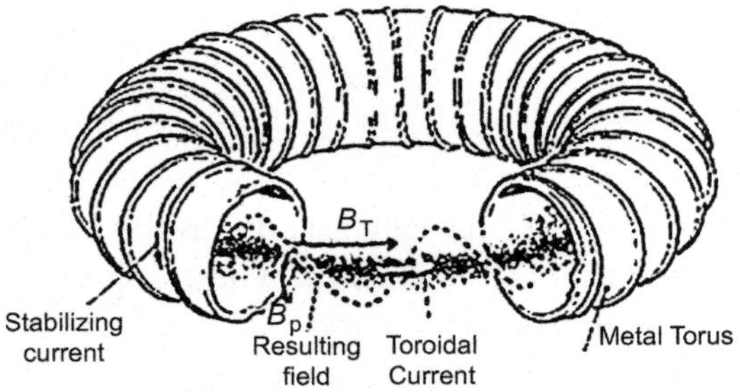

Figure 9.39 Schematic of the ZETA device and summary of experimental parameters. *Sykes, A., 2008. The Development of the Spherical Tokamak. Slide#5, UK AEA, Culham, Downloaded 9/7/15 from: http://www.sunist.org/shared%20documents/ISTW2008/TM/Talks/Sykes/Sykes%20talk1.pdf, with permission.*

The ZETA experiment did not confine the pinched column due to MHD instabilities. In later experiments, the addition of modest values of applied toroidal field were found to result in some gains in stability and longer confinement times; density and temperature of the confined plasma increased, but the improvement did not promise sufficient gains toward the goal of achieving fusion conditions. These results were reported at the First UN International Conference on Atomic Energy (Geneva, 1955), and the major findings are as follows.

Stability: Early linear pinch experiments demonstrated the presence of MHD disruptions (m = 0, sausage; m = 1, kink instabilities) (Haines, 2011). Later, in an attempt to define a stable configuration, an analysis of the fields and force balance for equilibrium was carried out by Kadomtsev (1966); this was based on the principle of magnetic flux distribution; it postulated a marginal stability criterion with magnetic energy conserved in boundary fluctuations and so did satisfy the adiabatic law: $pV^{\gamma} = \text{constant}$. This leads to the stability condition:

$$-\frac{r}{p}\frac{dp}{dr} = \frac{2\gamma}{1 + (1/2)\gamma\beta},$$

where:

$$\beta = \frac{2\mu_0 p}{B_0^2}.$$

Combining this with pressure balance leads to the equation:

$$1 - Z^{\gamma-1} = x^2 Z, \quad \text{with} \quad x = \frac{r}{r_0}$$

where r_0 is a length scale based on p_0 and γ. So a stable pressure–field distribution can be solved for as:

$$p(r) = p_0 \cdot Z^{\gamma},$$

where:

$$Z(r) = \frac{B_\theta(r)}{r}\left(\frac{r}{B_\theta}\right)_{r=0},$$

however, this unique configuration was not achieved in experiment.

Stabilizing axial current: It was found that some degree of stabilization could be achieved by applying an axial (or toroidal) magnetic field outside the plasma. Subsequently, to accomplish this, the plasma density had to be limited to values lower than desired (Haines, 2011). Also, stability could be enhanced by finite Larmor radius effects, sheared axial flow, and finite resistivity (Haines and Coppins, 1995). For the last of these, stabilization due to plasma resistance was found for conditions, where:

$$S = \frac{8.2 \times 10^{24} I^4 a}{Z(Z + T_i + T_e)^{1/2} A^{1/2} N_i^2 \ln \lambda_{ei}} < 100,$$
$$\text{for } Z \text{ (atom charge)} = 1 \text{ and } m_i = m_D.$$

A graphical presentation of the results of experiment and theory with respect to stabilization of the Z-pinch plasma column is shown in Figure 9.40. In the graph, ion Larmor effects are largely density dependent, while collision influences are seen with Hall effect, resistivity, and viscosity-related regimes whose boundaries are linear with increasing current and density. CGL refers to kinetic MHD model; IC and NRL to laboratory experiments.

Recent Applications: Dense (Linear) Z-Pinch X-ray Sources

The intense heating possible in the Z-pinch in the imploded plasma column on axis has been studied for neutron production as a part of fusion studies and also, independently, as an X-ray source (Haines, 2011). The neutron and X-ray production were enhanced by using the technology of Marx generator power supplies to provide more intense discharges. Recent research using peripheral wire arrays to initiate Z-pinch discharges resulted in significant improvements in X-ray production. In the wire array discharges, single wires carrying intense current were seen to create metal plasmas, which generated a plasma front that propagated toward the axis.

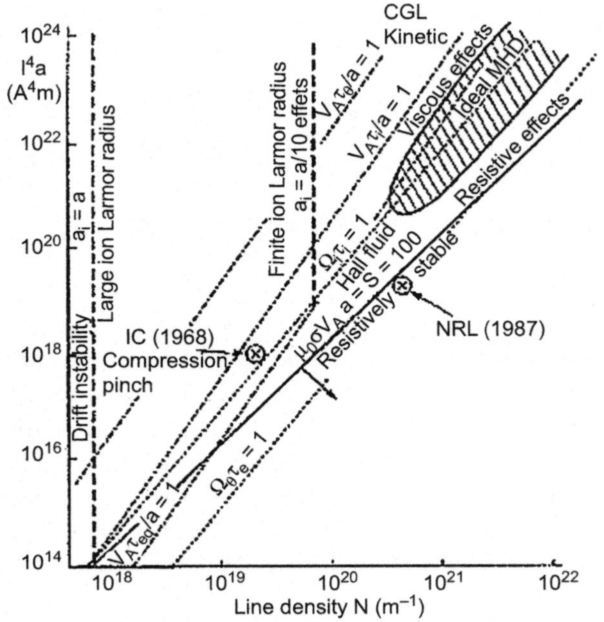

Figure 9.40 Stability regimes of the Z-pinch for D–T. *Haines, M.G., 2011. A review of the dense Z-Pinch. Plasma Phys. Controlled. Fusion, 53, 1. Figure 16, p. 33 with permission.*

The spacing of gaps between the wires can be made to form an effective channeling of plasma so that a smaller gap can intensify the implosion, and in that way create enhanced X-ray production (Beg, Lebedev et al., 2002; Cuneo et al., 2005). The dynamics of the plasma created by current carried by the wires has three components: precursor plasma (40–50%), main implosion (10–30%), and trailing mass (30–40%). The radial implosion trajectory corresponds to a partial (10%) entrapment snowplow of the plasma. Lebedev et al. (2001) showed that the radial distribution of mass at radius, r, and time, t, after initiation at R_0 is:

$$\rho(r,t) = \frac{\mu_0}{8\pi^2 R_0 r V_a^2}\left[I\left(t - \frac{R_0 - r}{V_a}\right)\right]^2,$$

where $I(\ldots)$ is the current at the time $\left(t - \frac{R_0 - r}{V_a}\right)$, with V_a being Alfven speed. The precursor plasma will stagnate on axis, but it is when the later-arriving current carrying main implosion arrives on-axis that the enhanced X-ray pulse occurs. The trajectory of these high-intensity

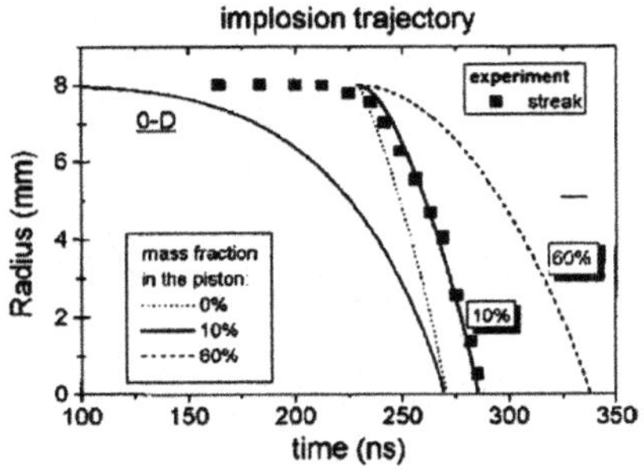

Figure 9.41 Implosion trajectory of 32-wire array. *Haines, M.G., 2011. A review of the dense Z-Pinch. Plasma Phys. Controlled. Fusion 53, 1. Figure 39, p. 69 with permission.*

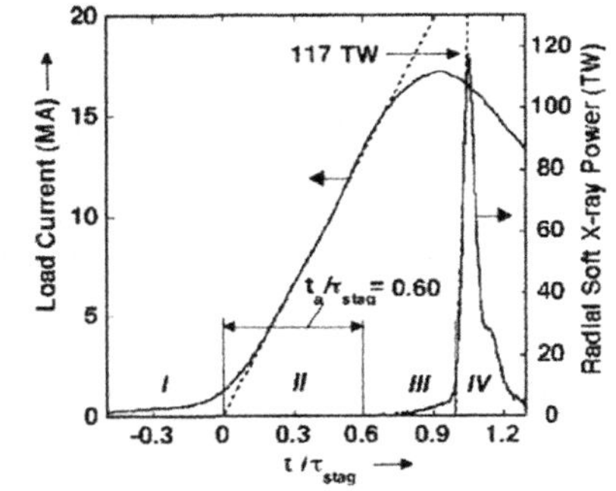

Figure 9.42 Current and power for a wire array showing four phases. *Haines, M.G., 2011. A review of the dense Z-Pinch. Plasma Phys. Controlled. Fusion 53, 1. Figure 39, p. 69 with permission.*

implosions is shown in Figure 9.41, and the history of the circuit current driving the implosion and the emission of X-rays is shown in Figure 9.42.

Of considerable significance in the energy deposition process is the ordering of accommodation times in the plasma species, with the equilibration time, $\tau_{eq}(i-e)$, being greater than the viscous heating time (Alfven transit time), which is greater than the ion–ion collision time, as:

$$\tau_{eq}(i-e) = \frac{3m_i(4\pi\varepsilon_0)^2}{8(2\pi m_e)^{1/2}e^{5/2}} \cdot \frac{T_e^{1/2}}{Z^2 m_e \ln \Lambda_{ie}} > \tau_{visc} = \frac{a}{V_a} \gg \tau_{ii},$$

where a is the column radius.

Summary: The study of pinch physics, pinch dynamics, heating, and confinement physics in pinch discharges has been fundamental to the progress toward controlled fusion. The sequence of different categories of physics contributions can be summarized, by topics, and is presented in Figure 9.43.

Applied Toroidal Field Configurations: Plasma Parameters and Device Development

Introduction

After the First United Nations (UN) Conference on the Peaceful uses of Atomic Energy in 1955, the Second UN Conference in 1958 (UN, 1958) provided a forum for detailed reporting on national programs for development of magnetic confinement devices; most significant was the

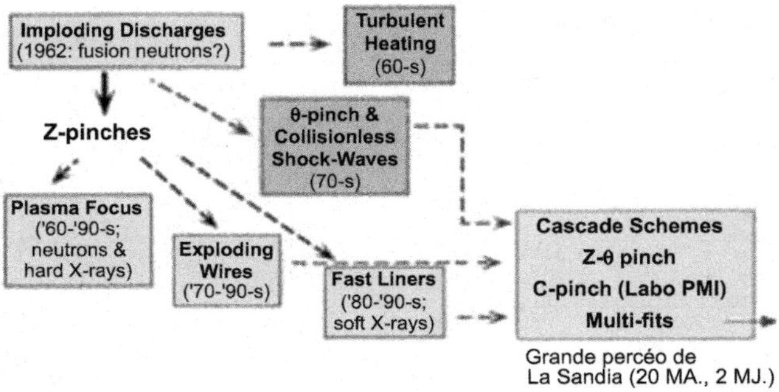

Figure 9.43 Outline of advances in imploding discharges. *Larour, J., January 13, 2011. Approach to Fusion with Z-Pinches: History and Recent Developments. Labratoire Phys. Plasmas, Palaiseau, FR. Downloaded from: http://www.lpp.fr, on 5/20/12. p. 30 in presentation with permission.*

inclusion of research results from the USSR. Fusion research in the Soviet Union generally pursued concepts similar to those in the West, but the results of their research were not known in the international community. This was significant, as earlier in this time period, in 1956, Kurchatov (USSR) had reported, in a lecture at UK AEA, Harwell, and a later publication (Kurchatov, 1956), that fusion conditions had not, in fact, been achieved in Z-pinch devices, contrary to Western research reports; that conclusion was based on neutron emission, which was later defined as due to instabilities. Research on controlled fusion gained importance; notably, the First International Conference on Controlled Fusion (1961) included presentations on minimum-B magnetic confinement configurations as being MHD stable. The Second International Conference (1965) included research results on tokamak magnetic configurations (toroidal chamber with axial magnetic field) TM-1, TM-2, T-2, T-3, and T-5. In a significant advance, theta pinch experiments were reported to have demonstrated kiloelectronvolt plasma temperatures. At the Third International Conference (1968), breakthrough research reported that longer (millisecond) confinement and kiloelectron volt temperatures had been achieved in the T-3 tokamak; these temperatures, $T_e > 1$ keV and confinement times, $\tau_E > 50\ \tau_{Bohm}$ (Bohm diffusion time) were later confirmed with data from definitive Western (Thomson scattering) diagnostic tests. The tokamak that achieved these advances, utilized the concept much stronger toroidal magnetic field to achieve stabilization, and that unique approach created a new framework for future fusion physics studies.

Tokamak

The overall toroidal geometry, the coordinates, and the functional variation of the toroidal magnetic field relative to the value on the major axis are shown in Figure 9.44. A confined plasma cross-section showing the radial distribution of plasma pressure, along with poloidal and toroidal magnetic field strengths in a tokamak confinement configuration, is shown in Figure 9.45.

The ultimate goal of the magnetic confinement configuration is to establish equilibrium, and in that context the plasma pressure and the effective pressure created by the magnetic fields must balance. The geometry of the tokamak fields can be described as follows (McMillan, 2012). The length of the toroidal field can be described as:

$$l_t = \int dl_t = \int \frac{B_t}{B_p} dl_p.$$

In one poloidal (p) turn, we have $l_t = 2\pi r_p \cdot (B_t/B_p)$, and the number of toroidal (t) turns in one poloidal turn is the safety factor, given by

$$q = \frac{l_t}{2\pi R_t} = \frac{r_p B_t}{R_t B_p}.$$

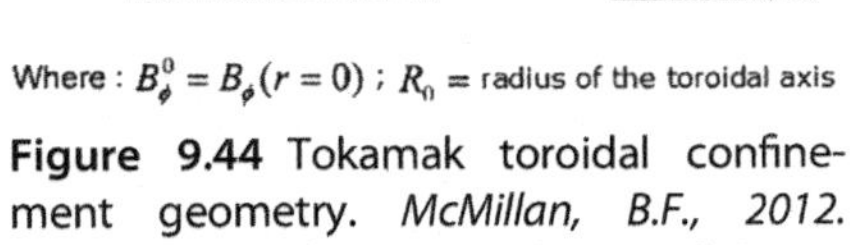

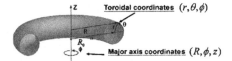

Figure 9.44 Tokamak toroidal confinement geometry. *McMillan, B.F., 2012. Lecture Slides for PX438 Physics of Fusion Power, Univ. Warwick, UK, http://www2.warwick.ac.uk/fac/sci/physics/current/teach/module_home/px438/ downloaded 4/5/2012, with permission.*

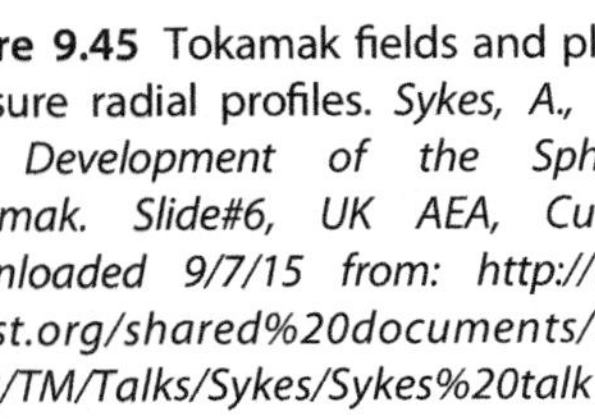

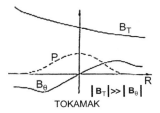

Figure 9.45 Tokamak fields and plasma pressure radial profiles. *Sykes, A., 2008. The Development of the Spherical Tokamak. Slide#6, UK AEA, Culham. Downloaded 9/7/15 from: http://www.sunist.org/shared%20documents/ISTW2008/TM/Talks/Sykes/Sykes%20talk1.pdf, with permission.*

This is an important parameter in defining stability of the confined plasma. Typically, for a value of $q = 3$, then:

$$\left(\frac{B_p}{B_t}\right) = \frac{r}{3R} \approx 0.1.$$

The specification of magnetic equilibrium in the toroidal tokamak configuration requires the definition of unique coordinates and functional forms (deBlank, 2006). Beginning (following Miyamoto, 1989) with cylindrical coordinates (r, ϕ, z), we define a magnetic surface, ψ, with:

$$\psi = rA_\phi(r, z),$$

where A_ϕ is the component of vector potential, and:

$$rB_r = -\frac{\partial\psi}{\partial z},$$

with:

$$rB_z = \frac{\partial\psi}{\partial r};$$

we also have:

$$p = p(\psi).$$

From the force and field equations for the plasma:

$$\nabla p = \overrightarrow{J} \times \overrightarrow{B}, \quad \nabla \cdot \overrightarrow{B} = 0, \quad \nabla \cdot \overrightarrow{J} = 0, \quad \nabla \times \overrightarrow{B} = \mu_0 \overrightarrow{J},$$

$$\text{then} \quad \overrightarrow{J} \cdot \nabla p = 0, \quad \text{and} \quad \frac{\partial p}{\partial r} \cdot \frac{\partial(rB_\phi)}{\partial z} + \frac{\partial p}{\partial z} \cdot \frac{2(rB_\phi)}{\partial r} = 0.$$

Then the toroidal field is:

$$rB_\phi = \frac{\mu_0 I(\psi)}{2\pi},$$

where $I(\psi) = I_p$ is the poloidal current within $\psi = rA_\phi$. The plasma–field radial equilibrium can be expressed as:

$$L(\psi) + \mu_0 r^2 \frac{\partial p(\psi)}{\partial \psi} + \frac{\mu_0^2}{8\pi} \frac{\partial[I(\psi)]^2}{\partial \psi} = 0,$$

where:

$$L(\psi) = \left[r\frac{\partial}{\partial r}\left(\frac{1}{r}\frac{\partial}{\partial r}\right) + \frac{\partial^2}{\partial z^2}\right]\psi,$$

which is the Grad–Shafranov equation for equilibrium. Taking p and I^2 as quadratic functions of ψ, $\psi_s = 0$ at the plasma boundary, and ψ_0 as the value at the magnetic axes, we can write:

$$L(\psi) + \left(\alpha r^2 + \overline{\beta}\right)\psi = 0,$$

where:

$$\alpha = \frac{2\mu_0\left(p_0 - p_s\right)}{\psi_0^2},$$

and:

$$\overline{\beta} = \frac{\mu_0^2}{4\pi^2}\frac{\left(I_0^2 - I_s^2\right)}{\psi_0^2}.$$

For the toroidal geometry we take (r, ϕ, z) for the major axis coordinates and (ρ, ω, ϕ) for the coordinates of the plasma and fields Figure 9.46.

When the plasma cross-section and the boundary wall at $(\rho = a)$ are circular, then at the plasma boudary (where R is the major axis of the torus):

$$\psi(\rho, \omega) = \frac{\mu_0 I_p R}{2\pi}\left(\ln\frac{8R}{a} - 2\right).$$

With this solution, equilibrium properties of plasma and magnetic field can be determined; this provides a fundamental description of the magnetic

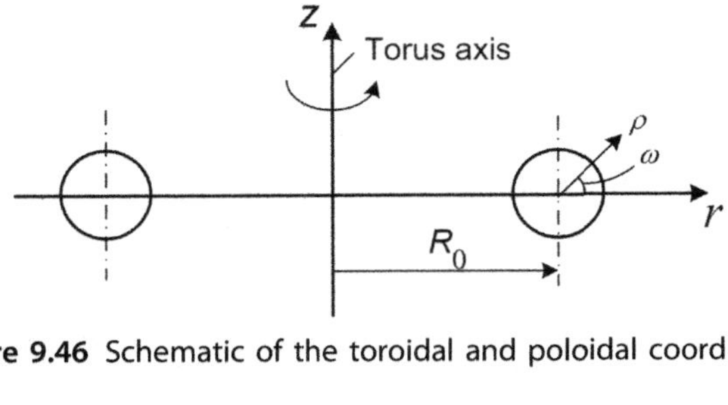

Figure 9.46 Schematic of the toroidal and poloidal coordinates.

confinement in a basic tokamak configuration. A more detailed discussion of various force components on the plasma related to tokamak magnetic fields has been presented in a later paper by deBlank (2006).

Reversed Field Pinch

In experiments with the early (1968) toroidal Z-pinch, ZETA, a relatively stable configuration of the confined plasma was discovered where there was some reversed magnetic field near the outer wall, but the physics of the occurrence was not then understood. Later, a self-organizing stable magnetic configuration was identified in which the poloidal field at large radius was reversed from that in the near-axis region. Taylor (1974) identified this as a "minimum energy" state, which evolved by relaxation of the plasma and fields under appropriate conditions. Subsequently, confinement devices conceived to utilize this magnetic field configuration were developed and appropriately named the reversed field pinch (RFP) (Bodin, 1990). Figure 9.47 shows the radial variation of the magnetic fields and plasma pressure in the RFP; the magnitude of the toroidal and poloidal fields are of the same order, with the safety factor, q, written as $q \sim B_z/B_\theta < 1$.

The RFP magnetic field configuration has a number of characteristics for confinement and heating of plasmas that are advantageous in a reactor configuration (Bodin et al., 1986), most notably including the effective confinement of high-β plasmas. The RFP is unique in that the stable configuration evolves (by MHD relaxation) from an initially unstable

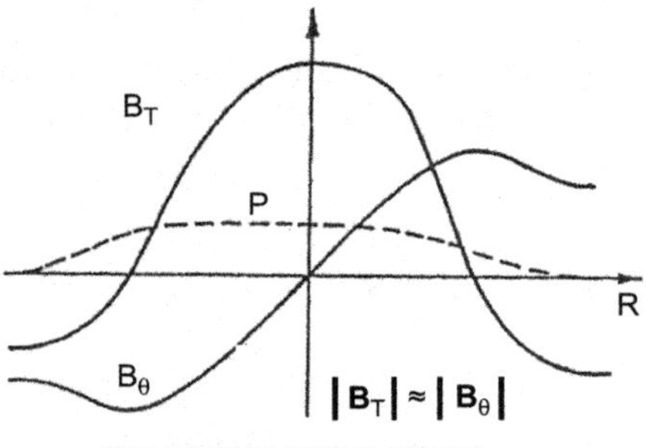

Figure 9.47 Radial profiles of magnetic field and fluid pressure in an RFP. *Sykes, A., 2008. The Development of the Spherical Tokamak. Slide#6, UK AEA, Culham. Downloaded 9/7/15 from: http://www.sunist.org/shared%20documents/ISTW2008/TM/Talks/Sykes/Sykes%20talk1.pdf, with permission.*

pinch configuration. Several basic aspects of the physics will be noted here.

The concept of relaxation to a minimum energy state was developed by Taylor (1974). In this theory, the plasma that is initially confined within magnetic fields undergoes a spontaneous change in magnetic field geometry to a more stable state by the action of finite resistivity and local reconnection of magnetic lines of force. We outline a basic analysis following Miyamoto (1989). The magnetic field is represented by a potential, $\vec{B} = \nabla \times \vec{A}$. The magnetic helicity is then defined as: $K = \int_{Vol} \vec{A} \cdot \vec{B} \, dVol$, over the volume surrounded by a magnetic surface. When the plasma is surrounded by a conducting coil, $(\vec{B} \cdot \vec{n}) = 0$ and $(\vec{E} \times \vec{n}) = 0$, where $\vec{n}$ is normal to the surface.

We can determine the rate of change of a variable helicity, K, to be:

$$\frac{\partial K}{\partial t} = -2 \int_{Vol} \vec{E} \cdot \vec{B} \, dVol,$$

and if we apply Ohms law, as:

$$\frac{\vec{J}}{\sigma} = \vec{E} + \vec{V} \times \vec{B},$$

then:

$$\frac{\partial K}{\partial t} = -2 \int_{Vol} \frac{\vec{J}}{\sigma} \cdot \vec{B} \, dVol.$$

Taylor assumed that the helicity for the whole plasma region, K_T, is constant, this allows for field reconfiguration; then:

$$\delta K_T = 2 \int_{Vol} \left(\vec{B} \cdot \delta \vec{A} \right) d\vec{r}$$

where, $d\vec{r}$ is a volume element.

Taking a force-free field $(\vec{J} \times \vec{B} = \nabla p = 0, \vec{J} \parallel \vec{B})$ plasma, the condition of minimum energy can be defined mathematically by the method of undetermined multipliers. The solution for the fields in cylindrical coordinates is:

$$B_r = 0, \quad B_\theta = B_0 J_1(\lambda_r), \quad B_z = B_0 J_0(\lambda_r),$$

where J_i are Bessel functions. Two parameters used to define RFP properties are the pinch parameter ($\overline{\theta}$) and the reversal ratio (F), which are expressed as:

$$\overline{\theta} \equiv \frac{B_\theta(a)}{\langle B_z \rangle} = \frac{\lambda a}{2}, \quad \text{and} \quad F \equiv \frac{B_z(a)}{\langle B_z \rangle} = \overline{\theta}\,\frac{J_0(2\overline{\theta})}{J_1(2\overline{\theta})},$$

where $\langle B_z \rangle$ is the volume average and $\lambda = \frac{\mu_0 \vec{J} \cdot \vec{B}}{B^2} = \frac{(\nabla \times \vec{B}) \times \vec{B}}{B^2} =$ Constant, in the Taylor model.

Another unique aspect of the RFP plasma is the experimental identification of regeneration of toroidal flux during the relaxation process (Baker et al., 1982). Research on this behavior has focused on fluctuations and anomalous transport to understand and analyze the MHD dynamo action.

Stellarators

The stellarator is a toroidal configuration that uses external coils outside the confined plasma to produce magnetic fields for stabilizing the plasma. This concept was one of the early (1951) attempts to control MHD instabilities with applied magnetic field configurations (Spitzer, 1958). In experiment, the C-Stellarator (1961, Princeton, USA) demonstrated limited confinement time and relatively low plasma temperatures compared to those being achieved in the tokamak (T-3). However, there is an inherent advantage in external stabilized configurations, as it is possible to achieve steady-state operation without plasma current instabilities. The magnetic field configurations in the stellarator are typically helical, as shown below (Figure 9.48).

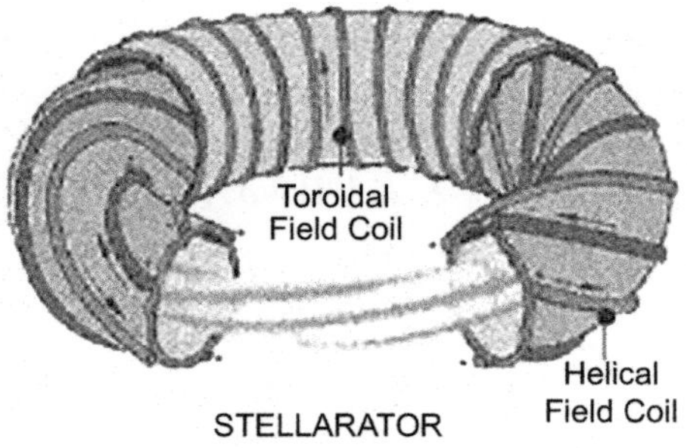

Figure 9.48 Schematic of stellarator magnetic confinement with exterior stabilizing coils. *McMillan, B.F., 2012. Lecture Slides for PX438 Physics of Fusion Power. Univ. Warwick, UK, http://www2.warwick.ac.uk/fac/sci/physics/current/teach/module_home/px438/ downloaded 4/5/2012, with permission.*

Again, we will here focus on the basic analysis and definition of the equilibrium of the magnetic confinement in the stellarator. We will identify steps in a brief outline for the confinement magnetic fields that have been presented (Miyamoto, 1989). In a cylindrical coordinate system (r, θ, z), we can express the field as (r, $\phi = \theta - \delta \propto z$), where, $\propto > 0$ and $\delta = \pm 1$. In terms of a field potential, $\overrightarrow{A}$, $\overrightarrow{B} = \nabla \times \overrightarrow{A}$, a magnetic surface can be defined as:

$$\psi = A_z + \delta \propto rA_\theta = \delta \propto rA_\theta = \text{Constant.}$$

A critical determination in the configuration is the rotational transform angle for the helical coil configuration. The geometry of the torus and helical coils is shown in Figure 9.49.

A line of magnetic force can be defined by:

$$\frac{dr}{B_r} = \frac{rd\theta}{B_\theta} = \frac{dz}{B_z},$$

and the rotational transform angle (i) is:

$$\frac{r \cdot (i)}{2\pi R} = \left\langle \frac{rd\theta}{dz} \right\rangle = \left\langle \frac{B_\theta}{B_z} \right\rangle,$$

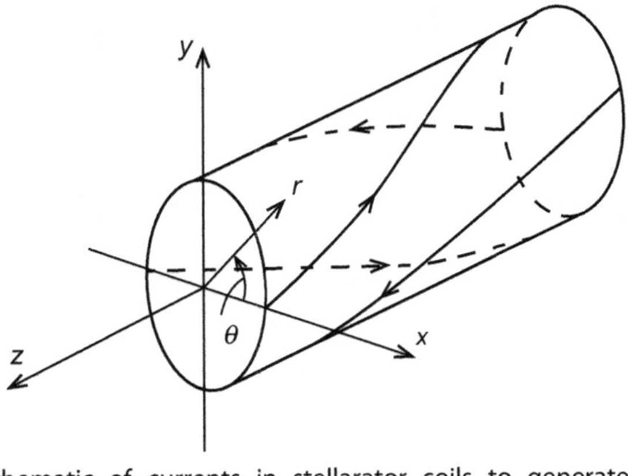

Figure 9.49 Schematic of currents in stellarator coils to generate the stabilizing fields. *Miyamoto, K., 1989. Fundamentals of Plasma Physics and Controlled Fusion, rev. ed. MIT, Cambridge, MA. Figure 17.2, p. 252 with permission.*

where the bracket, $\langle \ldots \rangle$, is an average over z. Using an expansion in modified Bessel functions:

$$\frac{(i)}{2\pi} = \delta\left(\frac{b}{B}\right)^2 \cdot \left(\frac{1}{2^l \cdot l!}\right) \cdot l^5 \cdot (l-1) \propto R\left[(l+r)^{2(l-2)} + \cdots\right],\ l \geq 2$$

where b is the field due to helical current, and l is a pole number, which defines the helical field ($l = 2$ represents the standard stellarator). The variation of magnetic fields along the line of force is shown (Figure 9.50). The behavior of ions in helical fields has been studied in detail, and ions have exhibited unique (banana) orbit behavior as shown in Figure 9.51.

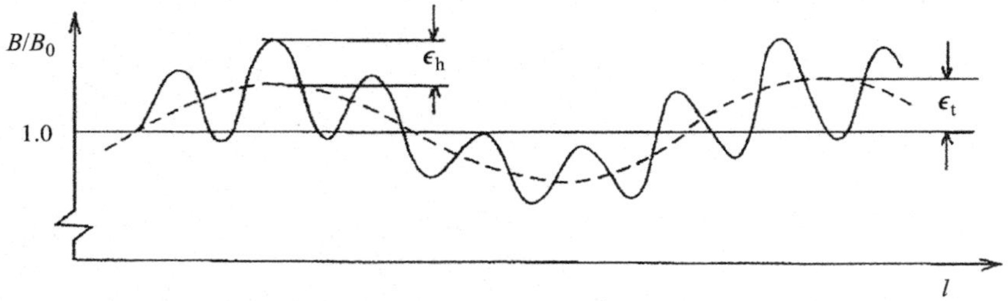

Figure 9.50 Variation of the magnitude of B along the length, l, of a line of magnetic force. *Miyamoto, K., 1989. Fundamentals of Plasma Physics and Controlled Fusion, rev. ed. MIT, Cambridge, MA. Figure 17.7, p. 257 with permission.*

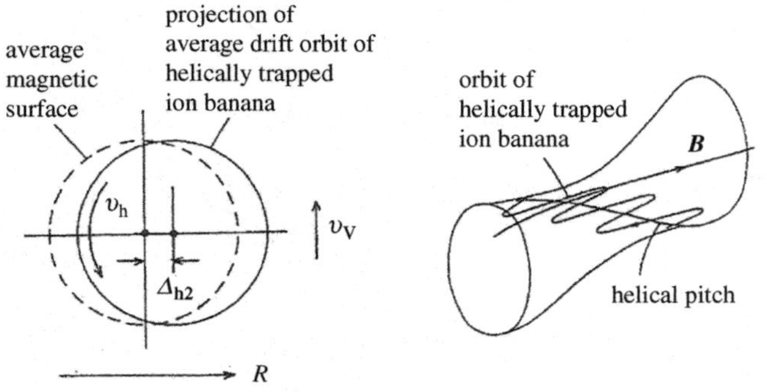

Figure 9.51 Orbits of ion trapped in the helical field (banana orbit). *Miyamoto, K., 1989. Fundamentals of Plasma Physics and Controlled Fusion, rev. ed. MIT, Cambridge, MA. Figure 17.8, p. 257 with permission.*

Compact Toroids: Plasma Parameters and Device Development

Introduction

The path for magnetic confinement and heating of plasmas to fusion conditions has primarily involved concepts and devices that required extensive external material and electrical structures. This has introduced inherent difficulties in size and scaling when considering reactor-related devices. Alternative devices that have simpler, less complex external structures have been developed in the course of physics studies and concept evaluation. In the early work on pulsed heating and compression dating back to the 1950s, some concepts and experiments were investigated that produced unique results, and their applications have now become important. Specifically, in experiments by Alfven et al. (1960) investigating coaxial plasma guns and Wells (1964) studying conical theta pinches, rings of magnetized plasma were identified, which evidenced self-conversion of toroidal and poloidal flux and the subsequent development of stable confining toroidal magnetic fields. Such behavior is related to the notions later expounded by Taylor (1986) that magnetic flux can undergo relaxation to minimum energy states that are longer lived.

The term, compact toroid, is used to refer to closed magnetic field geometries that have toroidal symmetry without compatible external structures (Wright, 1990). There are two prominent classes of devices of interest: (1) field reversed configurations (FRCs), most successfully evolved from trapped field theta pinch devices and (2) spheromaks, where the plasma is confined by relaxation of confining magnetic field configurations with a null toroidal magnetic field at the boundary conductor wall. The advantage of these classes of discharges is that the plasma can be created smaller in size (compact) and be magnetically enclosed so that it can be translated in space, which allows a less massive reactor to be envisioned.

Field Reversed Configurations

FRCs are compact toroids with negligible toroidal magnetic fields (B_θ) and a closed poloidal magnetic field structure (B_r, B_z) that is maintained by toroidal plasma current (Wright, 1990). One formation of a magnetically confined toroid was identified in studies of theta pinch discharges with trapped reversed magnetic fields, and it was described in a review by Intrator et al. (2004). This behavior was a natural consequence of the fact

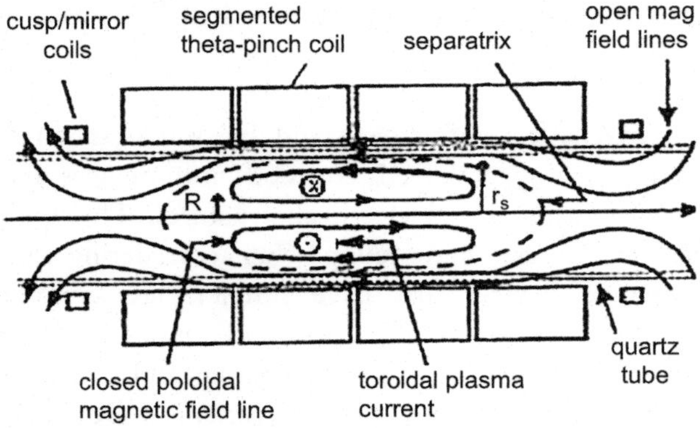

Figure 9.52 Schematic of cylindrical geometry showing poloidal and toroidal magnetic fields and cusp and mirror coils. *Intrator, T.P., et al., 2004. A high density field-reversed configuration plasma for magnetized target fusion. IEEE Trans. Plasma Sci. 32, 1. Figure 2, p. 3 with permission.*

that preionization plasma for theta pinch discharges will trap the preionization (PI) reversed field and allow for subsequent field reconnection at the coil axis (Figure 9.52).

A basic analysis of the plasma and field equilibrium has been presented (Intrator et al., 2008), and a brief review will now be described in terms of the above (Figure 9.52) geometry. The field reversed magnetic configuration has a safety factor value $(q = B_p/B_t) \gg 1$. The behavior of the plasma can be defined in terms of relative radii and length parameters: r_s is defined as the radius of the separatrix (i and o identify the inner and outer plasma radii, respectively), r_w is the wall radius, and l_s is the length of the plasma midradius location. The magnetic flux outside the separatrix is ψ_0 and inside the flux is ψ. With flux conservation, the flux tube volume inside a closed field line is:

$$V_\psi = \psi\left(\oint_c \frac{dl}{B}\right).$$

The radial force balance is written as:

$$p(\psi) + \frac{[B(\psi)]^2}{2\mu_0} = \frac{[B_w]^2}{2\mu_0},$$

and using the approximation:

$$\oint_c \frac{dl}{B} \approx \frac{l}{B},$$

we get:

$$\frac{l}{B} = \frac{2l}{\left[2\left(p_m - p(\psi)\right)\right]^{1/2}},$$

where p_m is a maximum pressure located on the magnetic axis (at $r_s/\sqrt{2}$). For the confined plasma in the compact toroid configuration, we can write an equilibrium expression for axial force balance as:

$$\left[\int_0^{r_s} 2\pi r\left(p - \frac{B^2}{2\mu_0}\right)dr + \int_{r_s}^{r_w} 2\pi r\left(-\frac{B_0^2}{2\mu_0}\right)dr\right]_{\text{midplane}} = \left[\int_0^{r_w} 2\pi r\left(-\frac{B_{ext}^2}{2\mu_0}\right)dr\right]_{\text{end}},$$

where B_0 refers to midplane conditions outside the plasma and B_{ext} refers to conditions at the end of the compact toroid. An expression for flux conservation can be written as:

$$\pi r_w^2 B_{ext} = \pi\left(r_w^2 - r_s^2\right)B_0.$$

With the introduction of the average ratio of fluid kinetic pressure and magnetic pressure, as:

$$\langle\beta\rangle = \left[\int_0^{r_s} 2\pi r p dr\right] \Big/ \left[\frac{2\mu_0}{B_0^2 \pi r_s^2}\right],$$

we can evaluate the average beta parameter to be:

$$\langle\beta\rangle = 1 - \left(\frac{r_s}{r_w}\right)^2 \frac{1}{2} = 1 - \frac{x_s^2}{2}.$$

So, smaller values of separatrix radii (r_s) result in smaller $\langle\beta\rangle$, and this is indicative of a geometry with more severe plasma property gradients that are related to greater energy and particle loss.

Following formation and stabilization of an FRC, achievement of fusion conditions involves translation of the plasma to achieve compression and heating. Appealing to force equilibrium and invoking inherently thermodynamic concepts, scaling relationships for the CT plasma properties have been evaluated (Intrator et al., 2008); these involve a profile index, ε, (*typical value* ≈ 0.25), to define trapped poloidal flux. For adiabatic flux compression, scaling of some representative parameters is as follows, with separatrix radius r_s related to the term, $x_s = \frac{r_s}{r_c}$.

CT Parameter	Scaling
$l_s = 2Z_s$	$\sim x_s^{2(4+3\varepsilon)/5} \langle\beta\rangle^{-(3+2\varepsilon)/5} r_c^{2/5}$
n_m	$\sim x_s^{-6(3+\varepsilon)/5} \langle\beta\rangle^{-2(1-\varepsilon)/5} r_c^{-12/5}$;
T	$\sim x_s^{-4(3+\varepsilon)/5} \langle\beta\rangle^{2(1-\varepsilon)/5} r_c^{-8/5}$.

It can be seen that the length, density, and temperature are critically related to x_s.

When the plasma toroid is first translated preliminary to application of heating or energy exchange processes, a toroid moving with velocity v_z has a total energy that is the sum of thermal and translational components as:

$$E = 5k_B T + m_i v_z^2,$$

where $T = T_e + T_i$.

Adiabatic relationships allow the velocity to be related to temperature and magnetic field energy changes as (Intrator et al., 2008):

$$v_z = \left[5\frac{k_B \Delta T}{m_i} + 2\frac{\Delta E_{BV}}{N m_i}\right]^{1/2},$$

where $E_{BV} = [B_{z0}^2/(2\mu_0)]\pi r_0^2 \cdot L$, with N being the number of ions in the volume with length L.

In MTF the FRC is translated from a formative region to a chamber for compression and heating to fusion conditions; this is accomplished in a converging and flux conserving configuration as shown in Figure 9.53.

When expansion and compression processes take place, there is inherent loss of magnetic flux, particles, and thermal energy according to well-defined physical laws. The compact toroid separatrix radius would behave as (Intrator et al., 2008):

$$x_s = (z_c)^{1/2}\left[\frac{\phi_p}{\pi r_c^2 B_l}\right]^{\frac{1}{(3+\varepsilon)}},$$

where ϕ_p is the poloidal flux. Typically, x_s was found to decrease by about 20% during translation and heating.

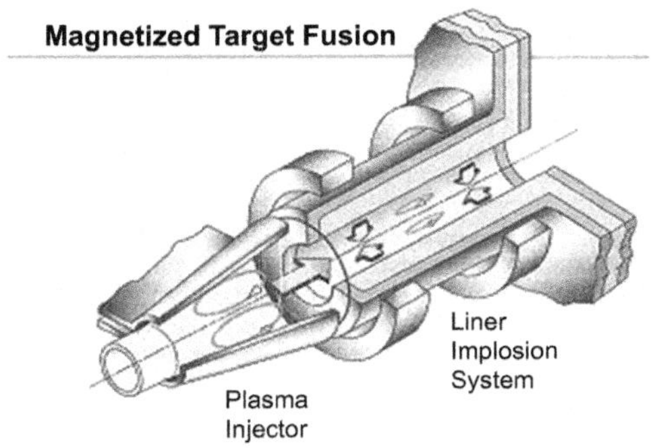

Figure 9.53 MTF schematic showing plasma formation region and liner implosion section. *Intrator, T.P., et al., 2004. A high density field-reversed configuration plasma for magnetized target fusion. IEEE Trans. Plasma Sci. 32, 1. Figure 1, p. 1 with permission.*

Spheromaks

The spheromak is a compact toroid of plasma with toroidal and poloidal magnetic fields of comparable strength (safety factor, $q \approx 1$) generated by currents flowing in the plasma and with no material linking the center of the torus (Bellan, 2000; Geddes et al., 1998; Bussac et al., 1978). Spheromaks result from self-organization and can evolve from different initially confined plasma structures; they are the result of an evolution toward a minimum energy state (Taylor, 1986).

The formation of a spheromak configuration in the CTX experiment (Wright, 1990) is shown schematically in Figure 9.54; this device utilizes the initiation of the process from a plasma gun geometry.

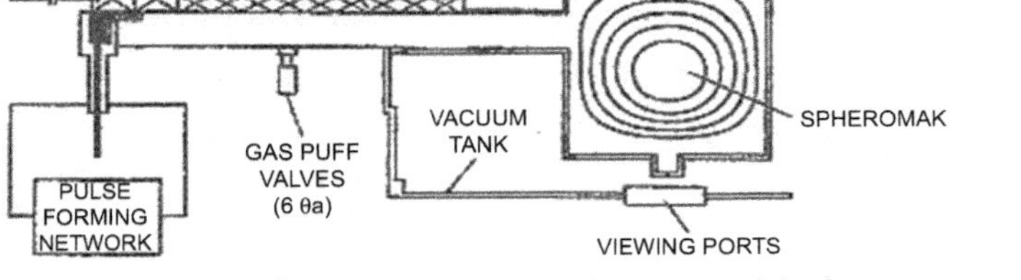

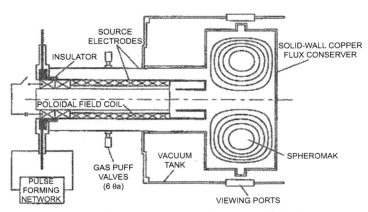

Figure 9.54 Schematic of CTX experiment using a coaxial plasma gun and flux conserver. *Wright, B.L., 1990. Field reversed configurations and Spheromaks. Nucl. Fusion 30, 1739. Figure 3, p. 1749 with permission.*

The basic aspects of the confined plasma and field geometry will be considered here (following Geddes et al., 1998). To form a spheromak, a threshold value of the inverse length parameter $\lambda_{th} = \frac{\mu_0 \cdot I_{gun}}{B_{gun} \cdot \pi r_{inner}^2}$ must be exceeded. The toroidal magnetic field in a spheromak vanishes at the boundary wall. The equilibrium is described by: $\nabla p = \vec{J} \times \vec{B}$, and if we specify a force-free state, $\vec{J} \times \vec{B} = 0$, we can determine that: $\nabla \times \vec{B} = \lambda \vec{B} = \mu_0 \vec{J}$. Here, λ is the inverse length parameter that describes the state of equilibrium (constant λ corresponds to the minimum energy state). For a closed perfectly conducting right circular cylinder, an analytic solution was reported (Geddes et al., 1998), with the following expression for the spatial variation of poloidal flux:

$$\psi = B_0 \frac{r}{k_r} J_1(k_r r) \sin(k_z z),$$

where B_0 is a constant and J_1 is the Bessel function. With the formalism, $\psi = \int \vec{B} \cdot \overrightarrow{dA}$, component B values can be determined directly. The field configuration is a function of the geometry and the poloidal flux with:

$$\lambda = \left[k_r^2 + k_z^2\right]^{1/2}, \quad \text{and} \quad k_z = \frac{\pi}{L}, \quad k_r = \frac{3.8}{R},$$

where R is the radius and L is the length of the conserver.

For more complex geometries, it is useful to examine computational results for specific experiments in order to get a better visualization of the fields and plasma (Okabayashi and Todd, 1980). Coordinates (x is radius) and regions are shown in Figure 9.55.

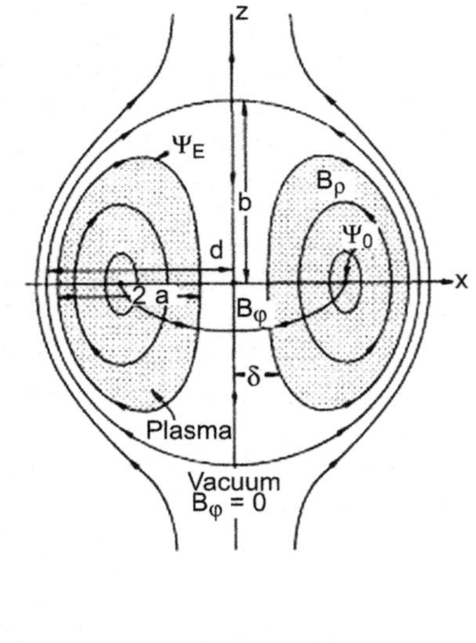

Figure 9.55 Spheromak geometry and coordinates. *Okabayoshi, M., Todd, A.M.M., 1980. A numerical study of MHD equilibrium and stability of the spheromak. Nucl. Fusion 20, 571. Figure 1, p. 571 with permission.*

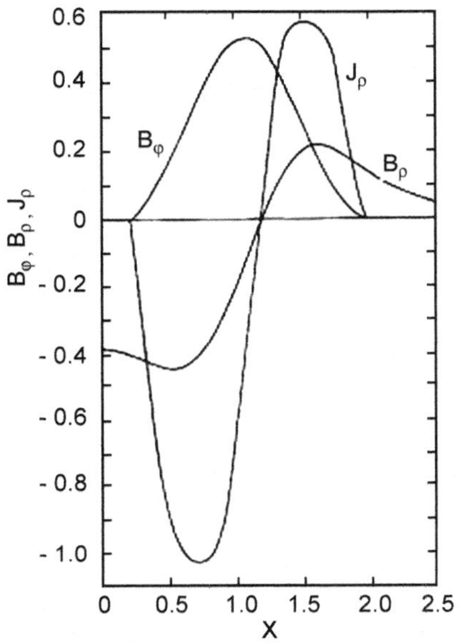

Figure 9.56 Radial variation of magnetic fields and current density in mid-plane. *Okabayoshi, M., Todd, A.M.M., 1980. A numerical study of MHD equilibrium and stability of the spheromak. Nucl. Fusion 20, 571. Figure 6(a), p. 574 with permission.*

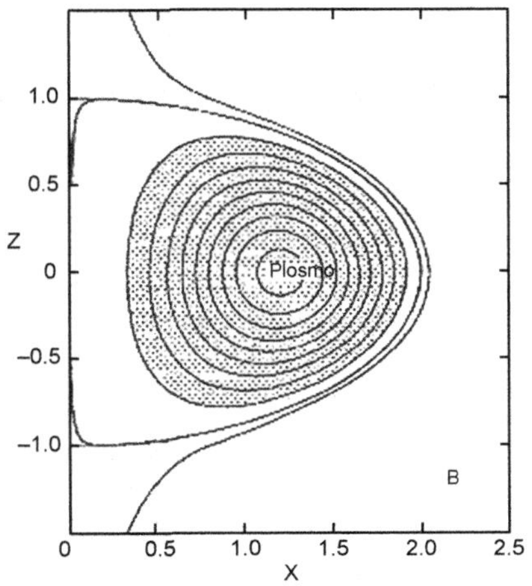

Figure 9.57 Magnetic flux contours in midplane. *Okabayoshi, M., Todd, A.M.M., 1980, A numerical study of MHD equilibrium and stability of the spheromak. Nucl. Fusion 20, 574. Figure 6(a), p. 574 with permission.*

For an optimized configuration of a slow formation experiment, Figure 9.56 shows the computational results for the radial variation of the poloidal and toroidal fields and the poloidal current. In Figure 9.57, the general topology of the solution with specific reference to the flux surfaces is shown; in particular, flux surfaces near the plasma axis are almost circular.

Reactor Concept Description (ITER)

Introduction

The goal of controlled magnetic fusion devices is the net release of useful energy from the fusion process that can be incorporated into a nuclear reactor device. The most easily achievable nuclear reaction involves deuterium and tritium, as:

$${}^{2}_{1}D + {}^{3}_{1}T \rightarrow {}^{4}_{2}\mathrm{He}(3.52\ \mathrm{MeV}) + {}^{1}_{0}n(14.08\ \mathrm{MeV}),$$

where the helium (α) particles' kinetic energy can be converted into useful (heat) form. As noted earlier, the parameter normally used to describe useful energy release is Q, the ratio of power from fusion to power loss, and then:

$$\frac{P_\alpha}{P_{Loss}} = \frac{1}{1 + \frac{5}{Q}}.$$

With $P_\alpha = P_{Loss}$, the system has reached ignition and is self-sustaining.

The International Thermonuclear Experimental Reactor (ITER) project (ITER, 2014) is a multinational (China, European Union, India, Japan, Russia, the United States) research and engineering endeavor to build an experimental tokamak reactor. The program agreement was approved in 2006, and the project is anticipated to last for 30 years. The facility is located in Caderache, France, with the construction phase scheduled to be completed by about 2020.

The ITER project has a number of programmatic aspects, and all are unique and challenging: physics, engineering, design, construction, and management. Some general discussion of the scope of these aspects has been presented in the literature (Holtkamp and ITER Project Team, 2007; NRC, 2009).

Burning Plasma Physics

The ITER experiment is designed to generate energy in excess of that required for operation, and it will consume reactants (burning) and produce products to operate with $Q \geq 10$, i.e., the energy produced is 10 times greater than that needed to reach fusion conditions. The toroidal reactor chamber is approximately 840 m^3, and it will produce 500 MW of power; operation is anticipated for a duration of 1000 s.

Table 9.3 Operating Parameters and Plasma Properties in ITER

Minor radius	2.0 m
Major radius	6.2 m
Toroidal magnetic field at R_0 from central axis (B_r)	5.3 T
Plasma current (I_P)	15 MA
Initial added plasma heating	73 MW
Electron density	1.1×10^{20} m^{-3}
Electron temperature	8.9 keV
Ion temperature	8.1 keV
Fusion power	500 MW
Pulse duration (induction mode)	400 s

There are significant plasma physics aspects to be studied in the operation of ITER (Green and ITER Team, 2003). There will be two different modes of operation of the experiment related to confinement and heating by the toroidal current: (1) inductively driven current for periods of 400 s and (2) steady-state burning with noninductive current drive (neutral beams and RF radiation). These new regimes of operation will allow improved understanding of the physics related to energetic particles and self-heating. One critical aspect of the ITER experiment is the incorporation of the divertor into the reactor; this is designed to remove impurities and replenish fuel. Basic parameters of the ITER device and plasma properties are presented in Table 9.3.

REFERENCES

Alfven, H., Lindberg, L., Mitlid, P., 1960. Experiments with plasma rings. J. Nucl. Energy, C: Plasma Phys. 1, 116.
Baker, D.A., et al., 1982. Performance of the ZT-40 RFP with an Inconel Liner. In: Proceedings of the Plasma Physics and Nuclear Fusion Research Conference, vol. 1. IAEA, pp. 587–597. IAEA-CN-4/H-2-1.
Beg, F.N., Lebedev, S.V., et al., 2002. The effect of current prepulse on wire array Z-pinch implosions. Phys. Plasmas 9, 375.
Bellan, P.M., 2000. Spheromaks: A Practical Application of Magnetohydrodynamic Dynamos and Plasma Self Organization. Imperial College, London.
Bennett, W.H., 1934. Magnetically self-focusing streams. Phys. Rev. 45, 890.
deBlank, H.J., 2006. Plasma equilibrium in tokamaks. Trans. Fusion Sci. Tech. 49, 111.
Bodin, H.A.B., Krakowski, R.A., Ortolani, S., 1986. The reversed field pinch: from experiment to reactor. Fusion Technol. 10, 307–353.
Bodin, H.A.B., 1990. The reversed field pinch. Nucl. Fusion 30, 1717.
Bussac, M.N., et al., 1978. Low-aspect-ratio limit of the toroidal reactor: the spheromak. In: Proc. Plasma Physics and Controlled Nuclear Fusion Research. Princeton Univ, Innsbruck. PPPL Rpt. 1472.
Chen, F.F., 1984. Introduction to Plasma Physics and Controlled Fusion, second ed. Plenum, New York.
Cuneo, M.E., et al., 2005. Characteristics and scaling of tungsten-wire-array z-pinch implosion dynamics at 20 MA. Phys. Rev. E 71, 046406.
Freidberg, J., 2007. Plasma Physics and Fusion Energy. University Press, Cambridge.
Geddes, C.G.R., Komack, T.W., Brown, M.R., 1998. Scaling studies of spheromak formation and equilibrium. Phys. Plasmas 5, 1027.
Glasstone, S., 1980. Fusion Energy. U.S. Department of Energy, Washington, D.C.
Green, B.J., ITER Team, 2003. ITER: burning plasma physics experiment. Plasma Phys. Controlled Fusion 45, 687–706.
Haines, M.G., 2011. A review of the dense Z-Pinch. Plasma Phys. Controlled Fusion 53, 1.
Haines, M.G., Coppins, M., 1995. A study of the stability of the Z-pinch. Phys. Rev. Lett. 5, 3285.
Holtkamp, N., ITER Project Team, 2007. An overview of the ITER Project. Fusion Eng. Des. 82, 427–434.
Intrator, T.P., et al., 2004. A high density field-reversed configuration plasma for magnetized target fusion. IEEE Trans. Plasma Sci. 32, 1.

Intrator, T.P., Siemon, R.E., Sieck, P.E., 2008. Applications of predictions for FRC translation to MTF. Phys. Plasmas 15, 042505.

ITER, 2014. https://WWW.ITER.org/sci.

Kodamstev, B.B., 1966. Hydromagnetic stability of a plasma. In: Reviews of Plasma Physics, 2. Consultants Bureau, New York, 153.

Kurchatov, I.V., 1956. The possibility of producing thermonuclear reactions in a gaseous discharge. At. Energ. 1, 76.

Larour, J., January 13, 2011. Approach to fusion with Z-Pinches: history and recent developments. Labratoire Phys. Plasmas, Palaiseau, FR. Downloaded from: http://www.lpp.fr. on 5/20/2012.

Lawson, J.D., 1957. Some criteria for a power producing thermonuclear reactor. Proc. Phys. Soc. B 70, 6.

Lebedev, S.V., et al., 2001. Effect of discrete wires on the implosion dynamics of wire array Z-pinches. Phys. Plasmas 8, 3734.

McMillan, B.F., 2012. Lecture Slides for PX438 Physics of Fusion Power. Univ. Warwick, UK. http://www2.warwick.ac.uk/fac/sci/physics/current/teach/module_home/px438/. Downloaded 4/5/2012.

Miyamoto, K., 1989. Fundamentals of Plasma Physics and Controlled Fusion, rev. ed. MIT, Cambridge, MA.

NRC, 2009. (National Research Council), Plasma Sciences Committee, a Review of the DOE Plan for US Fusion Community Participation in the ITER Project. National Academies, Washington, DC.

Okabayoshi, M., Todd, A.M.M., 1980. A numerical study of MHD equilibrium and stability of the spheromak. Nucl. Fusion 20, 571.

Spitzer, L.A., 1958. The stellarator concept. Phys. Fluids 1, 253.

Sykes, A., 2008. The Development of the Spherical Tokamak. In: 4th IAEA Technical Meeting on Spherical Tori, Frascati, Italy. UK AEA, Culham. Downloaded 9/7/15 from: http://www.sunist.org/shared%20documents/ISTW2008/TM/Talks/Sykes/Sykes%20talk1.pdf.

Taylor, J.B., 1974. Relaxation of toroidal plasma and generation of reversed magnetic Fields. Phys. Rev. Lett. 33, 1139.

Taylor, J.B., 1986. Relaxation and magnetic reconnection in plasmas. Rev. Mod. Phys. 58, 741.

Thomson, G.P., Blackman, M., 1947. Improvements In or Relating to Gas Discharge Apparatus for Producing Thermonuclear Reactions. UKAEA, UK. Patent application.

U N, 1958. In: Proceedings of the 2nd UN Conf. on Peaceful Uses of Atomic Energy (Geneva). UN, New York.

Wells, D.R., 1964. Axially symmetric force-free plasmoids. Phys. Fluids 7, 826.

Wright, B.L., 1990. Field reversed configurations and Spheromaks. Nucl. Fusion 30, 1739.

Space Plasma Environment and Plasma Dynamics

INTRODUCTION

The observable environment of the Earth includes our atmosphere, the Sun, the planets, and the stars. The Sun has manifested itself as the dominant source of light and energy, and, more benignly, the observable patterns of light emanating from the Sun, planets, and stars have given first clues to the complex makeup of the universe. More dramatically, the northern latitudes have given evidence of geoscale aurora light displays, which signal the existence of a dynamic physics beyond the Earth surface. In the mid-1800's a connection between dramatic solar activity, geomagnetic activity, and aurora displays was observed (Carrington) (Roberclauer and Siscoe, 2006). In the early 1900s, the transmission of radio waves between continents was (only later understood to be) possible because of the reflection of the electromagnetic waves from an electrically active layer above the Earth. The existence of a continuous stream of plasma particles outward from the Sun and incident upon the Earth was proposed (Parker, 1958) but not established by data until the 1950s (Luna 1 and Mariner 2) (Neugebauer and Snyder, 1962).

Since the time of the discovery of the solar wind, knowledge of the physics and significance of this phenomenon has been developing (Holzer, 2003). In the language of physics, regions with electromagnetic and fluid mechanical properties and the equations that describe their behavior have been identified. As the solar wind is a continuous stream of charged particles emanating from the Sun, such a flow will interact with the planets, and this interaction will differ depending upon the existence of an internally generated magnetic field. The Earth is one of the planets possessing such a magnetic field, and it is inherently dipole in structure. Our present understanding allows us to represent the Earth environment and flow dynamics as shown in the following schematic (Figure 9.58).

The figure shows a larger perspective, emphasizing the upstream compression and downstream extended tail of the magnetosphere structure. The interaction of the Solar Wind and the resulting near-Earth space environment is shown in Figure 9.59; the dominant feature here is the fluid mechanical bow shock wave formed about the magnetopause shape. Within the magnetosphere, the significant regions to be identified are the radiation belts, plasmasphere and the ionosphere nearest the Earth's surface.

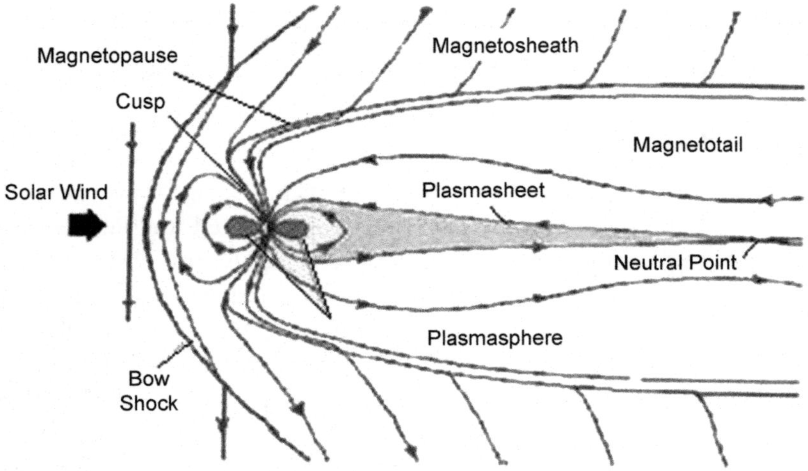

Figure 9.58 Schematic of solar wind interaction with the Earth's magnetic field. *NASA, 2014. The Earth's Magnetosphere, downloaded from web site: http://helios.gsfc.nasa.gov/magnet.html, 9/21/2014, Figure on p. 1 with permission.*

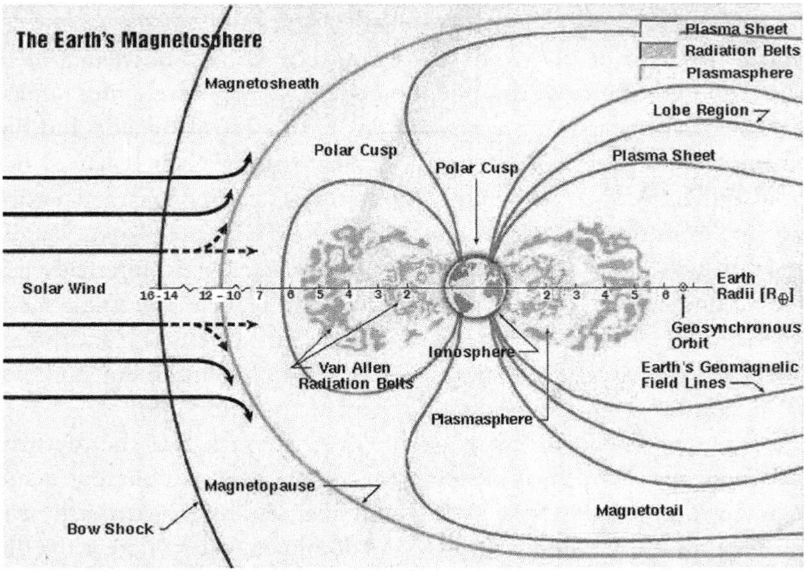

Figure 9.59 Schematic of near-earth magnetosphere environment. *Bailey, G., 2012. The Sun-earth Environment, Downloaded from web site: http://gbailey.staff.shef.ac.uk/researchoverview.html#space, 7/12/2012 with permission.*

The topics to be presented in this section are intended as an introduction to a broad subject that is treated in detail in other references (Holzer, 2003; Kivelson, 1995; Prolls, 2004; Gombosi, 1998; Parks, 2003). We will outline some of the fundamental aspects and governing relationships of the plasmas and phenomenology that are a part of the Earth's space plasma environment.

The Solar Wind and Geomagnetic Plasma Flow

The solar wind emanates from the Sun, a star with average, stable properties. It is the source of electromagnetic radiation and the flow of particles called the solar wind, which is, in fact, the flow field of the solar corona.

The physical makeup of the Sun can be defined by divisions within increasing radius from the core as shown in Figure 9.60; the physical properties of the Sun are presented in Table 9.4 (Prolss, 2004).

The most important aspect of the Sun is that it is the source of energy from thermonuclear reactions between hydrogen and hydrogen–deuterium in the core as:

$$\begin{aligned} \mathrm{H} + \mathrm{H} &\rightarrow 2\mathrm{H} + e^{+} + 1.2\ \mathrm{MeV} \\ 2\mathrm{H} + \mathrm{H} &\rightarrow 3\mathrm{He} + \gamma + 5.5\ \mathrm{MeV} \\ 3\mathrm{He} + 3\mathrm{He} &\rightarrow 4\mathrm{He} + 2\mathrm{H} + 12.9\ \mathrm{MeV} \end{aligned} .$$

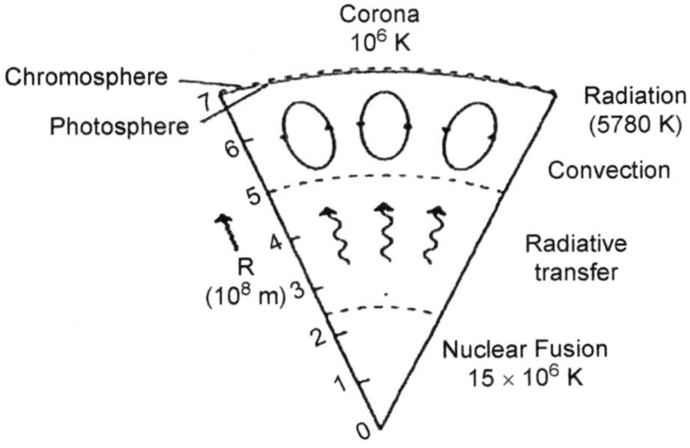

Figure 9.60 Cross-section through the Sun showing different regions of dominant activity. *Prolss, G., 2004. Physics of the Earth's Space Environment: An Introduction. Springer, Berlin. Figure 3.1, p. 81 with permission.*

Table 9.4 Physical Properties of the Sun

Radius (R_S)	6.96×10^8 m ($\simeq 109R_E$)
Mass (M_S)	1.99×10^{30} kg ($\simeq 333{,}000M_E$)
Composition (particle fraction)	91% H, 8% He, 1% ($M > 4$)
Composition (mass fraction)	72% H, 26% He, 2% ($M > 4$)
Mass density (mean)	1.41×10^3 kg/m^3 ($\simeq 0.25\ \rho_E$)
Mass density (at the center)	1.5×10^5 kg/m^3
Energy production rate (luminosity)	3.86×10^{26} W
Effective radiation temperature	5780 K
Distance from the Earth (mean)	149.6×10^9m = 1 AU
Solar constant	1.37×10^3 W/m^2 ($\pm 0.2\%$)
Age	4.6×10^9 years
Life expectancy (total)	10×10^9 years

Prolss (2004) with permission.

While the temperatures in the core are on the order of 10^7 K, the surface region most amenable to external examination, the photosphere, has temperatures on the order of 10^4 K, which is consistent with evidence and formulations for blackbody radiation. The solar atmosphere is composed of layers as shown in Figure 9.61, where the reference photosphere conditions are: $T = 5770$ K, $n = 10^{14}$ cm^{-3}, and weakly ionized plasma ($n_e/n_n = 10^{-4}$).

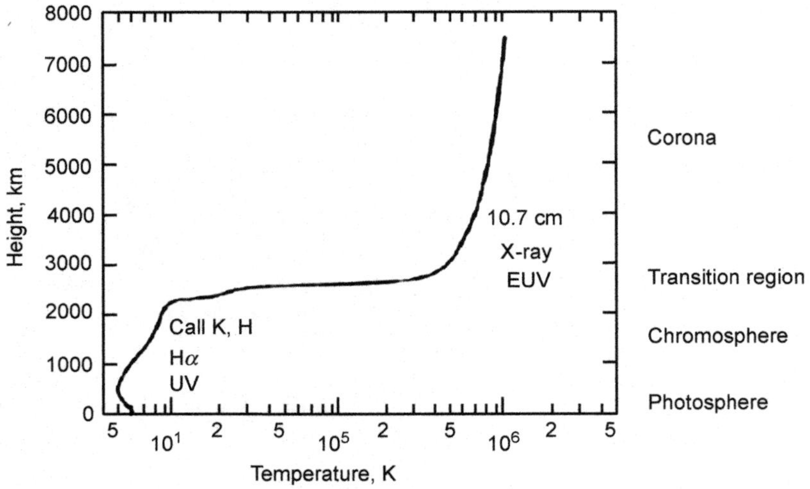

Figure 9.61 Temperature profile in the solar atmosphere and heights of radio emissions. *Prolss, G., 2004. Physics of the Earth's Space Environment: An Introduction. Springer, Berlin. Figure 3.3, p. 83 with permission.*

In a transition region (Figure 9.61), the density drops from 10^{14} to 10^8 cm^{-3} at about 3×10^3 km, while temperature increases from about 10^4 to 10^6 K. The gas is fully ionized plasma of protons and electrons. The process of plasma expansion and acceleration radially in all directions is the source of the interplanetary atmospheric solar wind.

Solar wind expansion process: The plasma environment near the Sun that is in aerostatic equilibrium extends out to about $3R_S$, with density and temperature variations with altitude consistent with that model. However, the plasma behavior beyond that radius, in interplanetary space, is not consistent with an aerostatic model. The relatively steady flow of plasma from the Sun was first established in the 1950s (Parker, 1958; Neugebauer and Snyder, 1962). Properties of what is termed the "solar wind" plasma are summarized in Table 9.5. The order of magnitude of the flow velocity, at about 500 km/s, was much higher than expected for a sound speed of about 50 km/s, thus indicating high supersonic (hypersonic) flow.

A theoretical model for this expansion of the corona plasma was first presented by Parker (1958); we will follow these calculations as outlined by Parks (2003, p. 215). We consider a plasma of protons and electrons. The equations are in spherical coordinates (r, θ, φ) as:

Mass:

$$\frac{1}{r^2}\frac{d}{dr}(r^2 \rho u) = 0;$$

Table 9.5 Mean Properties of the Solar Wind at Earth's Orbit

Composition	**($\simeq$96% H^+, 4% (0–20%) He^{++}, e^-)**
Density	$n_p \simeq n_e \simeq 6\ (0.1-100)\ \text{cm}^{-3}$
Velocity	$u_p \simeq u_e = u \simeq 470\ (170-2000)$ km/s
Proton flux	$n_p\, u \simeq 3 \times 10^{12}\ \text{m}^{-2}\text{s}^{-1}$
Momentum flux	$n_p m_H\, u^2 \simeq 2 \times 10^{-9}\ \text{N/m}^2$
Energy flux	$n_p m_H\, u^3/2 \simeq 0.5\ \text{mW/m}^2$
Temperature	$T \simeq 10^5\ (3500-5 \times 10^5)$ K
Plasma sound velocity	$v_{PS} \simeq 50$ km/s
Random velocity	$\bar{c}_P \simeq 46$ km/s
	$\bar{c}_e \simeq 2 \times 10^3$ km/s
Particle energy	$E_P \simeq 1.1$ keV (flow energy)
	$E_e \simeq 13$ eV (thermal energy)
Mean free path	$l_{P,P} \simeq l_{e,e} \simeq 10^8$ km
Coulomb collision time	$\tau_{P,P} \approx 30\ \tau_{e,P} > 20$ d

From Prolss (2004) with permission.

Momentum:

$$\rho u \frac{du}{dr} + \frac{dp}{dr} + \rho G \frac{M_{Sun}}{r^2} = 0;$$

Energy:

$$\frac{3}{2} u \frac{dp}{dr} + \frac{5}{2} p \frac{1}{r^2} \frac{d}{dr}(r^2 u^2) = 0,$$

where r is the distance from the center of the Sun, ρ is the density, u is the radial velocity, and p is the pressure, with

$$p = n_P k_B T + n_e k_B T.$$

We consider the radial flow as driven by the pressure difference, $\Delta p = p_{Sun} - p_{space} \approx p_{Sun}$ since $p_{Sun} \gg p_{space}$. By combination of energy and momentum equations, we get:

$$\frac{u^2 - a_s^2}{u} \frac{du}{dr} = 2 \frac{a_s^2}{r} - g_{Sun} \frac{R_{Sun}}{r^2},$$

where:

$$a_s^2 = \frac{5}{3} \frac{\rho}{p}.$$

For simplicity assume an isothermal expansion. Flows with expansion to vacuum will accelerate from static plasma to velocities that are supersonic. A mathematical solution can be expected to evidence singular points related to sound speed. If we assume that the solar gravity term is small, then:

$$\frac{1}{u} \frac{du}{dr} = \frac{1}{u^2 - a_s^2} \frac{2a_s^2}{r},$$

which has a singular point at $u = a_s$ and $du/dr > 0$ when $u > a_s$. So the flow expands subsonically ($u < a_s$) for $r < r_c$ and is supersonic ($u > a_s$) for $r > r_c$, where we have: $r_c = \frac{g_{Sun} R_{Sun}}{2a_s^2}$, for isothermal corona flow. For corona temperature, $T = 2 \times 10^6$ K, we get: $r_c \approx 6\ R_{Sun}$, where the location of the Earth is $r_{Earth} \approx 109\ R_{Sun}$. As we have made the assumption of isothermal plasma flow, the result is approximate. A model that would account for the temperature variation (Prolss, 2004) assumes:

$$T(r) = T(r_0) \left(\frac{r_0}{r} \right)^{\frac{2}{7}},$$

with:

$$T(r_0) = 10^6 \text{ K} \quad \text{at:} \quad r_0 = 3R_{Sun}.$$

Charged particle behavior for the plasma flow in the solar magnetic field does have an influence on flow interactions. Specifically, a more accurate model does distinguish between random energies (temperature) and energy components $\parallel$ and $\perp$ to the magnetic field. For the flow expansion in the interplanetary medium, for protons at 1 AU, $T_{\parallel}/T_{\perp} \doteq 1.5$, while electrons have almost equal component energies.

Geophysical Magnetosphere and Bow Shock

The most dominant feature of the space plasma flow environment of the Earth (Figure 9.58) is the occurrence of a solar-wind-induced bow shock wave about the magnetopause boundary. On the flow axis, the magnetopause (mp) forms where the solar wind (SW) dynamic pressure equals the magnetic pressure from the Earth's dipole $\left(\rho_{SW}u_{SW}^2 \approx \frac{B_{mp}^2}{2\mu_0}\right)$; this occurs at about: $R_{mp} \approx 10\, R_{Earth}$. The interaction of the solar wind with the geomagnetic field results in the formation of a magnetosphere cavity within the magnetopause.

A magnetopause current (Chapman–Ferraro current) separates the terrestrial dipole fields from interaction with the solar wind. This current can be determined by appealing to the MHD and Maxwell equations (Gombosi, 1998, p. 286), as:

$$\nabla \cdot \left(\rho \vec{u}\right) = 0;$$
$$\left(\rho \vec{u} \cdot \nabla\right) \vec{u} + \nabla p = \vec{J} \times \vec{B},$$

so,

$$\nabla(\rho u^2 + p) = \vec{J} \times \vec{B}.$$

If we neglect the magnetic field outside the magnetosphere and the plasma kinetic pressure inside the magnetopause, we have the balance at the magnetopause boundary as:

$$J_{mp} = \frac{\vec{B}_{Dipole}}{B_{Dipole}^2} \times \nabla(\rho u^2 + p) \approx \frac{1}{B_{Dipole}} \frac{\rho_{msh} u_{sh}^2}{\Delta_{mp}},$$

where ρ_{msh}, u_{msh} are the density and velocity in the magnetosheath (the region between the bow shock and the magnetopause) and Δ_{mp} is the thickness of the magnetopause. Typical values are: $B_{Dipole} \approx 3 \times 10^{-8}$ T, $\Delta \approx 500$ km, $n_{msh} \approx 8$ cm^{-3}, and $u_{msh} \approx 400$ km/s; so:

$$J_{mp} \approx 10^{-7} \mathrm{A/m^2}.$$

The supersonic solar wind–magnetosphere cavity interaction results in the formation of a geophysical (bow) shock wave and a following typical blunt body flow field with compression and expansion regions in the magnetosheath. Experimental evidence of the phenomenology is shown in Figure 9.62, and this flow field can be analyzed in detail with well-established fluid mechanics techniques. Specific measurements from satellites as they passed through the Earth's planetary bow shock wave are shown in Figure 9.63. Near-axis conditions are shown as the shock angle is near 90°.

The calculation of the quantitative changes that occur across the plasma shock wave can be carried out using MHD equations. We consider here the region near axis, which can be represented with normal shock assumptions for simplicity. The flow velocity and magnetic field across the shock wave can have both normal (n) and tangential (t) components; the flow equations can be written using the symbol [ ] to represent differences across the shock wave (2-1), as follows (Gombosi, 1998, p. 107):

Mass:

$$[\rho u_n] = 0;$$

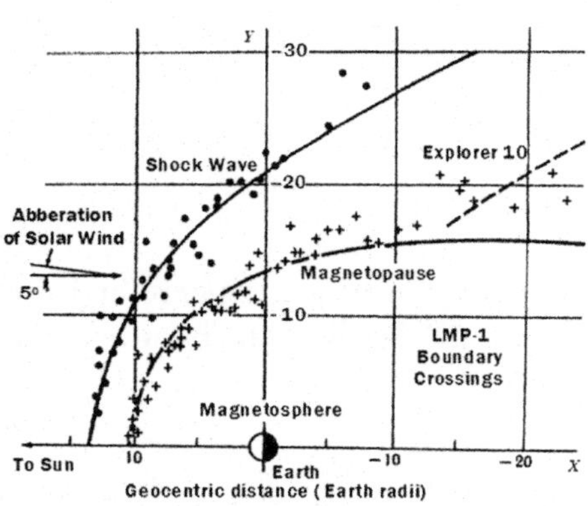

Figure 9.62 Magnetosphere shock wave data. *Ness, N.F., et al., 1964. Initial results of IMP-1 magnetic field experiment. J. Geophys. Res. 69, 3531. Figure 28, p. 3557 with permission.*

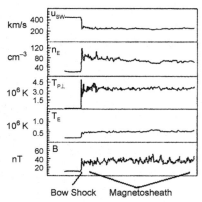

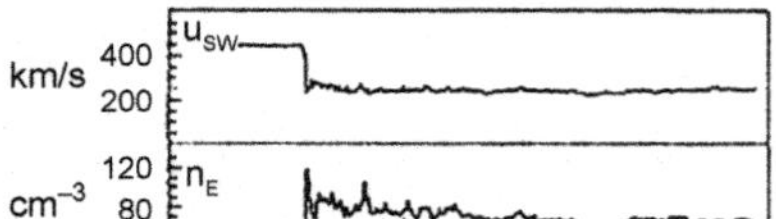

Figure 9.63 Discontinuous changes across magnetosphere shock wave (⊥ to B). *Prolss, G., 2004. Physics of the Earth's Space Environment: An Introduction. Springer, Berlin. Figure 6.31, p. 327 with permission.*

Momentum (t):

$$\left[\rho u_n \vec{u}_t - \frac{B_n \vec{B}_t}{\mu_0}\right] = 0;$$

Momentum (n):

$$\left[\rho u_n^2 + p + \frac{B_t^2 - B_n^2}{2\mu_0}\right] = 0;$$

Energy:

$$\left[\frac{1}{2}\rho\left(u^2 + u_t^2\right)u_n + \frac{\gamma}{\gamma - 1}\rho u_n + \frac{B_t^2}{\mu_0}u_n - \frac{B_n}{\mu_0}\left(\vec{B}_t \cdot \vec{u}_t\right)\right] = 0,$$

where γ is the ratio of specific heats. The Maxwell equations can be written as:

$$\left[u_n \vec{B}_t - B_n \vec{u}_t\right] = 0; \quad \text{and} \quad [B_n] = 0.$$

So:

$$\rho_1 u_{1n} = \rho_2 u_{2n};$$

$$\rho_1 u_{1n}[u_n] + [p] + \left[\frac{B_t^2}{2\mu_0}\right] = 0;$$

$$\left[\vec{u}_t\right] - \frac{B_n}{\rho_1 u_{1n} \mu_0}\left[\vec{B}_t\right] = 0;$$

$$\frac{1}{2}\rho_1 u_{1n}\left[u_n^2\right] + \frac{1}{2}\rho_1 u_{1n}\left[u_t^2\right] + \frac{\gamma}{\gamma - 1}[\rho u_n] + \left[\frac{B_t^2}{\mu_0}u_n\right] - \frac{B_n}{\mu_0}\left[\vec{B}_t \cdot \vec{u}_t\right] = 0;$$

$$\left[u_n \vec{B}_t\right] - B_n\left[\vec{u}_t\right] = 0.$$

It can be shown that the magnitude of B_t may change across the shock, but the direction will not change. The magnetic field remains in the same plane normal to the shock: $\vec{n} \cdot (\vec{B}_1 \times \vec{B}_2)$.

To simplify the general analysis of the bow shock jump conditions, we note that the flow speed is much greater than the sound speed or Alfven speed ($u \gg a_s,\ V_a$). We neglect incident solar wind magnetic pressure; downstream the kinetic pressure is greater than the tangential magnetic pressure component. The equations then become (Gombosi, 1998, p. 282):

Mass:

$$\rho_1 u_{1n} = \rho_2 u_{2n};$$

Momentum:

$$\rho_1 u_{1n}(u_{2n} - u_{1n}) + p_2 = 0;$$

$$u_{2n} B_{2t} - u_{1n} B_{1t} = 0;$$

$$u_{2t} - u_{1t} - \frac{B_{1n}}{(\rho_1 u_{1n})\mu_0}(B_{2t} - B_{1t}) = 0;$$

$$\frac{1}{2}\rho_1 u_{1n}\left(u_{2n}^2 - u_{1n}^2\right) + \frac{\gamma}{\gamma - 1}\rho_2 u_{2n} = 0.$$

We take $\gamma = 5/3$ as appropriate for the assumed fully ionized, monatomic plasma. So, we get for the jump conditions (in approximation):

$$u_{2n} = \frac{\gamma - 1}{\gamma + 1} u_{1n} \approx \frac{1}{4} u_{1n};$$

$$p_2 = \frac{2}{\gamma + 1}\rho_1 u_{1n}^2 \approx \frac{3}{4}\rho_1 u_{1n}^2;$$

$$\rho_2 = \frac{\gamma + 1}{\gamma - 1}\rho_1 \approx 4\rho_1;$$

$$B_{2t} = \frac{\gamma + 1}{\gamma - 1} B_{1t} \approx 4B_{1t};$$

$$u_{2t} - u_{1t} = \frac{2}{\gamma - 1}\frac{B_{1n}B_{1t}}{\mu_0 \rho_1 u_1} \approx 3\frac{B_{1n}B_{1t}}{\mu_0 \rho_1 u_1}.$$

The expected fluid mechanical effects of shock waves can be seen in the decrease of velocity and increase in density and pressure (and temperature) across the shock. However, the tangential magnetic field can be seen to increase significantly; since we know that $B_{1n} = B_{2n}$, then the magnetic field magnitude increases, and the magnetic pressure is increased in the magnetosheath.

A detailed study of the changes across the Earth's bow shock has been reported (Winterhalter et al., 1985); extensive data from satellite crossings were presented along with details of MHD shock calculations. The results documented the experimental findings and showed good agreement with theory. However, effects related to the angle between the magnetic field and the normal to the shock wave were significant, as predicted. A formulation was presented for the shape of the bow shock as an Earth-centered conic section with azimuthal symmetry about the Earth–Sun axis, as:

$$r = \frac{L}{1 + \varepsilon \cos \alpha},$$

where r is the radial distance to the shock wave, L is the semilatus rectum, α is the angle from the Sun–Earth axis, and ε is the eccentricity. Data indicated that: $\varepsilon = 0.7$ and $L = 23.5\ R_{Earth}$. So, along the Sun–Earth axis ($\alpha = 0$), the distance from the Earth to the bow shock is $r_s = 13.9\ R_{Earth}$.

Solar Wind and Magnetosphere Coupling

The physics and plasma interactions occurring in the magnetosphere cavity are varied and complex; regions are separated based on relative dominance of physical effects in momentum and energy of the plasma. These are shown in Figure 9.59. In order to develop an analytical framework to predict behavior, there have been physics-based models proposed for the magnetosphere: (1) an open model (Dungey, 1961), which allows connection of the Earth's magnetic field with the interplanetary magnetic field, and (2) the closed model (Chapman and Ferraro, 1930; Axford and Hines, 1961), which is based on a connection by diffusion across the magnetopause. The open model is generally accepted as more accurate, but the physics of the closed model allows understanding of local plasma processes.

The solar wind distends the downstream magnetic field into a long magnetotail (Figure 9.58), which extends tens of Earth radii from the planet's center. There are two distinctive regions in the Earth environment. The first is the innermost layer within 1000 km of the surface and is called the ionosphere. The second is the plasmasphere, a doughnut-shaped region surrounding the Earth with the central axis aligned with the planet's magnetic axis of the Earth's dipole field. The plasmasphere is located between 2–5 R_{Earth} and contains plasma with $n_e \approx 10^2\ \text{cm}^{-3}$ and

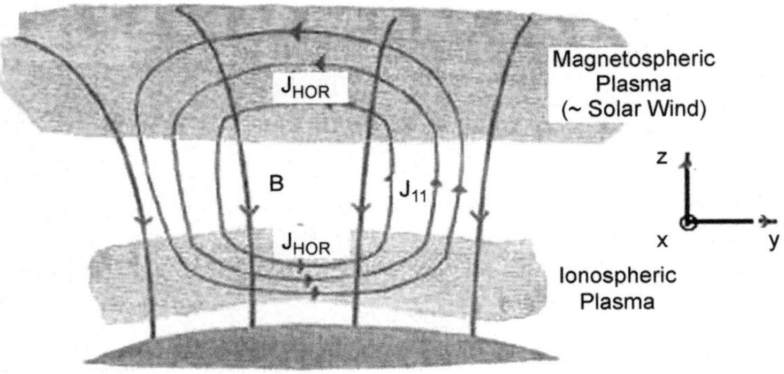

Figure 9.64 Schematic of ionosphere plasma linked to the magnetosphere by current streamlines. *Cravens, T.E., 1997. Physics of Solar System Plasmas. University Press, Cambridge. Figure 8.16, p. 371 with permission.*

$T_e = T_i \approx 1$ eV. Radiation (van Allen) belts have been identified near: $r_{inner} = 2R_{Earth}$ and $r_{outer} = 5R_{Earth}$. Plasmas in these belts are heated to megaelectronvolt levels by ring currents. The lower layers of the plasmasphere adjoin the outer (F) layer of the ionosphere.

An important effect in the magnetosphere is the coupling of the outer magnetosphere layers with the inner, near-Earth, ionosphere plasma by magnetic fields and the passage of a circuit current.

This phenomenon is shown in Figure 9.64. Field-aligned currents are called Birkeland currents.

We consider here the outline of a simple model (Cravens, 1997) of the physics of the interaction of plasmasphere and ionosphere. There must be stress (pressure) balance, which involves $\vec{J} \times \vec{B}$ forces. The current loop has components that are horizontal, j_{hor}, and vertical, which flows along the magnetic field line, so it is identified as: $j_{\|}$. As a current loop exists, we satisfy $\nabla \cdot \vec{J} = 0$; then:

$$\frac{dj_{\|}}{ds} = -\nabla \cdot \vec{J}_{hor},$$

where s is the distance along the magnetic field line. This defines that a current being driven in the horizontal direction (j_{hor}) will induce loop closure in the vertical direction, so:

$$\text{In the upper layer:} \quad \left(\vec{J} \times \vec{B}\right)_u = -j_{yu} B_0 \vec{y};$$

$$\text{In the lower layer:} \quad \left(\vec{J} \times \vec{B}\right)_L = -j_{yL} B_0 \vec{y},$$

where j_{yu} is in the $-y$-direction and B_0 is the dipole field ($+z$-direction). We write the momentum equations as:

$$\text{Upper:} \qquad \rho_u \frac{Du_u}{Dt} + \frac{\partial p}{\partial y} = -j_{yu} B_0;$$

$$\text{Lower:} \qquad \rho_L \frac{Du_L}{Dt} = -j_{yL} B_0 - \rho_L \nu u_L,$$

where ν is the ion-neutral collision frequency. In order to establish orders of magnitude, we make the assumption that $u_L \approx$ constant, then:

$$u_L = \frac{j_{yL} B_0}{\rho_L \nu}.$$

So the current density that is generated is the source of the flow speed.

We can determine the current magnitude by using the force balance relationship:

$$u_L = \frac{j_{yL} B_0}{\rho_L \nu} = \frac{j_{yu} B_0}{\rho_L \nu} = -\frac{1}{\rho_L \nu}\left[\rho_u \frac{Du_u}{Dt} + \frac{\partial p}{\partial y}\right],$$

which connects the top and bottom plasma regions. Using Ohm's law we get:

$$u_L = \frac{\Omega^2}{\nu^2 + \Omega^2} u_u,$$

where:

$$\Omega = \frac{eB_0}{m}.$$

With the current driven by $\vec{E} \times \vec{B}$ drift in the upper region, $u_u = E/B_0$, and so the flow speed in the lower region is smaller than that in the upper region. If we assume for the top region that:

$$\left.\frac{\partial p}{\partial y}\right|_u = 0,$$

we get:

$$\rho_u \frac{\partial u_u}{\partial t} = -\rho_L \nu u_L,$$

and using the above relationship of upper and lower speeds, we can solve for $u(t)$.

In this model the drift speed from $\vec{E} \times \vec{B}$ is the same in both regions. The current flow/vertical force phenomenology creates the connection between the magnetosphere and the ionosphere region close to the Earth's surface. The outer layer of the ionosphere (F region) does give evidence of plasma with $\vec{E} \times \vec{B}$ drift, which is diminished at lower altitudes.

Ionosphere Region of the Earth

Introduction

The ionosphere is the partially ionized region that forms due to the planet's atmosphere and ionization processes from electromagnetic radiation (photoionization) and energetic particles incident from external sources. So the ionosphere is fundamentally structured from lower altitude gases such as molecular nitrogen (80%) and oxygen (19%), carbon dioxide, argon, and water vapor. In meteorological terms the atmospheric layers are the troposphere (0–18 km), stratosphere (18–50 km), mesosphere (50–90 km), and thermosphere (90 km to near space). More importantly, the decreasing density with altitude assists the development of an electrical character in the atmosphere with altitude. A graph of the Earth's ionosphere showing electron density variation and the identification of different regions is presented in Figure 9.65.

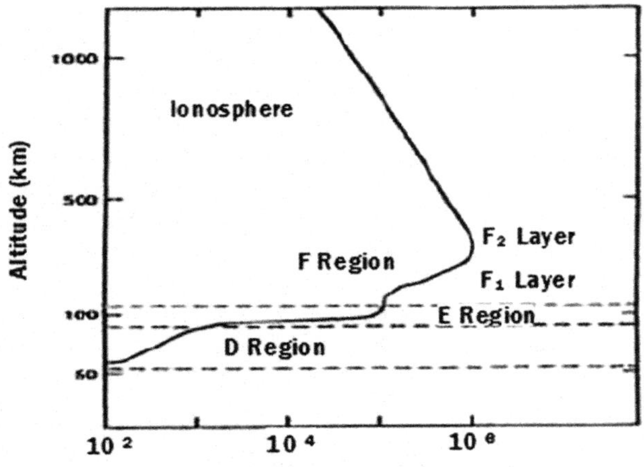

Figure 9.65 Electron density of the Earth's ionosphere as a function of altitude. *Kivelson, M., 1995. Introduction to Space Physics. Univ. Press, Cambridge. Figure 1.7, p. 11 with permission.*

Different regions have different electrical and physical properties; however, region identification is best understood from a historical perspective. The transmission of radio waves (megahertz) across the Atlantic Ocean in 1901 involved the reflection of the electromagnetic waves from a free electron layer, and this became known as the E layer. This layer occurs at about 100 km. In 1928, measurements indicated that higher electron density existed at altitudes above the E layer, and this region became known as the F layer. In daylight, this region can be separated into the F1 and F2 layers. Subsequently, a layer of weaker ionization was discovered below 90 km, and this was termed the D layer.

Ionization in the Ionosphere

A simple model of the Earth's atmosphere (basically the Chapman model) can be developed; here we follow the presentations of Hines et al., 1965; Papagiannis, 1972). We can express the decrease of number density with distance from the Earth's surface (h) as:

$$n(h) = n_0 e^{-h/H}$$

where H is a constant.

The primary source of ionization in the atmosphere is solar radiation. We can represent the absorption of radiation intensity (I) as:

$$dI = -\sigma_\nu I(h) n(h) dh,$$

where σ_ν is the coefficient (cross-section) of photoabsorption; we assume normal incidence on the atmosphere. If we take I_∞ as the incident intensity on the topside ionosphere, then:

$$I(h) = I_\infty \exp\left(-\sigma_\nu n_0 H e^{-h/H}\right).$$

Ionization is created by photoionization, which is expressed as:

$$G + h\nu \rightarrow G^+ + e^-,$$

where G represents moles of gas molecules of species G and h is the Planck's constant. Then the rate of production per unit volume of ion–electron pairs is:

$$\frac{dn}{dt}(h) \sim I(h) \cdot n(h).$$

So with density increasing with distance down from topside and intensity decreasing with distance from topside, there will be an altitude (h_m) where there will be maximum production rate and maximum electron density (n_m). We then write:

$$n(h) = n_m \exp\frac{1}{2}\left[1 - \left(\frac{h-h_m}{H}\right) - e^{-\left(\frac{h-h_m}{H}\right)}\right].$$

This functional form shows an electron density that increases exponentially from 0 at the Earth's surface to a maximum value and then an exponential decrease to a limiting value at topside, with $n/n_m \approx 1.5$. The details of radiation absorption will depend upon variations of density with altitude, resulting in preferential ionization reactions that change with altitude. This general behavior is shown in Figure 9.66; the resulting chemical composition is shown in Figure 9.67.

As can be seen in Figure 9.67, at lower altitudes species N and O are dominant, as would be expected. However, at altitudes above 1000 km, H^+ and He^+ become the dominant species. This behavior is related to the reaction:

$$O^+ + H \rightleftharpoons O + H^+.$$

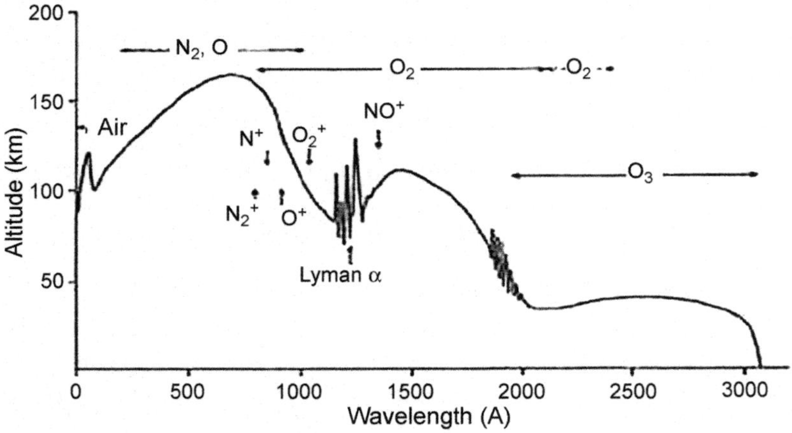

Figure 9.66 Altitude at which intensity of solar radiation (as a function of wavelength) drops to e^{-1} of its incident value. *Hines, C.O., et al., 1965. Physics of the Earth's Upper Atmosphere. Prentice-Hall, Englewood Cliffs, N J, Figure 2.2, p. 40 with permission.*

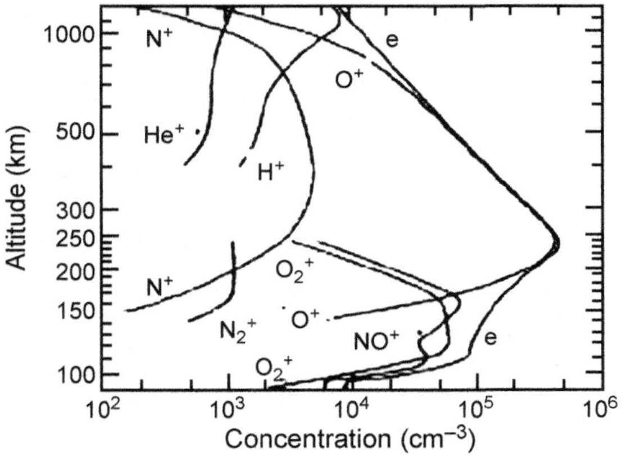

Figure 9.67 Composition of dayside ionosphere at solar minimum. *Gombosi, T.I., 1998. Physics of the Space Environment. Univ. Press, Cambridge. Figure 10.2, p. 181 with permission.*

If we assume that diffusion along the magnetic field lines is dominant in this topside region, we can solve for the distribution of H^+ in this plasma as:

$$n(H^+) = n_0 \exp\left(-\frac{s - s_0}{H_p}\right),$$

where s is the distance along magnetic field lines, subscript 0 refers to a lower (reference) altitude, and $H_p = k_B(T_e + T)/gm_H$ is the plasma scale height. This topside region plasma exists in the Earth's dipole magnetic field, and it reaches an equilibrium, stable spatial configuration. This region forms the transition boundary with the space environment.

Space Plasma Experiments

Introduction

Research programs have been established worldwide to gather data and to conduct theoretical research to model Earth's magnetosphere and the planetary system heliosphere. Such research is crucial to not only better understand the structure of our environment in the near-Earth regions but also establish the basic knowledge to enable future space exploration. As the focus of this work has been on plasma behavior and plasma dynamics, it is useful to present some descriptive information on space programs, space missions, and space experiments that have been and are being carried out to understand space plasma behavior. The experiments described here are largely US (NASA) supported and conducted.

Voyager (NASA Voyager, 2014): Interplanetary and Interstellar Plasmas Voyager 1 and 2 spacecraft were launched on September 5 and August 20, 1977, primarily to conduct close-up studies of Jupiter and Saturn and the larger moons of the planets. The mission was expanded to include Neptune and Uranus and 48 moons of the planets. While the spacecraft had a primary mission to gather planetary information, data were also gathered on interplanetary plasmas and magnetic fields, and plasma waves. Having been successful in measuring the solar wind and magnetic fields within the solar system, the mission have continued and has been making measurements in the interstellar medium. The spacecraft have passed through the "heliosheath," the region where the solar wind is slowed by the pressure of the interstellar plasma. The heliopause boundary defines the outer limits of the Sun's magnetic field and outward flow of the solar wind.

Instruments on the spacecraft measure the following:

1. Properties of the solar wind (ions: 10 eV–6 keV; electrons: 4 eV–6 keV)
2. Plasma waves (electric fields of plasma waves in the range: 10 Hz–56 kHz)
3. Magnetometers (magnetic field intensity in the ranges: 50,000–200,000 nT and 8000–50,000 nT).

Both spacecraft (2004 and 2007) found that the termination shock occurred at about 90 AU from the Sun. Voyager 1 entered interstellar space in August 2012, and as of September 2013 it was located at 125 AU from the Sun (Gurnett et al., 2013; Burlaga et al., 2013).

WIND (NASA WIND, 2014): Solar Wind Effects on Near-Earth Environment This spacecraft is part of a mission of the Global Geospace Science initiative, which is part of the Solar Terrestrial Physics program. The goal of these missions is to understand the solar terrestrial system so as to enable prediction of the Earth's magnetosphere and atmospheric reaction to changes in the solar wind. This specific (WIND) spacecraft was launched in November 1994; in its orbit at the L_1 Lagrange point about the Sun, it will have a lifetime of 60 years.

The experiment is designed to accomplish the following:

1. Gather complete information on plasmas and magnetic fields for magnetosphere and ionosphere studies.
2. Measure wave-plasma processes in the near-Earth solar wind.

The spacecraft has an array of instruments to measure particles, electromagnetic waves, gamma rays, as well as radio and plasma waves; it is specifically oriented to monitor for observations of solar flares and gamma ray bursts (Wilson, 2007).

THEMIS - (Time History of Events and Macroscale Interactions during Substorms): Magnetic Storm Effects This program is designed to study the onset of magnetic storms on the Earth's magnetosphere. Since its launch in 2007, five spacecraft have been able to monitor disturbances in space weather so as to provide understanding of transient and local plasma events. The satellites have actively monitored over 50 solar storms and found relationships to electromagnetic waves and perturbations in the relatively near-Earth van Allen radiation belts. Some of the specific findings are as follows:

1. The occurrence and behavior of plasma waves is linked to solar storms (Turner et al., 2013).
2. Perturbations in solar wind cause reactions in the Earth's magnetophase shock and correlate with ultralow-frequency waves (Hartinger et al., 2013).
3. Solar wind perturbations induce extreme variations in the Earth's bow shock and affect conditions throughout the magnetosphere (Korotova et al., 2009).

A related research program conducted by the European Space Agency involved four spacecraft to provide detailed maps of the magnetosphere. The spacecraft are at altitudes that penetrate the radiation belts and determined the populations of electrons of different energies. As a general observation, it was found that during quiet magnetic periods the plasmapause was located at about 6 R_E, but during active magnetic periods the plasmapause moved to about 4.5 R_E. The Earth's ionosphere influence was found to enter or exit this region in response to the dynamic events (Darrouzet et al., 2013).

REFERENCES

Axford, W.I., Hines, C.O., 1961. A unifying theory of high latitude geophysical phenomena and geomagnetic storms. Can. J. Phys. 39, 1433.

Bailey, G., 2012. The Sun-earth Environment. Downloaded from web site. http://gbailey.staff.shef.ac.uk/researchoverview.html#space, 7/12/2012.

Burlaga, L.F., et al., 2013. Evidence for a shock in interstellar plasma: Voyager 1. Astrophys. J. 778, 5.

Chapman, S., Ferraro, V.C.A., 1930. A new theory of magnetic storms. Nature 126, 129.

Cravens, T.E., 1997. Physics of Solar System Plasmas. University Press, Cambridge.

Darrouzet, F., et al., 2013. Links between the plasmapause and the radiation belt boundaries as observed by the instruments CIS, RAPID, and WHISPER on-board cluster. JGR: Space Phys. 118, 4176–4188.

Dungey, J.W., 1961. Interplanetary magnetic field and the Auroral Zones. Phys. Rev. Lett. 6, 47.

Gombosi, T.I., 1998. Physics of the Space Environment. Univ. Press, Cambridge.

Gurnett, D.A., et al., 2013. In Situ observations of interstellar plasmas with Voyager 1. Science 341, 1489–1492.
Hartinger, M.D., et al., 2013. The role of transient ion foreshock phenomena in driving Pc5 ULF wave activity. JGR: Space Phys. 118, 299–312.
Hines, C.O., et al., 1965. Physics of the Earth's Upper Atmosphere. Prentice-Hall, Englewood Cliffs, N J.
Holzer, T.E., 2003. Physics of the Solar Wind. Univ. Press, Cambridge.
Korotova, G., et al., 2009. THEMIS observations of compressional pulsations in the dawnside magnetosphere: a case study. Ann. Geophys. 27, 3725–3735.
Kivelson, M., 1995. Introduction to Space Physics. Univ. Press, Cambridge.
Ness, N.F., et al., 1964. Initial results of IMP-1 magnetic field Experiment. J. Geophys. Res. 69, 3531.
Neugebauer, M., Snyder, C.W., 1962. Solar plasma experiment. Science 138, 1095.
NASA, 2014. The Earth's Magnetosphere downloaded from web site. http://helios.gsfc.nasa.gov/magnet.html, 9/21/2014.
NASA-Voyager, 2014. http://voyager.jpl.nasa.gov/mission/.
NASA-WIND, 2014. https://heasarc.gsfc.nasa.gov/docs/heasarc/missions/wind.html.
Papagiannis, M.D., 1972. Space Physics and Space Astronomy. Gordon and Breach, New York.
Parker, E.N., 1958. Dynamics of the interplanetary gas and magnetic Fields. Astrophys. J. 128, 664.
Parks, G., 2003. Physics of Space Plasmas: An Introduction, second ed. Westview, Boulder, CO.
Prolss, G., 2004. Physics of the Earth's Space Environment: An Introduction. Springer, Berlin.
Robertclauer, C., Siscoe, G., 2006. The great historical geomagnetic storm of 1859: A modern look. Adv. Space Res. 38, 117.
Turner, D.L., et al., 2013. First observations of foreshock bubbles upstream of Earth's bowshock: Characteristics and comparisons to HFAs. JGR: Space Phys. 118, 1552–1570.
Wilson III, L.B., 2007. Waves in interplanetary shocks: a WIND/WAVES study. Phys. Rev. Lett. 99, 041101.
Winterhalter, D., et al., 1985. Magnetic field change across the earth's bow shock: comparison between observations and theory. J. Geophys. Res. 90, 3925.

Appendix A: Conversion between MKS and Gaussian System

		Units	
Physical Quantity	**Symbol**	**Rationalized mks**	**Gaussian**
Length	l	1 meter (m)	10^2 centimeters (cm)
Mass	m	1 kilogram (kg)	10^3 grams (g)
Time	t	1 second (s)	1 second (s)
Force	F	1 newton	10^5 dynes
Work Energy	W U	1 joule	10^7 ergs
Power	P	1 watt	10^7 erg/s
Charge	q	1 coulomb (coul)	3×10^9 statcoulombs
Charge density	ρ	1 coul/m^3	3×10^3 statcoul/cm^3
Current	I	1 ampere (coul/s)	3×10^9 statamperes
Current density	J	1 amp/m^2	3×10^5 statamp/cm^2
Electric field	E	1 volt/m	$\frac{1}{3} \times 10^{-4}$ statvolt/cm
Potential	Φ, V	1 volt	$\frac{1}{300}$ statvolt
Polarization	P	1 coul/m^2	3×10^5 statcoul/cm^2 (statvolt/cm)
Displacement	D	1 coul/m^2	$12\pi \times 10^5$ statvolt/cm (statcoul/cm^2)
Conductivity	σ	1 mho/m	9×10^9 s^{-1}
Resistance	R	1 ohm	$\frac{1}{9} \times 10^{-11}$ s/cm
Capacitance	C	1 farad	9×10^{11} cm
Magnetic flux	ϕ, F	1 weber	10^8 gauss cm^2 or maxwells
Magnetic induction	B	1 weber/m^2	10^4 gauss
Magnetic field	H	1 ampere-turn/m	$4\pi \times 10^{-3}$ oersted
Magnetization	M	1 weber/m^2	$\frac{1}{4\pi} \times 10^4$ gauss
Inductance	L	1 henry	$\frac{1}{9} \times 10^{-11}$

Continued

—Cont'd

Formulas

Quantity	Gaussian	mks
Velocity of light	c	$(\mu_0 \epsilon_0)^{-1/2}$
Electric field (potential, voltage)	$\mathbf{E}(\Phi, V)$	$\sqrt{4\pi\epsilon_0}\,\mathbf{E}(\Phi, V)$
Displacement	$\mathbf{D}$	$\sqrt{\frac{4\pi}{\epsilon_0}}\,\mathbf{D}$
Charge density (Charge, Current density, Current, Polarization)	$\rho(q, \mathbf{J}, I, \mathbf{P})$	$\frac{1}{\sqrt{4\pi\epsilon_0}}\rho(q, \mathbf{J}, I, \mathbf{P})$
Magnetic induction	$\mathbf{B}$	$\sqrt{\frac{4\pi}{\mu_0}}\,\mathbf{B}$
Magnetic field	$\mathbf{H}$	$\sqrt{4\pi\mu_0}\,\mathbf{H}$
Magnetization	$\mathbf{M}$	$\sqrt{\frac{\mu_0}{4\pi}}\,\mathbf{M}$
Conductivity	σ	$\frac{\sigma}{4\pi\epsilon_0}$
Dielectric constant	ϵ	$\frac{\epsilon}{\epsilon_0}$
Permeability	μ	$\frac{\mu}{\mu_0}$
Resistance (impedance)	$R\ (Z)$	$4\pi\epsilon_0 R(Z)$
Inductance	L	$4\pi\epsilon_0 L$
Capacitance	C	$\frac{1}{4\pi\epsilon_0} C$

Source: Jackson, J.D., 1962, Classical Electrodynamics, Wiley, New York.

Appendix B: Definite Integrals – Maxwellian Distribution Functions

Evaluation and values of definite integrals related to Maxwellian distribution functions

Source: Vincenti and Kruger (1965), *Introduction to Physical Gas Dynamics*, New York: Wiley.

A definite integral that appears frequently in kinetic theory and statistical mechanics is the following:

$$I_n(a) \equiv \int_0^\infty x^n e^{-ax^2}\,dx,$$

where $a > 0$ and n is a nonnegative integer. Values of this integral for specific values of n are

$$I_0(a) = \frac{1}{2}\left(\frac{\pi}{a}\right)^{\frac{1}{2}}, \quad I_1(a) = \frac{1}{2a},$$

$$I_2(a) = \frac{1}{4}\left(\frac{\pi}{a^3}\right)^{\frac{1}{2}}, \quad I_3(a) = \frac{1}{2a^2},$$

$$I_4(a) = \frac{3}{8}\left(\frac{\pi}{a^5}\right)^{\frac{1}{2}}, \quad I_5(a) = \frac{1}{a^3}, \text{ etc.}$$

Starting with I_0 and I_1, which will be established below, each integral can be found from one above it by means of the relation

$$I_{n+2}(a) = -\frac{dI_n}{da},$$

which follows obviously from the defining formula. The integral from $-\infty$ to $+\infty$ is twice the value given above if n is even and zero if n is odd.

Appendix C: Nomenclature

A	Area; arbitrary property in transport
c	Particle speed
C_p	Specific heat at constant pressure
C_V	Specific heat at constant volume
d	Molecular (particle) diameter
D	Diffusion coefficient
e	Energy per unit mass
E	Random thermal energy (per particle) (Ch. 2)
f(i)	Distribution function of property(i)
g	Degeneracy
G	Mass (identity) flux
h	Planck's constant
k	Boltzmann's constant
K	Coefficient of thermal conduction
m	Mass of individual molecule/atom particle
n	Number of particles per unit volume
n_i	Number of particles per unit volume that have the property, *i*
N	Total number of particles
q	Heat flux
Q	Energy partition function
R	Gas constant
$\overline{\mathbf{R}}$	Universal gas constant
S	Cross-section area for transport
S	Entropy
T	Temperature
V	Volume
x,y,z	Axis directions in rectangular coordinates
$\in, \overline{\in}, \in', \in'', \in'''$	Energy (molecular component) per particle (Ch. 3)
$\in$	Molecular (component) energy (Ch. 3)
θ_D	Characteristic energy of dissociation
υ	Molecule vibration frequency
α	Degree of ionization
p	Pressure (force/area, energy/volume)
u,v,w	Speed in *x*, *y*, *z* directions
$\overline{w}$	Average speed perpendicular to S
ω	Solid angle
Θ, ν_c	Collision frequency (collisions/time)
σ	Collision cross-section
ρ	Density (mass/volume)
ρ_D	Characteristic density of dissociation
ε_i	Energy in molecular component, *i*
ξ	Energy distribution function
φ	Speed distribution function

Γ	Flux per unit time
λ	Mean free path
τ	Stress
μ	Viscosity

SUBSCRIPT AND SUPERSCRIPT

i	Integer value
tr	Translation
rot	Rotation
vib	Vibration
el	Electronic
A	Atom species
AA, A_2	Molecule of same species
AB	Molecule with different species

Appendix D: Problems

CHAPTER 2

1. Newton had postulated that when particles collide with a surface, they lose their normal component of velocity. If this were true, and a gas in contact with a wall had equilibrium random motion of particles $\left(\frac{1}{2}m\overline{c^2} = \frac{3}{2}kT\right)$, determine the relationship of wall pressure with number density (n) and temperature (T).
2. Pressure is related to the flux of momentum to a wall (area). Suppose that a wall material in contact with a gas reflected the molecules with $\frac{1}{2}$ of the incoming normal ($\perp$) velocity component. Evaluate the energy flux (energy/time-area) to the wall as a function of temperature (T), number density, and molecule mass.
3. Consider a low-density gas in a long (length, L) thin (diameter, D) tube. Calculate the ratio of the number of collisions with the wall to the number of collisions that occur between the gas molecules, in terms of the mean free path (λ) and D.
4. Consider an equilibrium mixture of gas, species A, and molecular nitrogen (MW = 28). If the molecules of species A are found to collide with the wall twice as frequently as those of nitrogen, what is the molecular weight of species A?
5. Atomic oxygen (MW = 16, molecular diameter = 10^{-8} cm) at very low density (p = 10 N/m^2, T = 1200 K) is flowing (100 m/s) along a thin wall that separates it from gas at rest (T = 300 K). There is a hole in the wall, 1 cm diameter, through which there is flux. Calculate: (1) flux of x-momentum through the hole (in the y-direction) (kg-ms^{-1}/m^2-s), and (2) flux of thermal energy through the hole (in the −y-direction) (J/m^2-s).
6. Using mean free path methods and known value for Avogadro's number, calculate the diameter of an argon atom based on a value of viscosity 2.1×10^{-4} (dyne-s/cm^2) at 0 °C.
7. For a gas in an equilibrium velocity distribution, show that $\overline{c_i} = 0$ where c_i is the velocity in the i-direction.
8. With an equilibrium distribution for gas particles, the probability that a particle will travel a distance between x and x + dx before colliding with another particle is given by: $w(x) = Ae^{-\frac{x}{\lambda}}\,dx$, where λ is the

mean free path and A is a constant. (1) Evaluate A; (2) evaluate the probability of finding a particle that travels a distance $\geq 2\lambda$ before collision.

9. Consider a rarefied gas that is leaking slowly through a small hole of diameter, D, in a container with a vacuum on the other side. The particles in the container have mass, m, and temperature, T, and the mean free path $\lambda \gg D$. (1) Find the average speed of the escaping particles; (2) evaluate and compare the average root mean square speed of the gas in the container and the escaping gas.

10. In Problem 6 (above): (1) find total rate (particles/s) of escape; (2) find the ratio of the average (random) kinetic energy of the particles that escape to the average (random) kinetic energy of the particles in the container.

CHAPTER 3

1. Consider only random translational energy of particles. If it can be assumed that for three-dimensional motion the partition function is $Q_{tr} = Q_x Q_y Q_z$, where x, y, and z are different directions, then if there is a case where particles can move in only two directions, what is the partition function? Using that partition function derive the energy per particle for those particles that can move only in two directions. What is the pressure exerted on a wall by particles that move only in two directions, and how does it compare with particles that move in three directions?

2. In the atomic particle system, energies can be measured from several possible energy reference levels. Consider the possibility of measuring particle energies from an origin different from the original by a fixed amount, ϵ^*, so we have $\epsilon_i' = \epsilon_i + \epsilon^*$. If a new partition is then evaluated in terms of ϵ_i' rather than ϵ_i, then evaluate expressions for $E' = f(E)$ and $p' = f(p)$.

3. Consider a gas composed of two species whose molecules have diameters d_1 and d_2, and that can move only in two dimensions (individual molecules move in a plane). Calculate the collision frequency for these molecules and compare with the collision frequency for particles that move in three dimensions. Also, calculate the mean free path and compare with that for molecules that move and collide in three dimensions.

4. Carbon dioxide (CO_2) is found to have different specific heats at constant volume at different temperatures, as follows:

(C_V/R)	2.5	3.0	3.5
$(T\,\text{K})$	300	1000	5000.

For each temperature relate the specific heat values to the degrees of freedom of the molecule (*tr, rot, vib, el*) and calculate energy components that are contributing to the specific heat (do not calculate specific numbers for the *vib, el* energies).

5. Consider nitrogen gas at elevated temperatures; evaluate the specific heat at constant volume (C_V) for temperatures of 1000 and 3000 K; specifically calculate the component due to vibrational excitation. Calculate the ratio of specific heats $\left(\gamma = \frac{C_p}{C_V}\right)$ for each of these temperatures, assuming equilibrium. If the gas is expanded rapidly and the vibrational component of energy "freezes" at the value for 2000 K, what is γ in the "frozen" flow?
6. The ideal dissociating gas is categorized by the following relationships: $\frac{\alpha^2}{1-\alpha} = \frac{\rho_D}{\rho} e^{-\frac{\theta_D}{T}}$ and $e\frac{(energy)}{(mass)} = R_{A_2}[3T - (1-\alpha)\theta_D]$, where α is the degree of dissociation and ρ_D is a constant. Derive an equation for the specific heat at constant pressure, C_p, for this gas with $C_p = fn$ (constants, α, T).
7. Calculate the ionization rate from ground state due to electrons (in an equilibrium distribution) collisions (single ionization) for a 2 eV temperature for hydrogen, helium, neon, and argon.
8. Evaluate the degree of ionization and the number density of electrons in argon gas at a pressure of 1 atm and a temperature of 5000 K, assuming equilibrium distribution.
9. In the Saha equation for the degree of ionization of a gas in equilibrium, it is assumed that $\frac{2Z_+^{elec}}{Z_0^{elec}} = 1$. However, it is proposed that a more accurate calculation for high-temperature argon is $\frac{2Z_+^{elec}}{Z_0^{elec}} = \frac{(2)[4+2e^{-2062/T}]}{1}$; calculate this value for 10,000 K and compare it to the 1.0 value. Calculate the values of degree of ionization, α, from the Saha equation, using the two models, and 10,000 K and 10^1 N/m^2.

CHAPTER 4

1. Electrons are ejected from a source with velocity, V, in a horizontal direction. They enter into a region of length, L, (in the x-direction) in which they encounter an electric field of strength, E, in the y-direction and are deflected. Calculate the ratio of the angles of deflection for values of electric field E and 2E.
2. A long multiturn straight solenoid is a classic device for generating magnetic field. Neglecting end effects, derive an equation for the B field as a

function of I (current) and R and specifically identify the relationship $\vec{B}(r)$. For a diameter of 10 cm and a current of 100 A, what is the value of B induced?

3. A coaxial (z-direction) electrical discharge configuration has an anode central electrode (radius, R_A) and a cathode outer electrode (radius, R_C). Assume that a plasma fills the interelectrode region and that a radial discharge is uniform in z and θ. Determine the magnetic field as a function of current density and z, and also as a function of I and r. Determine the force/volume $(\vec{J} \times \vec{B})$ as a function of I(z), R_A, and R_C.
4. Use the law of magnetic induction, $\vec{B}(\vec{r})$, to show that $\nabla \times \vec{B} = \mu_0 \vec{J}$.
5. A cylindrical column of liquid sodium (a conductor: $\sigma = 10^6$ (ohm-m)$^{-1}$) is carrying an axial current, I, with uniform current density, $J = 10$ A/m^2. The sodium is flowing with a speed of 1.0 m/s, in a tube of 10 cm diameter, thus inducing a radial electric field because of the magnetic field induced by the axial current flow; there is no radial current. Find (1) the magnetic field, $B(j, r)$; (2) the radial electric field; and (3) the body force, $\vec{J} \times \vec{B}(r)$.
6. A wire, twisted pair magnetic probe has been designed to measure the magnetic field in a current sheet that sweeps by the probe. The probe diameter of the measuring area is 2 mm^2; the current sheet is 1.0 cm thick with uniform current density of 1000 A/cm^2 and sweeps by the sensor at 1 cm/μs. Using Maxwell's equations, determine the output signal induced at the ends of the wire leads of the twisted pair.
7. For the argon plasma conditions in Chapter 3, Problem 8, calculate the Lorentz (Spitzer) conductivity of the plasma (take $\ln \Lambda = 15$).
8. In the text material, a formula (Francis, 1960) was given for the rate at which electrons gain energy (i.e., power) per unit volume from an applied field, $E_0 e^{i\omega t}$, as related to the collision frequency, ν_c. From that expression, determine the high-frequency conductivity, $\sigma = j_e/E$, and compare it to the functional form for the steady drift motion of electrons in a DC field.

CHAPTER 5

1. Electrons are ejected from a source with velocity, V, in a horizontal direction. They enter into a region of length, L, (in the x-direction) in which they encounter a magnetic field of strength, B, which is in the (y-direction) direction normal to the velocity, and their trajectory

is altered. Calculate the ratio of the radius of the electron's orbit and the ratio of the cyclotron radius for values of the magnetic field B and 2B.

2. The ionosphere layers above the Earth's surface are known to reflect radio waves that are in the megahertz range. This occurs when the electron density in the ionosphere increases to the level where the radio frequency is on the order of the plasma frequency. The E layer of the ionosphere occurs at about 100 km, where the electron density is about 10^5 cm^{-3}, neutral density is about 10^{11} cm^{-3}, and $T = 500$ K. The F layer is about 200 km and the electron density is 10^6 cm^{-3}, neutrals are 10^9 cm^{-3}, $T_e = 2500$ K, and $T_i = 750$ K. Find the mean free paths, plasma frequencies, and Debye lengths at these two altitudes in the ionosphere.
3. Calculate the effect of collisions on the plasma frequency in the following way. If we derive an expression for ω_p with the added term $-mn\nu_e \vec{v_e}$ in the momentum equation, the solution of the momentum equation gives $v_e = e\vec{E}/m\nu_e$. Using mass, momentum, and Gauss equations for small perturbation, n', where $n = n_0 + n'$ and $n' = e^{i\omega_p t}$, calculate ω_p.
4. A spherical conductor of radius, a, is immersed in a plasma and charged to a potential, φ_0. The electrons are in a Maxwellian distribution but the ions are assumed to be static when a Debye sheath is formed about the sphere. Assuming $\varphi_0 \ll kT_e/e$, derive an expression for $\varphi(r)$ in terms of a, φ_0, and λ_D, assuming a solution in the form $e^{-Kr/r}$.
5. In a fully ionized nitrogen plasma at $n_e = 10^{16}$ cm^{-3} and $T_e = T_i =$ 2 eV, a long, thin wire (diameter, 0.005 in) Langmuir probe is introduced into the plasma, and the voltage is changed to collect electrons and then ions. Calculate: (1) the mean free paths, (2) the thickness of the Debye sheath, and (3) the voltage for electron saturation current collection and the voltage for ion saturation collection voltage. After determining the voltage for ion saturation current, double that voltage and estimate the thickness of the ion collection region and the increased current collected by the probe due to sheath growth and increased collection area.

CHAPTER 6

1. An ion source ejects two ions of the same mass but one is single charge (+) and the other is double charge. They travel in the x-direction and enter a region of length, L, which has uniform electric field, E, in the -y direction. The mass, speed, and field strength are such that the

deflections are small angle ($\theta < 20°$). Calculate the ratio of the deflection angles for the two ions.

2. In an electron bombardment ion thruster the ionization chamber is cylindrical with mercury gas (MW = 200.5) flowing in the axial direction. Electrons are introduced with 10 eV energy, perpendicular to the axis. The chamber is 10 cm in diameter, and to facilitate ionization collisions, an axial magnetic field is introduced in the chamber. What is the field strength needed to make the electrons spiral in a 5-cm-diameter orbit?
3. As in Problem 1, particles are injected in the x-direction into a region of (x-) length, L = 10 cm, which has a uniform electric field, E, in the −y-direction and a uniform magnetic field, B, in the −z-direction. If helium ions are injected into the region with 10 eV kinetic energy, what magnetic field strength is needed to cause the ions to have a 2-cm-diameter orbit around the field lines. Define all the motions of the ion in the E and B fields (specify directions). If an electron is injected in the x-direction with 10 eV kinetic energy, into the same fields as the ion, what is the cyclotron radius and frequency? What are the motions that the electron will follow (give the formulas for motion). Compare the ion and electron motions. If the magnetic field profile is changed, so that the uniform value (above) becomes the value at x = L/2 and there is linear variation from 0 at entry to the x = L location, define the drift that will be created for the ion and electron because of the field gradient.
4. A particle of charge, q, is located at a radial position, r, away from a long, straight wire. At time t = 0, current, I, is instantly created in the wire (inducing a magnetic field outside the wire). Also, a uniform electric field, E, is created parallel to the wire in the space around the wire. Derive equations for all the induced and drift velocity components that develop (indicate directions). Will the particle gain energy; if so, specify how?
5. Show that there is *no* mean drift velocity when charged particles move in the field $\overrightarrow{B} = (0, B_y(x), B)$, where B_y and $\frac{\partial B_y}{\partial x}$ are small.
6. A fully ionized hydrogen plasma with $n_e = 10^{12}$ cm^{-3} and $T = 100$ eV is to be contained in a magnetic mirror trap; assume no collisions. It is desired to trap 90% of the plasma at the ends; what mirror ratio $\left(R_m = \frac{B_m}{B_0}\right)$ will be required?

CHAPTER 7

1. Consider the additional term in the generalized Ohm's law related to Hall effect. Evaluate the dissipation due to this term as in $\vec{J}\cdot(\vec{E}+\vec{v}\times\vec{B})$.
2. Consider the effect on electrical current flow in a rectangular channel where plasma of scalar conductivity 20 (ohm-m)$^{-1}$ flows at 100 m/s along the x-axis and a magnetic field of strength 15.0 W/m^2 is applied along the y-axis (across the channel). The motion induces an electric field that drives currents in the plasma, which exhibits tensor conductivity. Evaluate the current density for the different directions (assuming current paths can be completed) for values of (1) $\omega\tau = 10$, (2) $\omega\tau = 100$, and (3) $= 1.0$.
3. The MHD approximations are based on the condition that $\frac{\varepsilon\omega}{\sigma} \ll 1$, the ratio of product of dielectric constant and frequency of electromagnetic fields divided by the conductivity of the plasma. By examining the regimes of plasma properties in different physical situations (devices in laboratories, space plasmas, etc.) define *one* situation where this is a bad assumption by defining plasma properties for the occurrence.
4. The pressure in a plasma with applied magnetic field can be broken into components $p_\perp$ and $p_\parallel$. Starting with the energy equation and momentum equation, and making an adiabatic assumption (infinite conductivity), show that $\frac{D}{Dt}\left(\frac{p_\parallel B^2}{\rho^3}\right) = 0$.
5. Two of the most important parameters that identify regimes of plasma behavior are the magnetic Reynolds number

$$\left(R_m = \mu_0\sigma_0 U_0 L_0 = \frac{\text{Induced B field}}{\text{Applied B field}} = \frac{U_0 L_0,\ inertia}{\mu_0\sigma_0^{-1},\ diffusion}\right)$$

whose key variable is the flow velocity (U_0) and the beta parameter

$$\left(\beta = \frac{nkT}{B_0^2/2\mu_0} = \frac{\text{Thermal energy density}}{\text{Magnetic energy density}}\right).$$

By reviewing the order of magnitude of typical device parameters, calculate:

a. Magnetic Reynolds number for an MPD (>1000 A) space thruster.
b. Tokamak fusion experiment or proposed reactor.

CHAPTER 8

1. Consider a linear pinch in equilibrium with a proton and electron plasma, with temperatures T_+, T_- and with a radial variation of density $n(r) = n_0\,[1 + f(r)]^{-2}$, where n_0 is the number density on-axis. (1) If the electron velocity (u) in the axial direction is uniform and much greater than the ion streaming velocity, show that $\frac{\mathrm{d}}{\mathrm{d}r}\left[\frac{r}{n}\frac{\mathrm{d}n}{\mathrm{d}r}\right] = -A\,r\,n$, and determine the functional relationship for A. (2) In the density distribution $n(r)$, solve for the functional form of $f(r)$.
2. Consider a plasma column of radius, a, in equilibrium with a coaxial magnetic field, B_0, and which has a pressure distribution $p = p_0\left(1 - \frac{r^2}{a^2}\right)$. Determine the current distribution, $J(r)$.
3. In the high-velocity flow of plasma being ejected in a plasma thruster, it is important to define the magnitude of diffusion across field lines relative to the convection that occurs along the field lines. For a nitrogen plasma with source density of 10^{15} cm^{-3} and temperature of 2.0 eV, in a magnetic field of 1.0 kG, calculate the convective flux (assuming sonic velocity) along the field lines and the diffusive flux across (straight) field lines (determine the dominant diffusive mode). Evaluate the magnetic Reynolds number.
4. It has been observed that some plasma waves demonstrate an equipartition of wave energy. For the case of an Alfven wave, show that the perturbations in flow velocity (u') and magnetic field (B') are related as $\frac{1}{2}\rho(u')^2 = \frac{B^2}{2\mu_0}$.
5. In the section on shock waves in plasma the solutions for several types of plasma shock waves were presented, including the jump conditions across a shock wave propagating perpendicular to a magnetic field (3). For the condition $\gamma < 2$, show that $\frac{v_2}{v_1} = r < r(B = 0)$, the shock strength is reduced with the existence of the magnetic field.

INDEX

Note: Page numbers followed by "f" and "t" indicate figures and tables respectively.

A

AC. *See* Alternating current
Accelerator grid, 238
Adiabatic invariants, 111–112
AFs. *See* Applied fields
Alfven velocity, 148
Alfven waves
  oblique, 170, 171f
  propagation of magnetic perturbations, 161
    in plasma, 162–164
  travel, 173
Alternating current (AC), 77
Ambipolar diffusion, 140
Ampere's law, 72, 138, 203, 213. *See also* Faraday's law
Anode, 246
Anomalous transport, 143
Applied fields (AFs), 196
  AF-MPD, 229–234
Applied Toroidal field configurations, 285–287
  RFP, 290–292
  stellarators, 292–294, 292f
  tokamak, 287–290, 287f

B

Beam ion production cost, 237
Birkeland currents, 316
Bohr model, 52f, 53
Boron nitride (BN), 245
Borosil (BN-SiO_2), 245
Bow shock, 311
  Earth's bow shock, 315
  magnetosphere shock wave, 312f
  Maxwell's equations, 313
  supersonic solar wind–magnetosphere cavity interaction, 312
Burning plasma physics, 302–303

C

Channel flow, 197
  with gas electromagnetic acceleration, 212
    1D channel flow, 215
    MPD thruster experiments, 213
    "sonic" transition, 212
    velocity gradient expression, 214
  isothermal assumption, 198
  isothermal flow, 199
  1D channel flow, 198f
  prediction of force interactions, 200
Chapman–Ferraro current, 311
Characteristic temperature, 37
Chemical kinetics, 49–50
Childs-Langmuir law, 97
Classical diffusion coefficient, 139
Closed-loop magnetic probes, 82–83
Cold plasma limit, 160–161
Collision mean free path, 87
Collisional shock waves, 51–52
Collisionless heating, 266
  distribution function, 267
  ion cyclotron, 269
  wave energy, 268
Collisionless sheath calculations, 95
  Childs-Langmuir law, 97
  effects, 98
  electrical discharge of plasma device, 97
  electron current with adverse potential gradient, 95
  ion current collection before saturation, 95
  Poisson's equation, 96–97
  presheath region with ion saturation bias, 96f
  saturation current, 95
    ion current, 95

Collisionless shocks, 189, 192f
Compact toroids, 295
FRCs, 295–298
spheromaks, 299–301
Conduction, 15
Conservation
of charge, 69–71
equations, 120–125
of mechanical energy, 126–127
Constraints, 26
Continuum plasma dynamics equations, 137–138
Convection current, 132
Coulomb logarithm, 76
Coulomb shielded potential, 90
Coupling plasma, 236
Current density, 73–74
Current sheet implosion, 251, 252f
average sheet velocity, 254
MHD model, 258
momentum balance for, 253
particle loss time, 257
plasma temperature, 255
sheet acceleration, 252
Cyclotron frequency, 92

D

DC. *See* Direct current
Debye length, 90
Definite integral, evaluation and values of, 327t
Dense Z-pinch X-ray sources, 283–285
Deuterium (D), 279
Device plasmas, 5
Diffusion, 15, 18–20
velocity, 144–145
Direct current (DC), 77
Discharge chamber, 236–238, 245
Discharge plasma, 236
Dispersive media, 158–159
Displacement current, 132
Dynamic pinch, 249, 274–277
current sheet implosion, 251–258
Kelvin-Stokes theorem, 250–251

E

Earth
magnetosphere structure, 3, 4f
rotation, 1
ECRH. *See* Electron cyclotron resonance heating
Electrical conductivity, 73–74, 142
evaluation, 74–76
Electromagnetic waves cutoff, 158–159
Electromagnetics, 65
Ampere's law, 72
atomic-scale particles, 65
conservation of charge, 69–71
electrostatics, 65
action at distance, 66
behavior of particles, 67
Coulomb's law, 66
divergence theorem, 67
electric field, 66–67
polarization of medium, 67
potential voltage difference, 68
volume of plasma, 66
Faraday's law, 71–72
forces and currents, 73–76
electrical conductivity and Ohm's law, 73–74
electrical conductivity evaluation, 74–76
plasma dielectric properties, 76
magnetostatics, 68–69
Maxwell's equations, 72–73
Electron cyclotron resonance heating (ECRH), 264
Electron saturation current, 94
Electrostatic body force, 133
Electrostatic(s), 65. *See also* Magnetostatics
action at distance, 66
behavior of particles, 67
Coulomb's law, 66
divergence theorem, 67
electric field, 66–67
particle collection
current and density behaviors, 94
Debye length, 93
densities and potential in sheath region, 93f

in Langmuir probes, 92–93
Maxwellian speed distribution, 94
plasma volume, 93
probe behavior, 94
polarization of medium, 67
potential voltage difference, 68
volume of plasma, 66
Energy
conservation, 137
density, 12
distribution function, 33–34
Energy transport. *See* Thermal conduction
Entropy, 24–26
Equilibrium kinetic theory, mathematical formulation of, 20
distribution function and average values, 20
average values of property calculation, 20–22
speed and velocity distribution functions determination
arrangements, 23t–24t
equilibrium distribution of molecular kinetic states, 22–23
probability considerations, 23–24
Equilibrium state analysis, 24
constraints, 26
energy distribution function, 27
entropy, 24–26
equilibrium velocity distribution function, 28
Lagrange multipliers method, 26
partition function, 26
Stirling approximation, 25
velocity distribution function, 27
Equilibrium velocity distribution function, 28
Exit beam neutralizer, 241

F

Faraday channel, 211
Faraday's law, 71–72, 138, 143, 202
Field reversed configuration (FRC), 295–296
MTF, 299f
plasma toroid, 298
radial force balance, 296–297
thermal energy, 298
Field-aligned currents, 316
Floating potential, 94
Fluid behavior of plasma. *See also* Fluid waves; Shock waves
continuum plasma dynamics equations, 137–138
hydromagnetic stability, 152–158
magnetic fields kinematics in, 143–147
magnetohydrostatics, 147–152
MHD, 137
plasma waves, 158–164
transport effects in plasmas and plasma devices, 138–143
Fluid mechanics, 156
Fluid waves. *See also* Shock waves
in compressible plasma medium, 164
directional orientation for wave propagation, 168f
fast and slow waves, 172
isentropic energy relationship, 167
Maxwell's equations, 165–166
oblique Alfven wave, 170, 171f
Fluid-plasma equilibrium configurations, 149–152
FRC. *See* Field reversed configuration
Frictional factor, 80
Friedrich's diagram, 172, 173f
Frozen flow, 47
reaction rate effect on specific heats, 47
gas constant for, 48
high-temperature state, 49
low-temperature state, 48–49
Mach number, 49
monatomic and diatomic species, 48
Fusion, 3. *See also* Magnetic—fusion plasmas

G

Gas
discharge plasma sources, 81, 81t
mixtures, 12–13
transient diffusion, 138
Generalized Ohm's law, 130
Geomagnetic plasma flow, 307–311
"Glow" discharge, 77–78

Grad-Shafranov equation for equilibrium, 288–289
Guiding center of particle orbit, 102

H

Hall parameter, 92, 243
Hall thruster, 241–242, 242f. *See also* Ion thruster
anode, 246
design and performance, 244
discharge chamber, 245
hollow cathode, 246
magnetic circuit, 245
particle interactions, 242–243
SPT-100, 244f
thruster components, 244–245
Hartmann flow, 201
Hartmann number, 205
velocity profiles for, 208f
High-frequency gas discharges, ionization in, 79
breakdown curves, 81, 81f
diffusion of electrons, 80
frictional factor, 80
gas discharge plasma sources, 81, 81t
medium or high pressures, 80
primary processes, 80
radio frequency gas discharges, 79–80
High-temperature air, equilibrium composition of, 40–41
compressibility factor, 42
disassociation reaction, 41
gas mixture, 42
ideal dissociating gas, 43, 45
numbers of species particles, 41–42
total mass of gas, 41
undissociated gas, 42
High-temperature gases, 45, 47
Hollow cathode, 246
Hydromagnetic channel flow, 200–201
Ampere's law, 203
Faraday's law, 202
Hartmann flow, 201
induced current density, 208f
Maxwell's equations, 209
MHD
channel interaction, 206
generator, 205f
power generation, 210–211
Ohm's law, 204
second-order *ODE*, 205
solutions applications, 209–210
velocity profile, 207
with viscous interactions, 201f
Hydromagnetic stability, 152–153
MHD
perturbations analysis, 154–158
stability physical considerations, 153
Rayleigh–Taylor instability
hydrodynamic, 156f
plasma, 156f

I

ICF. *See* Inertial confinement fusion
ICRH = Ion Cyclotron Resonance Heating, 264
Ideal dissociating gas, 43, 45
Inertial confinement fusion (ICF), 279
Internal cathode plasma, 236
Internal energy, 8–13
International thermonuclear experimental reactor (ITER), 279, 301–302
burning plasma physics, 302–303
operating parameters and plasma properties, 302t
Invariant of motion, 109
Ion optics, 238–239
Ion saturation current, 95
Ion thruster, 234–235. *See also* Hall thruster
discharge chamber, 236–238
ion accelerating grid region, 240f
ion optics, 238–239
Kaufman-type, 235f
thrust components, geometry, and potential arrangements, 235–236
Ion waves, 159
Ionization
atomic structure and electron arrangements, 52–53
Bohr's model of atom with electrons, 52f, 53
shell distribution of electrons, 53t

electron collisions, 55–56, 56f
excited atoms, 56
Maxwellian electron distribution, 57
single-electron impact, 56
in gases, 50–51
collisional shock waves, 51–52
creation and dynamics, 51
heavy particle collisions, 57
ionization (electron) loss mechanisms, 58–59
periodic table of elements, 55t
photoionization, 57–58
quantum model atom, 51
radiation (energy) loss from plasmas, 59–60
Saha equation, 52, 60, 63
loss mechanisms, 58
electron diffusion, 59
recombination, 58–59
potentials for gases, 53, 54t
Ionosphere, 315–316. *See also* Magnetosphere
ionization in, 319–321
region of Earth, 318
space plasma experiments, 321–323
ITER. *See* International thermonuclear experimental reactor

K

Keeper, 241
Kinetic theory of gases, 7
basic hypotheses, 7
control volume for molecule motion, 8f
diffusion, 18–20
gas mixtures, 12–13
particle motion along x axis, 9f
pressure, temperature, and internal energy, 8–13
secondary hypotheses, 8
thermal conduction, 18
and transport processes
particle collisions, 13–15
transport phenomena, 15–17
viscosity, 17–18

L

Lagrange multipliers method, 26
Landau damping effect. *See* Collisionless heating
Larmor radius, 91–92
Law of mass action, 49–50
Linear pinch stability, 155
Linear Z-pinch X-ray sources. *See* Dense Z-pinch X-ray sources
Lorentz conductivity, 75

M

Mach number, 186–187
Macroscopic equations of plasmas, 115
electromagnetic energy and momentum addition to, 115
conservation of momentum, 117–118
continuous distribution, 116
divergence of dyadic, 119
electromagnetic force, 119–120
magnetic and electric field density, 116–117
Maxwell's equations, 116
Poynting vector, 117
magnetofluid mechanics
conservation equations, 120–125
single fluid equations, 125–132
similarity parameters, 134–136
Magnetic
circuit, 245
compression and heating
dynamic pinch, 249–258
plasma flow within magnetic field lines, 258–262
pulsed magnetic fields, 249
fusion plasmas
applied toroidal field configurations, 285–294
compact toroids, 295–301
energy release, 279
ITER, 301–303
MTF, 279
Z-pinch, 280–285
helicity, 291
moment, 109–110

Magnetic (*Continued*)
pressure in plasma fluids, 148–149
probe response, 83–84
Magnetic field lines, plasma flow within, 258–259
collisional MHD model, 262
collisional plasma, 259
cusp velocity, 260
plasma properties, 260f
two-temperature plasma, 261–262
Magnetic field(s), 196
effects, 141–143
kinematics in plasma, 143
convection of B in plasmas, 145–147
diffusion of B in plasmas, 144–145
geometry of flux tubes, 146f
parameters for convection and diffusion, 147
related parameters
cyclotron frequency, 92
Hall parameter, 92
Larmor radius, 91–92
regimes of interaction, 92
transport properties—magnetic field effects, 141–143
Magnetized target fusion (MTF), 279
Magneto-plasma-dynamics (MPD), 223
AF-MPD, 229–234
arc configuration, 223
self-field, 225–229
Magnetofluid mechanics
conservation equations, 120–125
single fluid equations, 125–132
electron conservation of momentum, 127–128, 132
Magnetohydrodynamics (MHD), 132, 137
approximations, 132–134
generators, 132
perturbations analysis, 154–158
stability physical considerations, 153
thruster experiments, 213
Magnetohydrostatics, 147–148
Alfven velocity, 148
fluid-plasma equilibrium configurations, 149–152
linear pinch, 151f
magnetic pressure in plasma fluids, 148–149
Magnetosphere. *See also* Bow shock; Ionosphere
coupling, 315
drift speed, 318
ionosphere plasma, 316f
Ohm's law, 317
solar wind, 315–316
geophysical, 311–315
Magnetostatics, 68–69
Mass conservation, 137
Mass transport. *See* Diffusion
Maxwell's equations, 65, 72–73, 209, 313
illustrative applications of, 81–82
closed-loop magnetic probes, 82–83
Z-Pinch and magnetic probe response, 83–84
Maxwell–Boltzmann statistics, 89
Maxwellian distribution function. *See* Equilibrium velocity distribution function
Meter, kilogram, second (MKS), 65
and Gaussian system conversion, 325t
MHD. *See* Magnetohydrodynamics
MKS. *See* Meter, kilogram, second
Molecular energy, 33
calculations, 34–35
chemical kinetics, 49–50
energy distribution function, 33–34
evaluation, 35–36
frozen flow, 47, 49
high-temperature air, equilibrium composition of, 40–41, 45
high-temperature gases, 45, 47
partition function
analytic forms, 37, 39
and dissociation energy, 39–40
Molecule, 7
Momentum conservation, 137
Momentum transport. *See* Viscosity
MPD. *See* Magneto-plasma-dynamics
MTF. *See* Magnetized target fusion
Multicomponent species conservation equations, 115

N

National Aeronautics and Space Administration (NASA), 211
Normalized distribution function, 21–22

O

Ohm's law, 73–74, 137, 165, 178–179, 197–198, 204, 213, 291
Optical thickness of plasma, 59–60
Order of reaction, 50

P

Particle collisions, 13, 15
  model, 14f
  molecules of diameter, 13f
  number of molecules in volume, 14
  parameters, 87
  sphere of influence, 13–14
Particle motion
  with curvature of magnetic field lines, 107
  spatially varying (inhomogenous) magnetic fields, 105
    drift effect, 106
    Larmor orbit, 106–107
  time-varying magnetic field, 108
    invariant of motion, 109
    Larmor orbit, 108
    magnetic moment, 109–110
  uniform electric and magnetic fields, 102, 105
    behavior of free acceleration, 104
    equation of motion, 104
    geometry and plane motion, 103f
    particle drift velocity, 103f
Particle orbit theory, 99
  charged particle motion, 100
    in constant, uniform field, 102f
    coupled equations, 101
    guiding center of particle orbit, 102
    magnetic field direction, 100
    ordinary differential equations, 101
  equation of motion for particle, 99–100
Particle trapping in magnetic mirrors, 110–111
  magnetic field geometry, 110f
  particle motion, 111f
Partition function, 26
  analytic forms, 37
    electronic component, 37–38
    rotational component, 38
    translational component, 37
    vibrational component, 38–39
Photoionization, 57–58
Plasma, 1–2, 87. *See also* Space plasma environment
  density and temperature, 4f
  dielectric properties, 76
  dynamics, 195
  external parameters, 87
  ion production cost, 236–237
  laboratory/device applications
    device plasmas, 5
    general description, 3–5
  momentum and energy
    electromagnetic terms affecting, 217
    ion thruster, 234–239
    MPD and AF-MPD Arc, 223–234
    PPTs, 218–223
  in nature, 2
    solar plasma, 2
  oscillations and frequency, 90–91
  particle (collision) parameters, 87
  potential, 94
  property domains, 3f
  sheath
    Coulomb shielded potential, 90
    Debye length, 90
    formation and effects, 87–88
    Maxwell–Boltzmann statistics, 89
    perturbation potential, 89
    Poisson's equation, 89
    potential energy, 88–89
    separation length, 89
    shielding distance, 90
    work done, 88
  shock waves, 51–52, 174, 176
    field orientation for shock, 180f
    flow and magnetic fields, 180
    for Maxwell's equations, 177
    momentum and energy, 184–185
    momentum equation, 182
    Ohm's law, 178–179

Plasma (*Continued*)
plasma shock, 186–187
steady flow, 177
"switch-on" shock waves, 183
thrusters, 217
space charge limited current, 239–246
transport effects
ambipolar diffusion, 140
anomalous transport, 143
diffusion of particles, 138–139
species energy equilibration, 140–141
transport properties—magnetic field effects, 141–143
Plasma acceleration, 197
channel flow, 197–200
with gas electromagnetic acceleration, 212–215
and energy conversion, 197
flow control utilizing plasma interactions, 215–216
hydromagnetic channel flow, 200–211
Plasma behavior in gas discharges, 76–77. *See also* Electromagnetics
formation, 77–78
ionization
growth in electric fields, 78–79
in high-frequency gas discharges, 79, 81
Plasma waves, 158
Alfven waves, 161–164
dispersive media, 158–159
electromagnetic waves cutoff, 158–159
heating process, 264, 265f
collisionless heating, 266–269
heating by plasma waves, 264
plasma–wave coupling model, 266f
procedure for, 264
properties for heating, 265t
for space propulsion, 269–277
ion waves—propagation of magnetic perturbations, 159
longitudinal electron oscillations, 160
longitudinal ion oscillations, 160–161
Plasmas laser heating. *See* Dynamic pinch
Plume plasma, 236
Poisson's equation, 89
Poynting vector, 117
PPT. *See* Pulsed plasma thruster
Prandtl number, 19
Pressure, 8–13, 50–51
Pulsed plasma thruster (PPT), 217–218
discharge current history, 222f
optimization, 221
plasma interactions, 223
rectangular PPT geometry, 219f
resistance heating term, 219–220
solid Teflon ablation, 221–222
space thruster system, 220
thruster performance for water and Teflon, 224t

R

Radial force balance, 296–297
Radiation (energy) loss from plasmas, 59–60
Raman scattering, 276
Rayleigh–Taylor instability, 156
hydrodynamic, 156f
plasma, 156f
Reaction rate constant, 50
Reversed field pinch (RFP), 290–292

S

Saha equation, 52, 60, 63
Saturation current, 95
Screen grid, 238
Self-field, 196
MPD, 225–229
Sheath formation and effects, 87–88
characteristic of separation length, 89
Coulomb shielded potential, 90
Debye length, 90
Maxwell–Boltzmann statistics, 89
perturbation potential, 89
Poisson's equation, 89
potential energy, 88–89
shielding distance, 90
work calculation, 88

Shielding distance, 90
Shock waves
collisionless shocks, 192f
dispersion, 215
magnetic Reynolds number parameter, 188
in ordinary fluid flow, 174–175
plasma shock wave physics extended reviews, 188–192
plasmas, 174, 176
field orientation for shock, 180f
flow and magnetic fields, 180
for Maxwell's equations, 177
momentum and energy, 184–185
momentum equation, 182
Ohm's law, 178–179
plasma shock, 186–187
steady flow, 177
"switch-on" shock waves, 183
structure, 187
Similarity parameters, 134–136
Single fluid equations, 125–132
Snowplow model, 251
Solar plasma, 2, 4f
Solar wind (SW), 307, 311
charged particle behavior, 311
corona plasma, 309
expansion process, 309
and magnetosphere coupling, 315–318
mean properties of, 309t
solar gravity, 310–311
Sun physical properties, 308t
Solid Teflon ablation, 221–222
"Sonic" transition, 212
Sound waves. *See* Ion waves
Space charge limited current, 239
exit beam neutralizer, 241
Hall thruster, 241–246
ions in grid region, 240
Space plasma environment, 307. *See also* Plasma
geomagnetic plasma flow, 307–311
geophysical magnetosphere and bow shock, 311–315
ionosphere region of Earth, 318–323
near-earth magnetosphere environment, 306f
solar wind, 307–311
interaction with Earth's magnetic field, 306f
and magnetosphere coupling, 315–318
Sun, 305
Space plasma experiments, 321–323
Space propulsion
plasma wave heating for, 269–270
plasma end loss in collisionless magnetic expansion, 271–273
plasmas laser heating, 274–277
variable specific impulse magnetoplasmadynamic rocket, 270–271, 270f
thrusters, 5
Space thruster system, 220
Species energy equilibration, 140–141
Speed and velocities, average values of, 28, 31
average speed, 30, 31f
mean square speed, 30
speed distribution function, 29–30
velocity space, 29, 29f
Speed distribution function, 29
Sphere of influence, 13–14
Spheromaks, 299–301
Standard temperature and pressure (STP), 76–77
Stationary plasma thruster (SPT), 244
Steady state diffusion, 139
Stellarators, 292–294, 292f
Stirling approximation, 25
STP. *See* Standard temperature and pressure
SW. *See* Solar wind
"switch-on" shock waves, 183

T

TAL. *See* Thruster with anode layers
Temperature, 8–13
characteristic, 37–38
of plasma devices, 4f
specific heat of air, 46f
Sun, 1
thermonuclear fusion of hydrogen to helium, 3

Thermal conduction, 18
Thermal diffusivity, 19
Theta pinch. *See* Dynamic pinch
Thruster with anode layers (TAL), 244
Tokamak, 287–290, 287f
Townsend (growth) coefficient, 78–79
Tritium (T), 279

V

Velocity distribution function, 27
Vibrational component, 38–39
Viscosity, 15, 17–18
Viscous diffusivity, 19
Vlasov equation, steady-state, 272

Z

Z-pinch, 83–84, 280
  axial current stabilization, 283
  dense X-ray sources, 283–285
  heat loss to electrodes, 281
  linear pinch experiments, 282
  magnetic fusion conditions, 281
  toroidal configuration, 281
  ZETA device, 282f

Printed in the United States
By Bookmasters